Modelling of agricultural production: weather, soils and crops

Simulation Monographs

Simulation Monographs is a series on computer simulation in agriculture and its supporting sciences

Modelling of agricultural production: weather, soils and crops

H. van Keulen and J. Wolf

Pudoc Wageningen 1986

CIP

ISBN 90 220 0858 4

Printed in the Netherlands

CONTENTS

PREFACE

This book introduces the reader into the quantitative aspects of agricultural production, as influenced by environmental conditions and management practices. The aim is to familiarize the reader with the subjects in such a way that first estimates of agricultural production potentials in situations relevant to him can be made. For that purpose many exercises and examples have been included in the text to facilitate direct application of the theory presented.

The approach presented in this book is developed by the Centre for World Food Studies (SOW), an interdisciplinary research group working on problems related to world food supply and agricultural production potentials and limitations.

The direct motive for publishing this Simulation Monograph was an international course on the same object, organized in Wageningen by dr. J.H. de Ru of the Foundation for Post-Graduate Courses of the Agricultural University in Wageningen. The course was organized in close cooperation with dr. D.A. Rijks of the Applications Program of the World Meteorological Organization in Geneva and was financially supported by the Dutch Directorate General for International Cooperation (DGIS), the European Community (EC), and the Food and Agricultural Organization of the United Nations (FAO). The book has been edited on the basis of the lectures presented during the course with ample cooperation of the authors and invaluable advice of prof. dr. C.T. de Wit. The contributions of ir. D.M. Jansen and Mrs. H.H. van Laar during the course and during the editing stage of the book were of great help. Ing. P.W.J. Uithol is gratefully acknowledged for his accurate work on the list of references. Many thanks are due to Mrs. R. Helder, who skillfully and enthousiastically typed the first versions of most of the contributions, to Mrs. M.A. Boss, who performed the task of finalizing the manuscript and Mr. G.C. Beekhof for his punctual drawings.

H. van Keulen
J. Wolf

CONTRIBUTORS

Berkhout, J.A.A., Centre for World Food Studies, c/o C.A.B.O., P.O. Box 14, 6700 AA Wageningen, the Netherlands.

Driessen, P.M., Department of Soil Science and Geology, Agricultural University, P.O. Box 37, 6700 AA Wageningen, the Netherlands.

Faber, D.C., Centre for World Food Studies, c/o C.A.B.O., P.O. Box 14, 6700 AA Wageningen, the Netherlands.

Heemst, H.D.J. van, Centre for Agrobiological Research, P.O. Box 14, 6700 AA Wageningen, the Netherlands.

Keulen, H. van, Centre for Agrobiological Research, P.O. Box 14, 6700 AA Wageningen, the Netherlands.

Laar, H.H. van, Department of Theoretical Production Ecology, Agricultural University, P.O. Box 430, 6700 AA Wageningen, the Netherlands.

Rijsdijk, F.H., Fyto consult bv, Boeslaan 21, 6703 EN Wageningen, the Netherlands.

Schouten, H., Directie Organisatie en Efficiency, Ministerie van Landbouw en Visserij, P.O. Box 20401, 2500 EK Den Haag, the Netherlands.

Wit, C.T. de, Department of Theoretical Production Ecology, Agricultural University, P.O. Box 430, 6700 AA Wageningen, the Netherlands.

Wolf, J., Centre for World Food Studies, c/o C.A.B.O., P.O. Box 14, 6700 AA Wageningen, the Netherlands.

1 INTRODUCTION

1.1 Introduction

C.T. de Wit

Agriculture may be defined as the human activity that produces useful organic material by means of plants and animals, with the sun as the source of energy. The minimum required number of resources is small: labour and land, with some sun and rain. For many soil types and climates, farming systems have been developed that enable subsistence in food, clothing, shelter and fuel, provided sufficient land is available. Unless conditions are very favourable, these farming systems do not produce much more than bare necessities. However, man is an animal species that thrives on brick and concrete and the development of civilisation is very much intertwined with that of urban life. To sustain a substantial non – farming population, the productivity of the farming population has to be much higher than its subsistence level. This is only possible if the non – farm sector produces industrial means of production for the farmers within an economic structure that provides sufficient incentives for their use.

Although a sharp distinction is not possible, these means of production may be classified as labour saving, yield increasing and yield protecting such as machines, fertilizers and pesticides, respectively. Only the yield protecting inputs require little energy for their manufacture and use, although their development would hardly have occurred independently of the chemical industry. With some exaggeration, modern agriculture could therefore be defined as the human activity that transforms inedible fossile energy (mineral oil and natural gas) into edible energy through plants, animals and the sun.

Up to World War II, the emphasis in agriculture in the U.S.A. was on mechanization. Horses were replaced by tractors, so that land that was used to grow food for horses could be used to cultivate crops for other purposes. In this way, the agricultural output of the nation as a whole increased considerably. The yield increases per hectare were, however, small: for wheat only about 3 kg ha^{-1} yr^{-1} as is seen in Figure 1. In Europe in the same period, more emphasis was given to increasing the productivity per unit of land. However, the results were not impressive: ranging from about 4 kg ha^{-1} yr^{-1} in the United Kingdom to 18 kg ha^{-1} yr^{-1} in the Netherlands.

A few years after World War II, the annual yield increase suddenly improved, reaching 50 – 80 kg ha^{-1} yr^{-1} as is illustrated in Figure 1 for the United States and the United Kingdom. In general terms, this persistent yield increase may be attributed to the simultaneous effect of soil amelioration, the use of fertilizers and the control of diseases, as well as to the introduction of varieties that were able to make good use of these increased inputs. In many regions, wheat yields are still so low that an absolute yield increase of 50 – 80 kg ha^{-1}

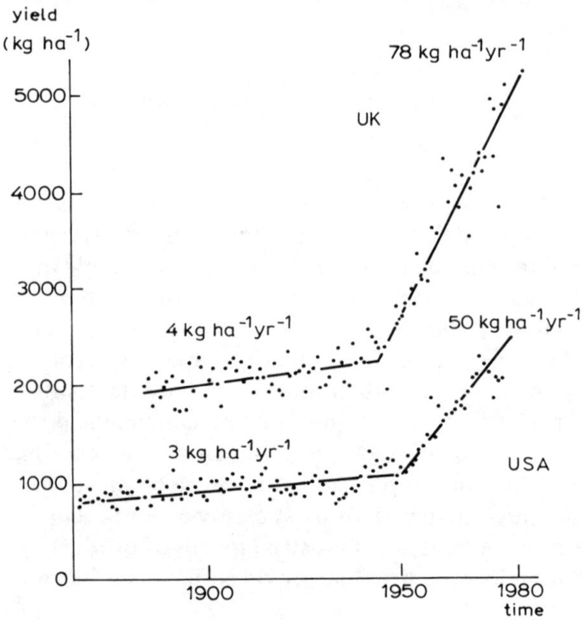

Figure 1. Average wheat yields in the United Kingdom and the United States over the last century.

yr^{-1} represents a relative increase of over 2 percent per year. The situation is not much different for other crops. Such yield increases outstrip the growth of the population in the industrialized countries. Any slack that is created in this way appears to be taken up by increased use of grain and land for milk and meat production and by taking land out of production.

In contrast, the annual yield increases in Africa, South America and Asia appear to be on an average 10, 19 and 25 kg ha^{-1} yr^{-1}, respectively (Figure 2). This is slightly higher than in the industrialized part of the world before World War II, which indicates that some of the knowledge and means of production are trickling down from North to South. However, this occurs at a rate that is too low to prevent hunger and malnutrition. For instance, in Africa with an average grain yield of 1000 kg ha^{-1}, the increase of 10 kg ha^{-1} yr^{-1} amounts to only 1 percent per year and even this may be too high an estimate for the last ten years. This growth in yield is far less than the relative growth rate of the population, which is 2 – 3 percent per year. Up to now, the difference has been more or less made up for by cultivating larger areas, but land that can be reclaimed by simple means within the social – economic framework of the family or the village is becoming scarce, so that more advanced technology is indispensable for further reclamation. Hence, to improve the food situation,

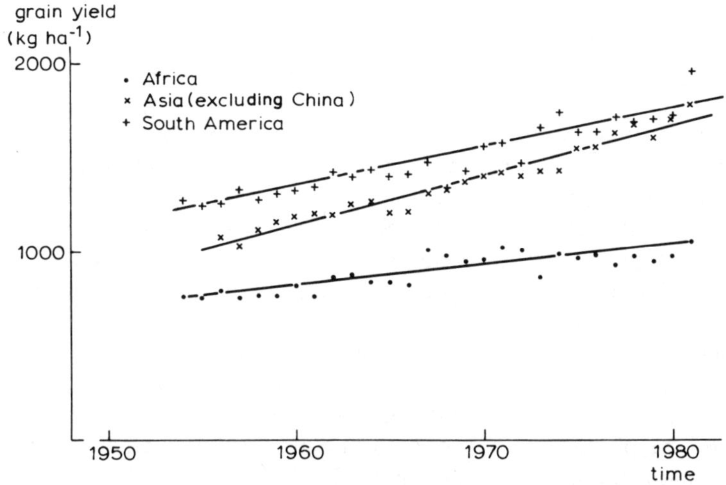

Figure 2. Average grain yields from 1954-1980 for Africa, Asia and South America.

either more machinery has to be used to extend the area under cultivation or more inputs, e.g. fertilizers, to increase the yield per unit area. Both of these paths require an open economy in which the farmer receives sufficient money for his agricultural products to pay for the necessary means of production. Too often the terms of trade are not so favourable, but claims that it is possible to improve substantially the food production in the world without these technical means are not justified in view of what is known about the agricultural production process.

Average yields, as used here for illustration, hide many differences. Some developing countries are reasonably well endowed with resources and have an economic structure that promotes agricultural development. Their price levels are such that it pays, at least for some farmers, to improve the soil, to apply fertilizers, to practise disease control and to use the proper plant varieties. However, policies that enable the poorest segments of the population to purchase the bare necessities may be lacking, so that hunger and malnutrition continue to exist. This is the case even in the richer countries of the world. There are also poor countries with infertile soils and an unfavourable climate, often with few other endowments and landlocked and with a demographic structure that results in rapid population growth. Such countries can only import the necessary agricultural inputs if they can export cash crops. Transport costs for both their import and export are often so high that even this path to increasing production is blocked. Then progress depends on political and economic solidarity that exceeds national boundaries.

The prospects for improvement of the food situation have, therefore, also national and international political and economic dimensions. International policy agreements aimed at stabilization of the prices on the world market at a fair level and at promotion of the opportunities of developing countries to penetrate the markets of the rich countries, may create a more favourable position for developing and poor countries. Such agreements must then be complemented by national development strategies that enable farmers to increase their output and, in particular, to improve the production opportunities for the poor. This broad range of problems has been the focus of research undertaken by the Centre for World Food Studies, which is situated in Amsterdam and Wageningen. For this research, national economic models with emphasis on the agricultural sector are being developed and linked to a global model to analyse and improve the policies of national governments and international agencies. These national economic models contain agricultural production modules that account for the possibilities of production and can distinguish between regions and commodities. That part of the work of the Centre focuses on the physical and agronomic factors that determine agricultural production and is the subject treated in this book.

The main purpose pursued is to familiarize the reader with the processes that govern the technical possibilities for agricultural production in a region in such a way that quantitative estimates can be made of the yield levels of the main crops under various constraints and of the inputs that are needed for their realization. The approach is necessarily simplifying, so the quantitative estimates should not be considered as the final answer, but rather as a framework for further analysis of possibilities and constraints that are based on factual knowledge, which can only be obtained by fieldwork.

For this approach, a hierarchical procedure is adopted which is in a schematic way presented in Figure 3. The rectangles in the second row represent the factors that ultimately determine the production potential. Climate and soil are fixed properties for a given region and, in combination with the level of reclamation, characterize the land quality level. The characteristics of agricultural crops may be changed by breeding, the scope for improvement in this respect being reasonably well – defined. For a given land quality level, the yield potential is therefore fixed for a fairly long period of time, and may, therefore, be calculated with reasonable accuracy.

In the further analysis, the goal is not to define a production function describing the relationship between the yield and all possible combinations of growth factors, because, by the nature of the agricultural production process, no unique solution to such a production function exists. Instead, a reasonable combination of growth factors should be established that will result in the yield level that is in accord with the land quality level. Thus, the yield level is considered concurrently as a dependent variable, determined by crop characteristics and land quality level, and as an independent variable, dictating the required input combination for its realization. This is reflected in the direction of the arrows in Figure 3: towards the yield level as well as away from the yield level.

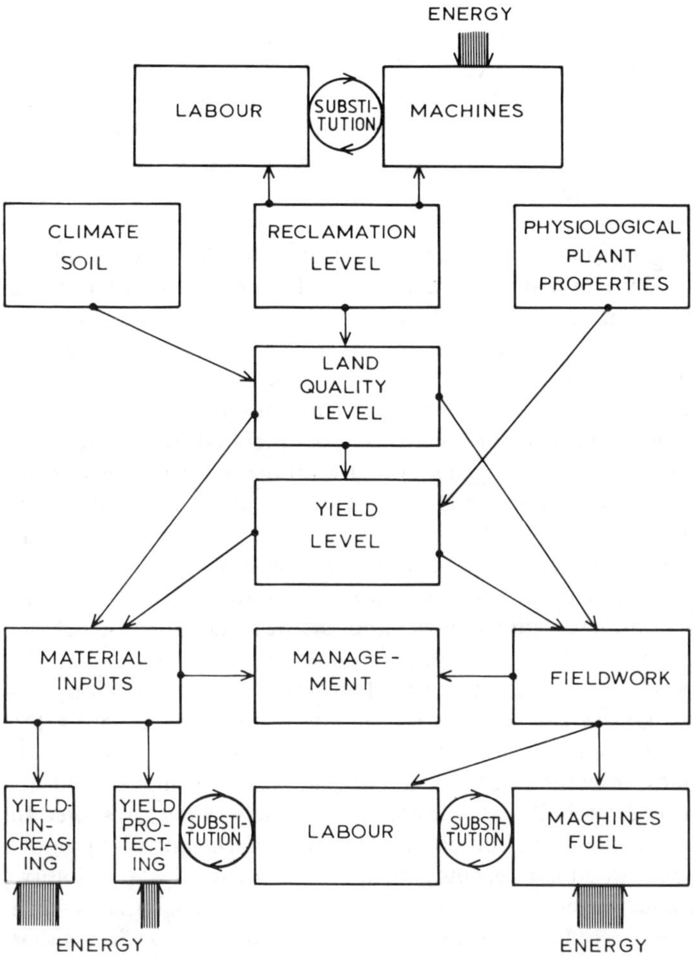

Figure 3. Schematic representation of the hierarchical analysis procedure.

With respect to the required inputs, a distinction is made between field work and material inputs. The necessary field work can be described in physical terms, for example, the requirements for ploughing, harrowing, weeding, the length of supply and transport lines, etc. The time required for these activities is to a large extent independent of the yield level as they must be done anyway. The total length of the period available to do the work depends, however, strongly on soil type and weather conditions. In performing field work, considerable substitution is possible between manual labour and activities relying on heavy mechanical equipment and their associated fossil energy requirements. The material inputs can be classified into yield – increasing inputs and yield protecting inputs. The required amounts of yield – increasing inputs such as water, minerals and nitrogen, are directly influenced by the required yield level, soil type and weather conditions. Characteristic for these inputs is that they cannot be substituted for by labour. This is in contrast to the yield protecting inputs, biocides, for which alternatives are available, i.e. labour – intensive weeding versus the use of herbicides and manual insect eradication versus insecticides.

The land quality level is, on the one hand, determined by intrinsic soil properties and the prevailing weather conditions and, on the other hand, by the level of reclamation. In a schematized set up, four levels of reclamation may be distinguished (Section 7.1). The lowest level refers to land in an almost virgin condition upon which agricultural activitities are limited to food gathering, fuelwood collection and extensive grazing. In the next level of reclamation, the land is cleared to allow more permanent agricultural use, with or without fallow periods. The moisture regime is fully dictated by weather conditions. Flooding, if any, can only be avoided by simple modifications of the topography or by building simple dams. The next level pertains to land upon which improvements have been carried out, such as levelling, simple terracing and the construction of open ditches to control excess water. The highest reclamation level refers to land in a favourable condition for crop growth, well levelled, with complete water control and the necessary infrastructure. Sufficient water is available for irrigation as required.

In addition to defining the present status of the land in a given region, it is also important to quantify the reclamation activities necessary to bring the land to a higher land quality level. This applies especially to the amount of vegetation and stones to be removed, the amount of soil to be moved and the infrastructure that must be built. This aspect of the analysis is represented in the first row (Figure 3). Reclamation can be carried out with manual labour. However, as has been stated above, this is often only a theoretical possibility, because most of the acreage that could easily be reclaimed has already been developed. Even in the People's Republic of China, it has been concluded that it is almost inevitable to resort to the use of mechanical means. The activities to be performed are therefore better defined for various technological levels in terms of the available equipment.

In the further analysis four hierarchically ordered production situations are distinguished. For the highest hierarchical production situation water, minerals, and nitrogen are in optimal supply. Crop yield is then only determined by the type of crop, the prevailing level of irradiance, and the temperature regime. Models to calculate potential growth (Section 2.1) and development (Section 2.2) of healthy closed green crop surfaces are available. Other models provide potential transpiration rates, so that the total water requirement may be obtained (Section 3.1). For most regions, sufficient experimental data are available to judge the feasibility of growing the major crops and to define so – called cropping calendars, which stipulate time of sowing, emergence, flowering, ripening, etc. (Section 6.1). Theoretical considerations and field data may be combined to develop simple calculation models for the relevant crops, yielding the time course of dry matter production and transpiration and the economic yield as outputs (Section 2.3).

For the second hierarchical production situation, it is assumed that the supply of nitrogen and minerals is still optimal, but the influence of moisture availability on transpiration and crop production is taken into account (Section 3.3). Water supply to the canopy is dependent mainly on rainfall and sometimes on supplementary irrigation. Water consumption is mainly determined by environmental conditions and the cover of the soil by the crop. The physical properties of the soil and the climatic conditions are now of major importance. From these data, the water balance is calculated. (Section 3.2). This enables determination of periods with excess or shortage of water, which will result in reduced growth rates compared with those for the first hierarchical level (Section 3.4). The models also enable the calculation of the number of workable hours in the field: an important parameter in the analysis of farming systems (Section 6.1).

For the third hierarchical production situation, apart from water and irradiance, the plant nutrients nitrogen and phosphorus may at times limit growth. Special emphasis is given to the use of nitrogen fertilizer, because of the large quantities required each year, it's cost and the mobility of nitrogen in the soil – plant – atmosphere system. The effect of nitrogen on production and the amount of nitrogen needed to achieve the production level of the second hierarchical production situation is determined by considering separately the relation between the amount of nitrogen that is taken up by the crop and the yield, and the relation between this uptake and the amount of fertilizer applied, or its recovery (Section 4.1). Phosphorus is treated in a similar fashion.

The recovery of nitrogen from fertilizer depends on the relative importance of the processes of uptake by the plants, mineralization from organic material, immobilization by soil – microbes, leaching and denitrification (formation of gaseous nitrogen). These processes are being modelled, but for the time being it is still necessary to rely also on the results of fertilizer experiments. This also holds for the determination of the amount of nitrogen that is available from natural sources. The recovery of phosphorus from fertilizer depends on such factors as the presence of soil constituents as aluminium and calcium in forms that render the phosphorus unavailable for the plants. The interaction between nitrogen and phosphorus fertilizer is treated by considering the P/N ratio in the plant tissue (Section 4.2). Whether minerals such as potassium, calcium and magnesium, are in sufficient supply and if the pH is in the proper range is most conveniently evaluated by means of soil analysis.

Subsistence farming is a concrete example of the fourth hierarchical production situation, i.e. hardly any external inputs are used. A generalized treatment may be based on the concept that under these conditions any farming system moves towards an equilibrium for the input and output of the main limiting growth factors (Section 6.2). The yield level at this equilibrium must be above a certain minimum to make the effort of farming worthwhile. Plant nutrients, especially nitrogen, ultimately limit the production in many situations. This implies that the effect of improved cultivation practices has to be judged on the basis of their temporal or permanent effects on the uptake of the limiting element. Since the possibilities to improve this situation are very restricted under conditions of low soil fertility, the effects of improved crop husbandry practices other than fertilizer application, cannot be cumulative, whether this concerns improved varieties, better cultivation or pest and disease control.

Of course, at any production level pests, diseases and weeds may interfere. Their effect is treated by making a distinction between diseases that are of special importance in high yield situations and those of special importance in low yield situations (Section 6.3). Different types of damage may be distinguished, which are evaluated in terms of damage levels.

This book is to a large extent based on the results of elaborate simulation models that have been developed in the past decade. Many of them have been described in other volumes of the Simulation Monograph series. For the present purpose these models have been simplified to such an extent that all calculations can be done with a simple scientific pocket calculator. For extended use this may be too cumbersome and for that reason a FORTRAN program is presented (Chapter 9), which allows the user to do the calculations on most home computers.

Any model, and certainly the computational schemes presented in this book are simplified representations of the complex real world. Therefore the results obtained should always be critically examined in the light of the practical experience and the results of field experiments, the more so if the crop is grown at the extremes of the range of conditions under which it is normally grown.

2 POTENTIAL CROP PRODUCTION

2.1 Physiological principles

H.D.J. van Heemst

In agriculture, solar energy is conserved for future use via its fixation in biomass by the process of photosynthesis. In this process CO_2 from the air is converted into carbohydrates $(CH_2O)_n$ according to the overall reaction:

$$CO_2 + H_2O + \text{solar energy} \rightarrow CH_2O + O_2 \quad (1)$$

This process is also called CO_2 assimilation. Part of the carbohydrates produced is used as building material for structural plant dry matter, as cellulose, proteins, lignin and fats and part is used as a source of energy for plant processes. The release of energy from carbohydrates produced during the assimilation process is described by the equation:

$$CH_2O + O_2 \rightarrow CO_2 + H_2O + \text{chemical energy} \quad (2)$$

This process is called respiration. About 40% of the weight of the carbohydrates formed during the assimilation process is lost by respiration. Subtraction of the rate of respiration from the assimilation rate gives the rate of increase in plant dry weight, i.e. the growth rate. In Figure 4, the time course of growth rate and total dry matter accumulation is in a schematic way presented for a summer wheat crop. The growth rates are obtained from the dry matter accumulation curve by determining at each point the slope of the curve.

With respect to the growth rate three phases may be distinguished: (i) during the first phase, the crop consists of individual plants that do not shade each other and the growth rate increases; (ii) in the second phase the crop covers the soil completely and the growth rate is constant; (iii) in the third phase the crop is maturing and the growth rate is decreasing..

In the first phase the major part of the assimilates is invested in leaf growth. This increase in leaf area is accompanied by a proportional increase in energy interception, because neighbouring plants are so small that mutual shading hardly plays a role. Individual plant weight increases by a constant proportion per day, thus leading to exponential growth. After a closed crop surface has been formed, more leaf growth does not lead to more light interception, hence the growth rate remains constant and total plant weight increases linearly. In the last phase leaf senescence leads to a decrease in the growth rate.

The major part of the total dry matter accumulation is achieved during the second phase. Total dry matter production of the crop is thus largely determined by the magnitude of the growth rate during the linear phase and the duration of that phase.

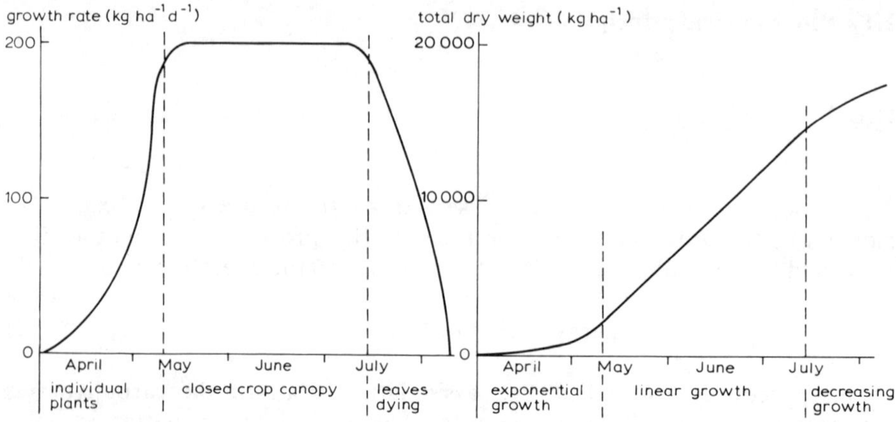

Figure 4. Schematized course of growth rate and total dry weight of summer wheat in time.

The duration of the period of linear growth is species and cultivar specific and, moreover, is influenced by environmental conditions (Section 2.2). The actual growth rate is predominantly influenced by environmental conditions, such as solar radiation and temperature, the supply of nutrients and water, and the occurrence of weeds, pests and diseases.

With an optimal supply of water and nutrients and in the absence of weeds, pests and diseases, the growth rate is determined by solar radiation and temperature and is referred to as the potential growth rate. Such conditions are supposed to prevail when discussing the basic processes of plant growth. A simple model for the calculation of potential dry matter production will be presented that may be applied to various crops at different locations.

2.1.1 CO_2 assimilation of a single leaf

In the leaves of a plant the photosynthetically active radiation is absorbed by green chlorophyl and other pigments and is used for the reduction of CO_2. Not all radiation of the sun is photosynthetically active, but only the visible radiation in the wavelength range from 400 to 700 nm, which represents about 50% of the total global radiation (Figure 5).

The rate of CO_2 assimilation of a leaf can be measured by enclosing a leaf in a so called leaf – chamber and analysing the CO_2 concentration of the incoming and the outgoing air, that passes the leaf at a known flow rate. When the assimilation rate is determined at various radiation intensities, a light response curve can be constructed as illustrated in Figure 6 for leaves of plant species referred to as C_3 and C_4 types. The main parameters characterizing these

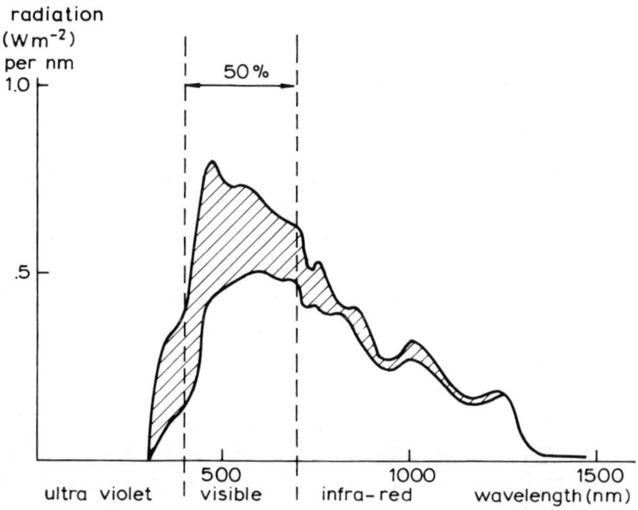

Figure 5. Spectral distribution of total solar radiation (upper curve) and direct solar radiation (lower curve). Solar elevation is 30° and precipitable water in the atmosphere is 21 mm. (Source: Monteith, 1973)

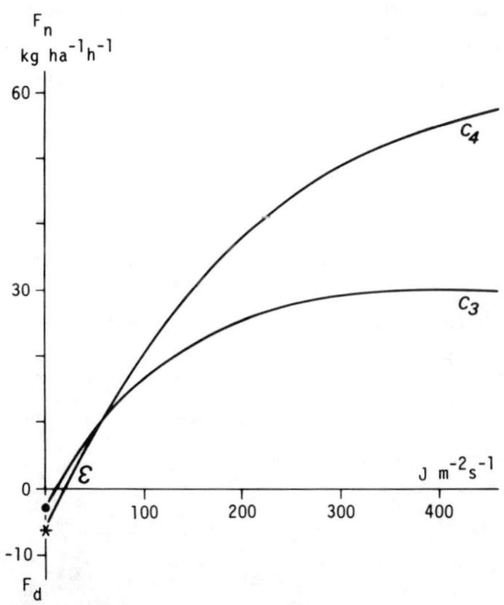

Figure 6. Characteristic net CO_2 assimilation functions for individual leaves of C_3 and C_4 plant species.

curves are the initial light use efficiency, ε, the respiration rate in the dark, R_d, and the maximum rate of net CO_2 assimilation at high light intensity, F_m. The latter ranges from 30 – 90 kg ha^{-1} (leaf) h^{-1} for C_4 type plants and from 15 – 50 kg ha^{-1} (leaf) h^{-1} for C_3 type plants, depending on environmental conditions. The gross rate of CO_2 assimilation, F_g, is the sum of the net rate and the concurrent dark respiration. The dark respiration is at normal temperatures roughly one – ninth of the maximum net assimilation rate.

The maximum net assimilation rate and the dark respiration rate are much more affected by temperature than the initial light use efficiency. The effect of temperature on the maximum assimilation rate is illustrated in Figure 7 for a C_3 and a C_4 type plant. However, these temperature responses were obtained with plants grown under controlled conditions at a temperature close to the optimum found in Figure 7. Under field conditions where plants are subjected to fluctuating temperature conditions, there appears to be adaptation of the photosynthetic apparatus. It was found that for such plants the maximum leaf assimilation rate was practically independent of temperature above about 13 °C for C_4 species and above 8 °C for C_3 species.

The difference in initial light use efficiency between the C_3 and C_4 types of photosynthesis is small, but the assimilation rate at light saturation is for the C_4 type plants generally higher. The names C_3 and C_4 refer to the length of the C skeleton of the first stable product in the photosynthetic process. Several characteristics of these two plant types are different (Gifford, 1974), such as: (i) the main carboxylating enzyme in the C_4 photosynthetic pathway has an affinity to CO_2 that is about twice as high as that in the C_3 photosynthetic pathway; (ii) in the C_3 type plants a respiratory process takes place in the light

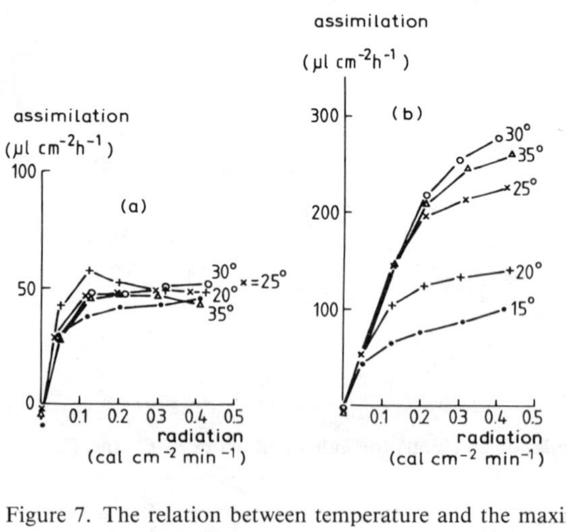

Figure 7. The relation between temperature and the maximum rate of CO_2 assimilation for a C_3 (a) and a C_4 (b) crop species.

which results in a dependence of assimilation rate on the oxygen concentration in the ambient air, whereas that process is absent in C_4 species; (iii) under conditions where the CO_2 concentration in the intercellular space is regulated over a wide range of external CO_2 concentrations and light intensities through adaptation of stomatal aperture, the level at which the internal concentration is maintained in C_4 types is about half of that in C_3 types (Raschke, 1975; Goudriaan & van Laar, 1978b). This last characteristic will be discussed in detail in Section 3.3.

Examples of species having the C_3 type of assimilation, which prevail in the temperate zones, are small grains, including rice. Species that are of the C_4 type, which are more abundant in subtropical and tropical regions, are maize, sorghum, millet, sugar cane and most tropical grasses. Extensive lists of C_4 species have been compiled by Downton (1975) and Raghavendra & Das (1978).

2.1.2 Canopy CO_2 assimilation

The rate of CO_2 assimilation of a crop depends on incoming visible radiation in the same way as that of an individual leaf. Suppose for simplicity a crop with a horizontal layer of large leaves, forming a closed surface. This layer acts as one big leaf, and knowing the light intensity, the rate of CO_2 assimilation can be read from Figure 6, taking into account that 10% of the incoming visible radiation is reflected, 10% is transmitted through the leaves, 10% is absorbed by pigments not contributing to photosynthesis, and that only the remaining 70% is absorbed by the chloroplasts. At an incoming visible radiation intensity of 300 J m^{-2} s^{-1} this crop, if it was a C_3 species, would have a CO_2 assimilation rate of about 25 kg ha^{-1} h^{-1}. Such a crop has a leaf area index (LAI) of one, because there is 1 m^2 of leaf area per m^2 of soil surface area. When another layer of such big leaves is situated under the first one, the crop has a LAI of 2, because there is 2 m^2 of leaf area per m^2 of soil surface area. The incoming radiation intensity in the second layer is equal to the light transmitted through the first layer, thus 10% of 300, or 30 J m^{-2} s^{-1}, resulting in an additional assimilation rate of about 3 kg ha^{-1} h^{-1}. The result is a small increase in assimilation rate for the two layer crop. Adding more layers under the second one will not substantially increase the assimilation rate of such a crop with layers of large horizontal leaves.

In reality, a crop does not consist of horizontal layers of large closely fitting leaves, but the leaves of a crop are spread in every direction and the light is therefore more evenly distributed over the leaves. The light extinction in a canopy can be experimentally determined by measuring the light intensity at different levels in the crop, while at the same time measuring the cumulative leaf area at the same levels. The result of such an experiment is presented in Figure 8, which depicts the relation between the relative light intensity and the cumulative LAI, counting the leaf area from the top of the canopy down-

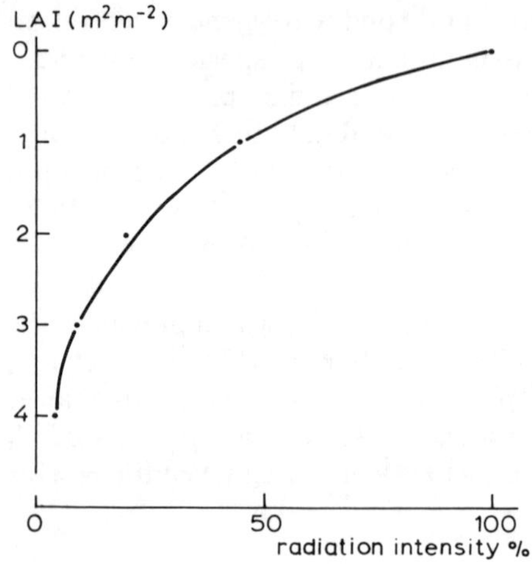

Figure 8. Extinction of radiation in a crop canopy.

wards. The extinction of the light is exponential for an increasing number of leaf layers. For any LAI the proportion of absorbed radiation can be read from Figure 8. In combination with Figure 6 this yields an estimate of the assimilation rate of the crop. For an LAI of four, the CO_2 assimilation rate is about 39 kg ha^{-1} h^{-1}, or about one and a half times that of the crop with layers of large horizontal leaves. The reason for this is that in a real crop the light intensity distribution over the leaves is more even and therefore more leaves are exposed to light intensities in the linear part of the light response curve.

The procedure just outlined is a schematized way of calculating the rate of CO_2 assimilation of a crop. Reality is more complicated, as the influence of direct and diffuse light, total leaf area, leaf angle distribution, leaf optical properties and solar height on the light distribution within the canopy have to be taken into account. The problem has been tackled with computer models (de Wit 1965; Duncan et al., 1967; Goudriaan, 1977) which calculate the assimilation rate of a canopy at any moment of a day in response to the incoming photosynthetically active radiation, which is dependent on solar height and the degree of cloudiness of the sky.

In a schematized set up, two situations are considered: a completely clear sky and a completely cloudy sky. Integration of the instantaneous rates yields the daily total amount of CO_2 fixed. In Tables 1 and 2 these daily totals are presented as a function of geographical latitude for both completely clear and completely overcast days, under the assumption of zero respiration and an LAI of five, for two maximum rates of gross CO_2 assimilation of a single leaf at high light intensity, F_g : 40 kg ha^{-1} (leaf) h^{-1}, typical for a C_3 type of plant,

Table 1. Calculated gross CO_2 assimilation rate (kg ha^{-1} d^{-1}) of a closed canopy with a spherical leaf angle distribution, for clear (F_{cl}) and overcast (F_{ov}) days, and a maximum leaf CO_2 assimilation rate, F_g, of 40 kg ha^{-1} d^{-1}.

Date		15	15	15	15	15	15	15	15	15	15	15	15
Northern Hemisph.		Jan.	Feb.	Mar.	Apr.	May	June	July	Aug.	Sep.	Oct.	Nov.	Dec.
Southern Hemisph.		July	Aug.	Sep.	Oct.	Nov.	Dec.	Jan.	Feb.	Mar.	Apr.	May	June
Latitude													
0°	F_{cl}	728	753	768	761	737	720	727	752	768	760	736	720
	F_{ov}	306	320	328	324	311	302	306	319	328	324	311	302
10°	F_{cl}	652	701	748	779	786	784	785	784	765	720	667	638
	F_{ov}	270	295	319	334	336	333	335	336	327	305	277	262
20°	F_{cl}	562	634	713	783	820	834	829	802	745	665	583	542
	F_{ov}	226	261	300	334	351	356	355	343	316	276	236	216
30°	F_{cl}	454	549	659	768	839	869	858	804	708	591	481	429
	F_{ov}	175	219	271	324	357	371	366	341	295	239	187	163
40°	F_{cl}	333	445	586	737	843	892	873	788	652	497	364	304
	F_{ov}	120	169	233	304	354	377	368	329	264	193	133	107
50°	F_{cl}	202	324	491	686	833	904	877	757	574	384	234	172
	F_{ov}	63	114	187	275	343	375	363	307	224	140	77	52
60°	F_{cl}	68	191	375	615	813	915	875	708	474	255	102	39
	F_{ov}	15	57	132	236	323	368	351	277	175	83	25	8
70°	F_{cl}	0	46	240	527	798	967	896	649	353	114	0	0
	F_{ov}	0	10	73	189	302	369	341	240	118	27	0	0

(Source: Goudriaan & Van Laar, 1978a)

and 70 kg ha^{-1} (leaf) h^{-1}, typical for a C_4 type of plant. On the basis of such tables, which for various maximum rates of CO_2 assimilation at high light intensity can be found in Goudriaan & Van Laar (1978a), potential crop assimilation can be calculated for any date, given the type of crop (C_3 or C_4), the latitude of the location and the fraction of the time the sky is clouded.

Crop type determines which table is used; given the latitude and the date, the assimilation rate of a closed canopy for a clear and an overcast day is obtained by interpolation. The assimilation rate for partially overcast days is obtained form the formula:

$$F_{gc} = f_o \cdot F_{ov} + (1 - f_o) \cdot F_{cl} \qquad (3)$$

where

F_{gc} is the gross canopy CO_2 assimilation rate (kg ha^{-1} d^{-1})

f_o is the fraction of the day the sky is overcast (f_o is 0 for completely clear days, f_o is 1 for completely overcast days)

Table 2. Calculated gross CO_2 assimilation rate (kg ha^{-1} d^{-1}) of a closed canopy with a spherical leaf angle distribution, for clear (F_{cl}) and overcast (F_{ov}) days, and a maximum leaf CO_2 assimilation rate, F_g, of 70 kg ha^{-1} d^{-1}.

Date		15	15	15	15	15	15	15	15	15	15	15	15
Northern Hemisph.		Jan.	Feb.	Mar.	Apr.	May	June	July	Aug.	Sep.	Oct.	Nov.	Dec.
Southern Hemisph.		July	Aug.	Sep.	Oct.	Nov.	Dec.	Jan.	Feb.	Mar.	Apr.	May	June
Latitude													
0°	F_{cl}	959	995	1017	1007	973	947	958	993	1018	1007	971	947
	F_{ov}	326	341	350	346	331	321	325	340	351	346	331	321
10°	F_{cl}	852	922	989	1032	1039	1035	1037	1038	1012	949	873	832
	F_{ov}	285	313	340	357	358	356	357	359	349	324	294	277
20°	F_{cl}	726	827	937	1035	1086	1103	1097	1062	983	870	755	698
	F_{ov}	237	276	319	356	375	381	379	366	336	292	248	226
30°	F_{cl}	577	707	860	1011	1109	1149	1134	1060	927	765	613	542
	F_{ov}	182	229	287	345	381	396	391	363	313	251	195	170
40°	F_{cl}	410	562	755	962	1108	1175	1150	1033	845	633	452	372
	F_{ov}	123	176	245	322	377	402	392	349	278	201	138	110
50°	F_{cl}	236	397	620	885	1086	1183	1145	982	733	477	278	198
	F_{ov}	65	117	194	289	362	398	384	324	234	145	78	53
60°	F_{cl}	71	220	460	779	1046	1182	1129	905	591	301	109	40
	F_{ov}	15	58	136	246	340	388	369	290	181	85	25	8
70°	F_{cl}	0	47	277	649	1006	1222	1132	810	421	121	0	0
	F_{ov}	0	10	74	195	314	385	356	249	120	28	0	0

(Source: Goudriaan & Van Laar, 1978a)

F_{ov} is the gross CO_2 assimilation rate on completely overcast days (kg ha^{-1} d^{-1})

F_{cl} is the gross CO_2 assimilation rate on a perfectly clear day (kg ha^{-1} d^{-1})

The fraction of the day the sky is overcast is obtained from the measured actual daily global irradiation and the daily global irradiation on a perfectly clear day, which is tabulated in Table 3.
Daily global irradiation on a completely overcast day may be approximated by multiplying the value for a perfectly clear day with 0.2. Thus:

$$f_o = (H_g - H_a)/(H_g - 0.2 \cdot H_g) \qquad (4)$$

where

H_g is total global irradiation on a perfectly clear day (J m^{-2} d^{-1})
H_a is measured total global irradiation (J m^{-2} d^{-1})

If the canopy does not form a closed cover, as at the beginning and the end of the growth cycle, not all incoming radiation is intercepted, and CO_2 assimilation is reduced relative to that of a closed canopy. The reduction is estimated from the fraction of the incoming radiation intercepted by the crop, as discussed earlier:

$$f_h = (1 - e^{-k_e \cdot LAI}) \qquad (5)$$

where

f_h is the fraction of light intercepted by the crop
k_e is the extinction coefficient for visible light, the value being between 0.5 and 0.8, depending on crop geometry

---

Exercise 1
Calculate the daily gross CO_2 assimilation for the middle of each month of the year for a completely clear and for a completely overcast sky at your own location, assuming a closed canopy, for both a C_3 and a C_4 type of crop.

Exercise 2
Repeat Exercise 1 assuming LAI = 1.5

The rate of CO_2 assimilation has been expressed sofar in amounts of CO_2. The absorbed CO_2 is reduced in the crop to carbohydrates or sugars $(CH_2O)_n$. To get an assimilation rate expressed in CH_2O, the rate in CO_2 is multiplied by 30/44 (the ratio of their molecular weights).

2.1.3 Respiration

The sugars produced in the assimilation process may be converted into structural dry matter, they may be accumulated and temporarily stored as reserves, or they may be used as a source of energy. The plant needs energy for two processes. On the one hand for maintenance of ionic gradients and resynthesis of degrading structural proteins; on the other hand for the conversion of primary photosynthetic products into structural plant material. In these processes CO_2 is produced, thus they are respiratory processes: the first one is

Table 3. Total global radiation, H_g, (10^6 J m^{-2} d^{-1}) for a standard clear day.

Date	15	15	15	15	15	15	15	15	15	15	15	15
Northern Hemisph.	Jan.	Feb.	Mar.	Apr.	May	June	July	Aug.	Sep.	Oct.	Nov.	Dec.
Southern Hemisph.	July	Aug.	Sep.	Oct.	Nov.	Dec.	Jan.	Feb.	Mar.	Apr.	May	June
Latitude												
0°	28.00	29.44	30.32	29.90	28.52	27.54	27.94	29.36	30.34	29.88	28.46	27.54
10°	24.34	26.88	29.34	30.86	30.96	30.68	30.82	31.02	30.18	27.90	25.10	23.60
20°	20.00	22.46	27.36	30.76	32.44	32.92	32.76	31.68	28.96	24.98	21.00	19.06
30°	15.18	19.30	24.42	29.62	32.90	34.24	33.74	31.28	26.74	21.34	16.34	14.10
40°	10.12	14.60	20.64	27.48	32.36	34.58	33.72	29.86	23.60	16.80	11.34	9.00
50°	5.22	9.60	16.14	24.40	30.88	34.02	32.82	27.50	19.60	11.92	6.38	4.22
60°	1.22	4.68	11.16	20.50	28.62	32.86	31.20	24.30	14.94	6.84	2.00	0.64
70°	0.00	0.76	5.96	15.98	26.12	32.18	29.70	20.56	9.78	2.20	0.00	0.00
80°	0.00	0.00	1.26	11.32	25.74	33.44	30.48	17.62	4.44	0.00	0.00	0.00
90°	0.00	0.00	0.00	9.72	26.04	33.98	30.94	17.46	0.38	0.00	0.00	0.00

(Source: Goudriaan & Van Laar, 1978a)

called maintenance respiration, the second growth respiration.

Maintenance respiration

The proteins in the plant, especially in the leaves, consist mainly of enzymes, which have only a limited life span. They deteriorate at a relative rate of about 0.1 per day at a temperature of 20 °C, and have to be resynthesized. The rate of protein turnover is temperature dependent with a Q_{10} of about 2 (Penning de Vries et al., 1979). This means that the rate of protein turnover doubles for temperature increases of 10 °C.

The concentration of ions in the vacuoles of plant cells is higher than in the surrounding tissue, which causes leakage of ions from the vacuoles. To maintain the desired internal concentration, the ions have to be taken up against a concentration gradient. That requires an active transport through cell membranes, which demands energy.

Although accurate data on maintenance requirements are scarce, reasonable estimates of the relative maintenance respiration rate can be made on the basis of the composition of the biomass present. Such estimates are given in Table 4 for four groups of crops, each group having approximately the same chemical composition.

Growth respiration

The conversion of primary photosynthates into structural plant material as cellulose, proteins, lignin and fats requires substrate for building materials and energy for synthesis of the end product, the transport of sugars and the uptake of nitrogen and minerals. Therefore, part of the sugars assimilated is respired to provide energy for the synthesis of new plant components. Another part is lost as refuse in the process of synthesis. The magnitude of growth respiration is determined by the composition of the end product formed. Thus the weight efficiency of conversion of primary photosynthates into structural plant material varies with the composition of that material. Fats and lignin are produced at high costs; structural carbohydrates and organic acids are relatively cheap. Proteins and nucleic acids form an intermediate group (Table 5).

Table 4. Relative maintenance respiration rate, R_m, at 20 °C (kg $kg^{-1}d^{-1}$), and conversion efficiency, E_g, (kg kg^{-1}).

Crop group	R_m	E_g
Root/tuber crops	0.010	0.75
cereals	0.015	0.70
protein-rich seed crops	0.025	0.65
oil-rich seed crops	0.030	0.50

Table 5. Efficiency of conversion, E_g, of substrate (sugars) into plant constituents (kg kg^{-1}).

Compound	E_g
Carbohydrates	0.826
Nitrogenous compounds (normal mix of amino-acids, proteins and nucleic acids) from NO_3^-	0.404
from NH_4^+	0.616
Organic acids	1.104
Lignin	0.465
Lipids	0.330

(Source: Penning de Vries, 1975)

For the same groups of crops distinguished above, the conversion efficiencies are tabulated in Table 4. At higher temperatures, the rate of conversion of primary photosynthates into structural plant material changes, but the conversion efficiency remains constant, because the biochemical pathway is not affected by temperature. Conversion of primary photosynthates into structural plant material occurs to a large extent at night. Low night temperatures may hamper this conversion to such an extent that not all the assimilates formed during the day can be converted into structural material. As a result, carbohydrates and starch accumulate in the plant and eventually this may affect the assimilation rate, either through a biochemical feedback or through physical damage to the chloroplasts. Under such conditions the assimilation rate is virtually determined by the capacity of the plant to convert the assimilation products.

2.1.4 Dry matter accumulation

On the basis of the processes presented in this section, the daily rate of increase in structural dry weight of a crop surface may be approximated by the formula

$$\Delta W = E_g \cdot (F_{gs} - R_m \cdot W) \quad (6)$$

where

ΔW is the rate of increase in structural dry weight (kg ha^{-1} d^{-1})
E_g is the conversion efficiency of carbohydrate into dry matter (kg kg^{-1}); see Table 4
F_{gs} is the gross rate of crop assimilation expressed in carbohydrates (kg ha^{-1} d^{-1})

R_m is the relative maintenance respiration rate (kg kg^{-1} d^{-1}); Table 4
W is the total dry weight of the live parts of the crop (kg ha^{-1})

In a temperate, humid climate e.g. in the Netherlands, the potential growth rate, as calculated by Equation 6, appears to be about 200 kg ha^{-1} d^{-1} during the growing season (Table 6). Experimental evidence confirming these estimates is given by Sibma (1968), who calculated growth curves for a number of field crops growing under near – optimal conditions, as shown in Figure 9. The main agricultural crops in the Netherlands all appear to have practically the same slope. That the C_4 type crop maize shows the same slope is because in the Netherlands it is grown at the limit of its temperature range.

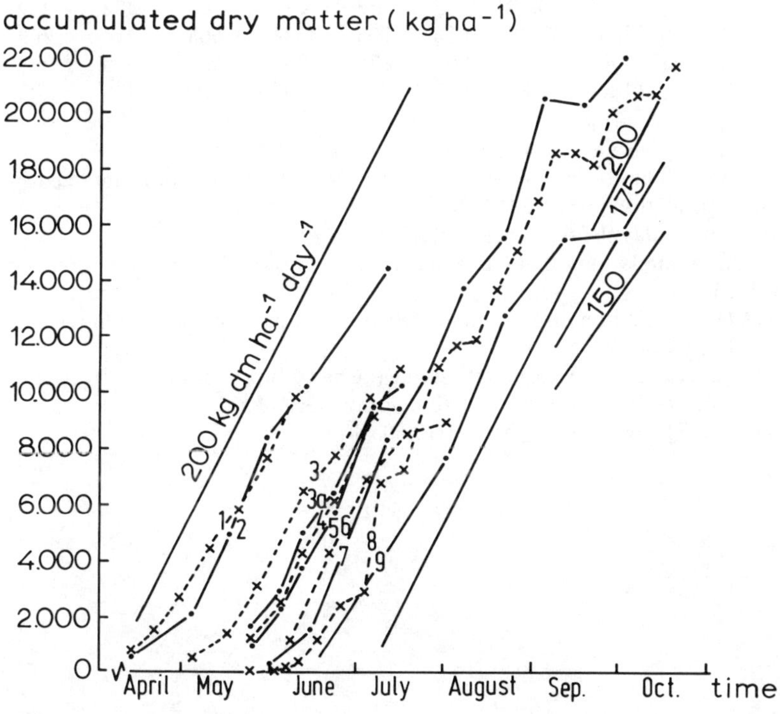

Figure 9. Growth rates of the main agricultural crops in the Netherlands under (near)-optimal growth conditions compared to growth rates of 200, 175 and 150 kg ha^{-1} d^{-}, respectively. 1. grass 2. wheat 3. oats + barley 3a. oats + peas 4. oats 5. peas 6. barley 7. potatoes 8. sugar beets 9. maize. (Source: Sibma, 1968)

Exercise 3

Calculate the potential growth rate per month of a C_3 crop for your own location, following the scheme presented in Table 6.

Estimate the fraction overcast from your own experience, if no data on radiation are available (heavy clouds: $f_o = 1$; clear skies prevailing: $f_o = 0$).

Table 6. Example of calculation scheme for the potential growth rate at De Bilt, the Netherlands (52 °N) assuming the overall loss by respiration to be 40%.

Month	H_a	H_g	f_o	F_{cl}	F_{ov}	F_{gc}	F_{gs}	ΔW
May	16.92	30.43	0.55	829	339	560	382	229
June	18.60	33.78	0.56	906	374	608	414	249
July	16.45	32.50	0.62	877	361	557	380	228
August	14.57	26.86	0.57	747	301	493	336	202

H_a = long term average actual global radiation (10^6 J m^{-2} d^{-1})
H_g = total global radiation on a clear day at 52° N.L. (10^6 J m^{-2} d^{-1}) (obtained by linear interpolation in Table 3)
f_o = fraction of the day the sky is overcast (Equation 4)
F_{cl} = gross CO_2 assimilation rate on completely clear days (kg ha^{-1} d^{-1}) (interpolation in Table 1 or 2)
F_{ov} = gross CO_2 assimilation rate on completely overcast days (kg ha^{-1} d^{-1}) (interpolation in Table 1 or 2)
F_{gc} = actual gross canopy CO_2 assimilation rate (kg ha^{-1} d^{-1}) (Equation 3)
F_{gs} = gross canopy assimilation rate in carbohydrates (30/44 x F_{gc})
ΔW = the potential growth rate (kg ha^{-1} d^{-1}) (0.60 x F_{gs})

2.2 Crop phenology and dry matter distribution

H.D.J. van Heemst

2.2.1 Introduction

Figure 4 shows the time course of above ground dry matter accumulation of a summer wheat crop during its growth cycle. With respect to the growth rate, three growth stages can be distinguished (Section 2.1). However, a crop not only accumulates weight, it also passes through successive phenological development stages: after sowing or planting, a cereal crop first forms roots, leaves and stems during the pre-anthesis phase, subsequently it flowers, and the seeds set and fill and the crop matures in the post-anthesis phase. These phenological stages are schematically illustrated for a rice crop in Figure 10.

Recognizing the distinction between growth and development, growth is defined as the increase in weight or volume of the total plant or the various plant organs, and development is defined as the passing through consecutive phenological phases; it is characterized by the order and rate of appearance of vegetative and reproductive plant organs. The two processes, growth and development, are often strongly interrelated, which is probably the reason why the term development is used often when growth is meant.

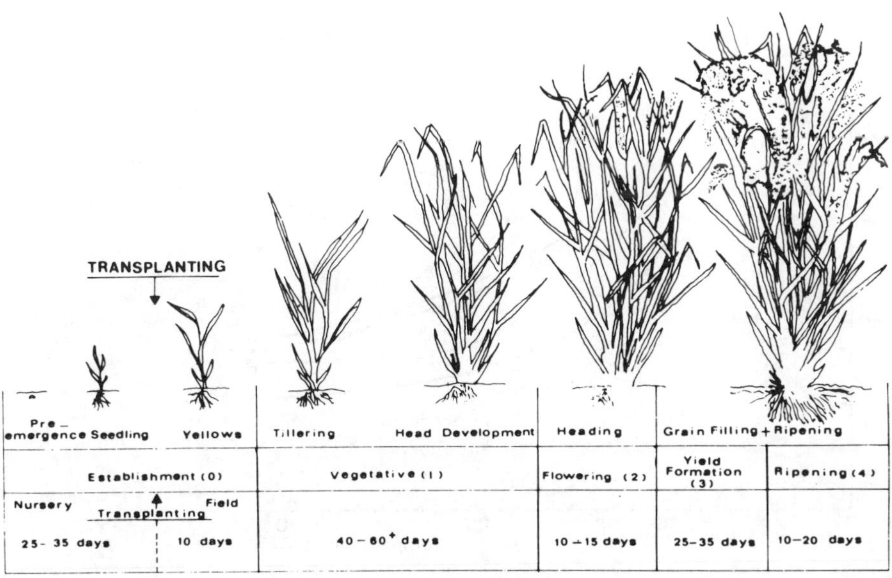

Figure 10. Developmental phases of rice. (Source: Doorenbos & Kassam, 1979)

The order of appearance of the various organs is a species characteristic, it may vary among species and is almost independent of the circumstances. The timing and rate of organ appearance, however, is dependent on environmental conditions and is, consequently, highly variable. Important events in the development of cereals are for instance, emergence, floral initiation, terminal spikelet formation, the moment of flowering (anthesis) and the beginning and end of grain filling. For tuber crops, the onset of tuber bulking is also such an event.

The major environmental conditions influencing phenological development are temperature and day length. Many plant species or cultivars need a period of low temperature to induce flowering, for example winter wheat, winter rye and sugar – beet. The process taking place during this period is called vernalization or jarowization. Summer crops in temperate climates and tropical crops do not need a period of low temperature to induce flowering. For winter crops the low temperature requirements first must be satisfied. For all crops, higher temperatures generally shorten the length of a given phenological phase. Van Dobben (1979) collected data on the length of the period from emergence to anthesis for a number of crop species, grown at various constant temperatures (Figure 11). The shape of the curves relating the number of days to anthesis to temperature suggests a constant product of days and temperature. This product is the temperature sum or the so called Thermal Unit (TU, expressed in units of day degrees). The most common method of obtaining TU values for the duration of a phenological phase is to add average daily temperatures above a threshold value . The range of threshold temperatures

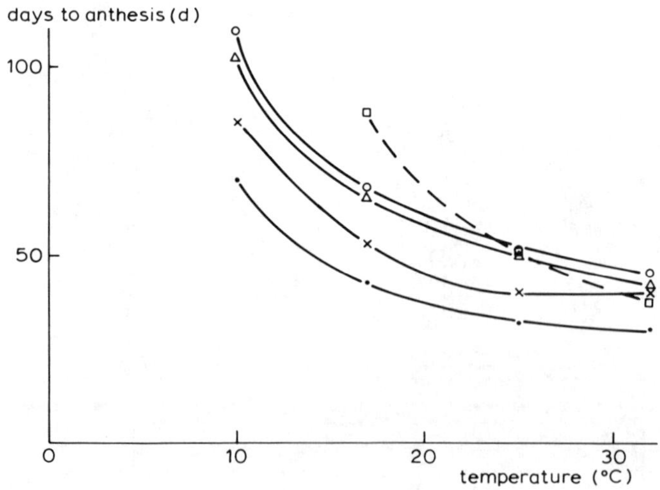

Figure 11. The influence of temperature on the length of the pre-anthesis phase for various field crops. (●) rye; (○) wheat; (Δ) flax; (□) maize; (x) peas. (Source: van Dobben, 1979)

varies between 0 and 10 °C for different species or varieties (Table 7). Sometimes an optimum temperature exists. In that case, temperatures exceeding the optimum, are replaced in the calculation by the optimum temperature itself.

Therefore the higher the temperature, the shorter the length of the total growing period of a crop or, in other words, the higher its rate of development. If development is expressed on a numerical scale, that ranges from 0 to 2, with 0 being emergence, 1 anthesis and 2 maturity, then the development rate is defined as that part of the scale that is accumulated per unit time. Generally a grain crop does not flower in the middle of its growing period. Consequently, the development rate during the pre – anthesis phase differs from the development rate during the post – anthesis phase at the same temperature. If, for example, the time lapse between emergence and anthesis for a certain crop variety in a specific environment is 50 days and between anthesis and maturity 25 days, then the average development rate during the pre – anthesis phase is 1/50 or 0.02 d^{-1}, and during the post – anthesis phase it is 1/25 or 0.04 d^{-1}. The numerical values between 0 and 2, obtained by adding the daily development rates are defined as the development stage.

Exercise 4

Transform the graphs for wheat and maize in Figure 11 into curves of development rate versus temperature. What do you notice about the curves?

For some species or cultivars the effect of temperature on development rate is modified by the influence of the length of the day, or. in fact, the length of the dark period. This effect is called photoperiodism. With regard to this mechanism, plants may be classified into three groups: (i) day – neutral plants, for which development rate is insensitive to day length; (ii) long – day

Table 7. Indicative threshold values, T_o (°C) used for the calculation of TU values for different species.

Crop	T_o
Maize	10
Soybean	10
Sorghum	7-10
Pea	4
Chickpea	4
Wheat	0-5
Rice	0-10

plants, for which anthesis is induced by the occurrence of long days (and therefore short nights); (iii) short – day plants, for which anthesis is induced by the occurrence of short days (and therefore long nights). The reaction to day length may be an important characteristic when a new species or cultivar is introduced in a region, even when it originates from a region at about the same latitude and – consequently – the same photoperiod. The reason is that the growing seasons at the two locations may not coincide due to differences in rainfall pattern. The effects of day length are not treated quantitatively here, because we assume that in each region species with the proper day – length reaction are cultivated.

Although the basic processes governing phenological development and biomass production act independently, both phenomena are strongly interrelated. If the rate of development is high, total biomass production will be low, because the period of linear growth will be short (Section 2.1). Moreover, crops are generally not grown for total biomass, but for their storage organs, such as tubers, grains or pods. These storage organs grow only during the latter part of the growth cycle, after roots, leaves and stems have been produced. A short growing period, resulting in a low vegetative biomass, especially of leaves responsible for light interception, leads inevitably to a poor crop. On the other hand, too much biomass invested in vegetative organs may lead to a relatively low production of storage organs, because in that case the maintenance requirements are high. Therefore, not only total biomass production is of interest, but also its distribution over the various plant parts. The actual proportion of leaves, stems, roots and storage organs in the total biomass at a certain moment depends on the preceding growth rates, which are governed by the weather and the leaf area index in the past, and the partitioning of that dry matter increase over the various plant parts. A fixed distribution pattern, for instance partitioning factors defined as a function of development stage, does not necessarily lead to a constant ratio of various organs. A simple example will illustrate this.

Suppose there is at a certain moment a crop in the field, that comprises 1000 kg ha^{-1} leaves and 400 kg ha^{-1} stems. During the following 10 days, the average growth rate is 200 kg ha^{-1} d^{-1}; the partitioning factors for leaf and stem being 0.6 and 0.4, respectively. In the subsequent 10 day period, the growth rate is only 100 kg ha^{-1} d^{-1} because of a much lower energy availability, and the partitioning factors have changed to 0.3 and 0.7 for leaf and stem, respectively. At the end of the second period the weight of leaves is $1000 + 0.6 \times 200 \times 10 + 0.3 \times 100 \times 10 = 2500$ kg ha^{-1} and the weight of stems is $400 + 0.4 \times 200 \times 10 + 0.7 \times 100 \times 10 = 1900$ kg ha^{-1}. This results in a leaf – stem ratio of 1.32. Now suppose the growth rate in the first period to be 100 kg ha^{-1} d^{-1} and in the second period 200 kg ha^{-1} d^{-1}. Assume too, an identical distribution pattern. Then the weight of leaves at the end of the second period is $1000 + 0.6 \times 100 \times 10 + 0.3 \times 200 \times 10 = 2200$ kg ha^{-1} and the weight of stems $400 + 0.4 \times 100 \times 10 + 0.7 \times 200 \times 10 = 2200$ kg ha^{-1}.

Then the leaf – stem ratio is 1.00.

In this example the development pattern and the time course of partitioning factors were assumed to be identical. However, development is not identical each year, as it responds to differences in environmental conditions. It is therefore not possible to relate the distribution pattern to crop age. Usually the partitioning of the current assimilate supply over the various plant parts is expressed in a distribution pattern in dependence of the development stage of the crop. Such configurations are characteristic for each crop. The effects of environmental conditions other than temperature on the distribution pattern are often very small, especially in the potential production situation; they are therefore not taken into consideration here.

In the next part of this section the partitioning of newly formed dry matter over the various plant organs is treated in relation to the development stage of the crop. Examples will be given for the crops rice, maize and cassava.

2.2.2 Development and dry matter distribution in rice

The total growing period of rice from transplanting to maturity generally varies from 90 to 150 days, depending on the variety, temperature and sensitivity to day length (Figure 10). Short duration varieties are in general day – neutral, long duration varieties are short – day plants.

Rice is a species with a terminal inflorescence, and such determinate species show a fixed development pattern: after anthesis leaf, stem and root formation stops and the only organ increasing in weight is the ear. The relation between development rate and temperature is a variety – specific characteristic, which needs to be established experimentally (Section 5.4). As an example, the TU values for two cultivars, the medium – duration breeding line B9C/Md/3/3 and the short duration cultivar IR5, for the pre – anthesis period from transplanting till anthesis are 2700 and 2150 d °C. respectively, taking into account a threshold temperature of 0 °C (van Keulen, 1976). The duration of those periods at 20 °C can be calculated as 2700/20, or 135 days, and 2150/20 or 108 days, respectively. The associated development rates (the inverse of the duration in days) at this temperature are 1/135, or 0.0074 and 1/108, or 0.0093 d^{-1}, respectively. The TU value for the post – anthesis period appeared to be equal for both cultivars, 650 d °C.

Exercise 5

Calculate the duration of the post – anthesis period and the development rate for both cultivars at 20 and 25 °C, respectively.

The similarity of the TU values for the post – anthesis phase for both cultivars is not a coincidence. In wheat too, differences in growth duration between cultivars grown under identical environmental conditions are mainly due to differences in the length of the period from emergence to anthesis (Nuttonson, 1955, 1953).

For actual field situations, the development rate may be deduced from the dates of emergence, anthesis and maturity, and air temperature data from the nearest meteorological station, which should not be too different from the experimental site.

Exercise 6

In Chiang Mai, a province in Northern Thailand, the upland rice breeding line Khichang x RDI was planted at two locations: Phabujom (800 m altitude, average maximum temperature 27 °C, average minimum temperature 20 °C); and at Chang Khian, (altitude 1200 m, average maximum temperature 24 °C, average minimum temperature 17 °C).

Emergence at Phabujom and Chang Khian occurred on 8 June and 5 June, anthesis on 16 September and 1 October, and maturity on 13 October and 2 November, respectively.

Calculate the average development rates for the pre – and post – anthesis phases at both locations, and the TU values for the two phases. A threshold temperature of 0 °C may be assumed.

Construct the relation between average temperature and development rate for this cultivar.

The fraction of the total dry matter increase apportioned to root and shoot, respectively, as a function of the development stage of a rice crop for the period between transplanting (development stage 0) and anthesis (development stage 1) is given in Figure 12a. The partitioning of shoot dry matter increase between leaf blades and 'stems', as a function of the development stage of a rice crop for the period between transplanting and anthesis is given in Figure 12b. It is assumed that after anthesis all dry matter increase benefits the reproductive organs.

Thus, if the growth rate is 200 kg ha^{-1} d^{-1} at development stage 0.5, a fraction of 0.92 of the total dry matter increase, or 184 kg ha^{-1} d^{-1} is apportioned to the shoots. Of that portion, a fraction of 0.50, or 92 kg ha^{-1} d^{-1} is apportioned to the leaf blades and 92 kg ha^{-1} d^{-1} to the 'stems'.

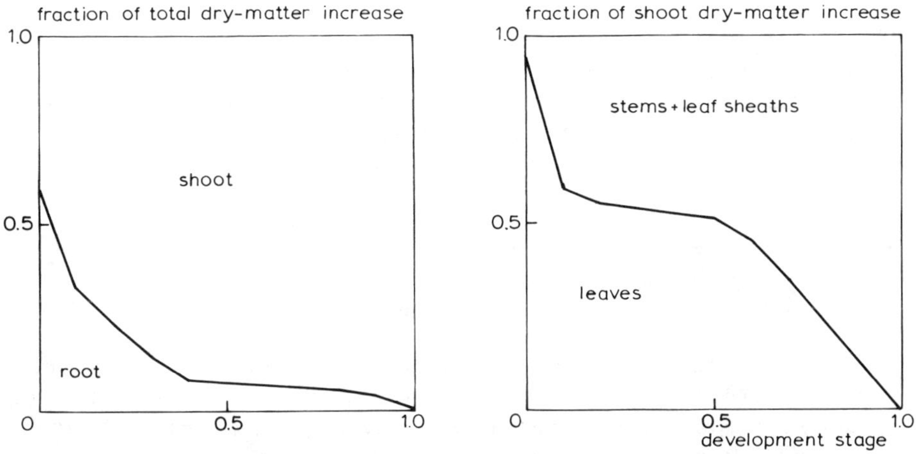

Figure 12. Partitioning factors for plant parts of rice in the course of development.

Exercise 7

Two upland rice varieties are grown at the same location. They are harvested periodically at ten – day intervals. The dry weight of the above ground biomass is given below for the two varieties A and B.

Variety A		Variety B	
harvest day	biomass, kg ha^{-1}	harvest day	biomass, kg ha^{-1}
0	50	0	50
10	100	10	100
20	200	20	200
30	400	30	400
40	800	40	800
50	1400	50	1400
60	2200	60	2200
70	3300	70	3300
80	4600	80	4600
90	6300	90	6300
100	8400	100	8100
110	10500	110	9500
120	11900		
125	12500		

Day 0 is emergence; anthesis for variety A is on day 95 and for variety B on day 80 after emergence.
Calculate for both varieties the dry weight of leaves, stems and grains at the various harvest times. Use for the calculation the partitioning factors given in Figure 12. Assume a constant temperature regime.
(Note: The partitioning factors in Figure 12 are functions of development stage, not age)

2.2.3 Development and dry matter distribution in maize

For maize, the total growing period from emergence to maturity varies from 80 to 110 days for short duration varieties, and from 110 to 140 days for medium duration varieties, when average daily temperatures are above 20 °C. Under cooler conditions, maize is mostly grown as a forage crop because the associated extended length of the growing period does not permit timely maturation, due to the low temperatures, especially towards the end of the growing season.

Maize is considered to be either day – neutral or a short – day plant. The flower has separate male and female parts. The male flowers are in the tassel at the top of the plant; the female flowers are in cobs at nodes along the middle of the stem (Figure 13). The pre – anthesis period ends at silking. The

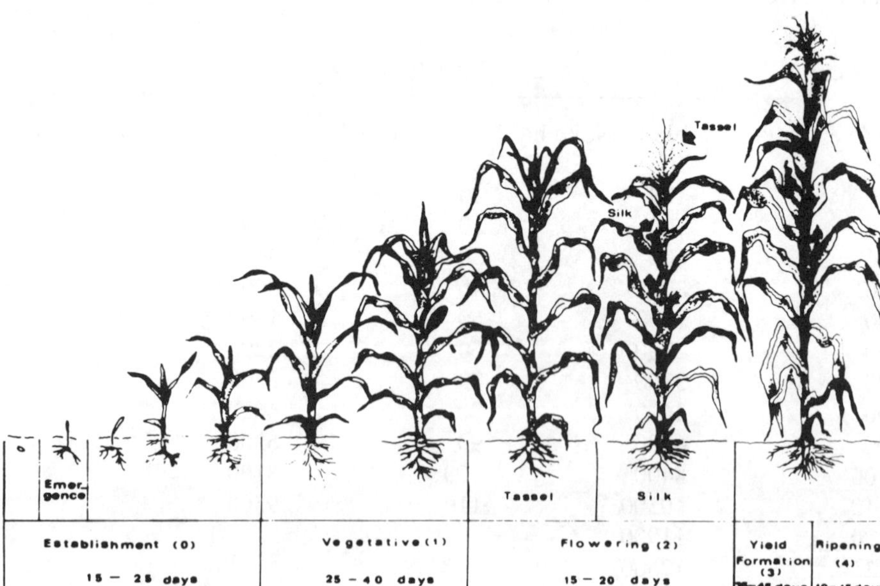

Figure 13. Developmental phases for maize. (Source: Doorenbos & Kassam, 1979)

duration of the interval between emergence and silking is affected by both genetic factors and environmental conditions. For most common cultivars, the time from silking to maturity under normal environmental conditions is identical, an average 50 – 55 days.

The most frequently used method for determining the temperature requirement from emergence to silking is a direct summation of average daily temperatures, taking into account a base temperature of 10 °C. Mederski et al. (1973) established TU values of 625, 640, and 755 d °C for the period between emergence and silking, and 650, 655, and 635 d °C for the post – silking period for the varieties Ohio 401, DeKalb XL45 and Pioneer 3306, respectively. The latter values may vary with the definition of maturity. In this case, it is defined as the moment that a black layer develops at the base of the kernel, which marks the end of the period of effective grain filling. Similar to rice, the TU value for the post – silking period is almost identical for the three varieties at 650 d °C. The average duration of that period at 20 °C is 650/(20 – 10) = 65 days; at 25 °C it becomes 650/(25 – 10) = 43 days (the base temperature is 10 °C!). The corresponding development rates are 1/65, or 0.0154 d^{-1}, and 1/43, or 0.0233 d^{-1}, respectively.

Exercise 8

A maize variety is grown at a location (air temperature given in Table 8), for which a TU value of 760 d °C has been established for the period between emergence and silking, and one of 660 d °C between silking and maturity. For both periods a threshold value of 10 °C may be taken into account. The crop emerged 1 June.

Calculate the dates on which development stages 0.2, 0.4, 0.6, 0.8, 1.0, 1.5, and 2.0 are reached.

What are these dates if emergence takes place on 15 June.

Calculate the average development rates for the pre – silking and post – silking periods for both emergence dates.

Exercise 9

A maize crop is harvested periodically at ten – day intervals. The harvested plants are separated into leaves and stems. The dry weights of the plant parts are given below:

Date	leaf weight ($kg\ ha^{-1}$)	stem weight ($kg\ ha^{-1}$)
10 June	200	0
20 June	1400	300

30 June	2800	800
10 July	4500	2000
20 July	5700	3800

Emergence date is 1 June; silking date is 5 August.
Draw a graph of the fraction of the weight increment allocated to the leaves as a function of the development stage of the crop. Assume a constant temperature regime.

Table 8. Average air temperatures (°C) to be used in Exercise 8.

	Month			
Date	June	July	August	September
1	16	20	24	33
2	15	22	26	30
3	14	21	25	30
4	16	19	26	29
5	18	18	27	28
6	17	18	24	30
7	16	20	25	29
8	13	19	24	28
9	10	21	26	29
10	9	21	27	28
11	12	22	28	27
12	16	23	30	30
13	20	30	31	26
14	23	20	28	25
15	22	24	31	26
16	19	24	30	25
17	18	23	29	24
18	18	22	28	27
19	16	23	31	24
20	20	20	32	25
21	22	22	31	24
22	21	23	30	22
23	24	24	29	23
24	21	23	29	22
25	20	24	30	23
26	19	26	26	22
27	20	25	33	24
28	19	26	28	25
29	23	25	29	24
30	18	24	26	23
31		25	33	

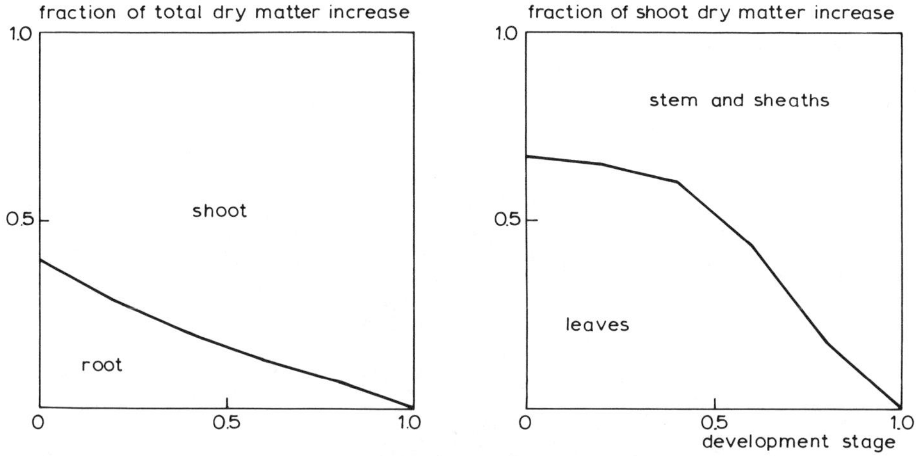

Figure 14. Partitioning factors for plant parts of maize in the course of development.

The distribution pattern for maize (Figure 14) is constructed from experimental data in the same way as is done for a part of the pre – silking period in Exercise 9.

2.2.4 Development and dry matter distribution in cassava

Cassava is a perennial plant that, for agricultural production, is propagated almost exclusively through stem cuttings. Many cultivars exist, their development pattern varying according to ambient climatic conditions. Short – duration cultivars are harvested 6 – 12 months after planting; long duration cultivars are often left in the field for periods of two years or more. Unlike rice and maize, the economic product is not the grain, but the storage root, which consists predominantly of starch. If environmental conditions are conducive for initiation storage roots are initiated about eight weeks after planting, which is mainly under photoperiodic control. Under short – day conditions, storage root initiation occurs readily, but it is delayed when day length exceeds 10 – 12 h, and consequently yields are lower. For this reason, cassava is most productive between 15° N and 15° S latitudes.

In the development of cassava, three phases may be distinguished:

– establishment,

– early growth and foliage formation,

– simultaneous formation of foliage and starch accumulation in the storage roots. The last phase may continue for more than a year. In Figure 15, dry – matter accumulation for various plant parts of cassava is shown for a period of two years. The graph shows that starch accumulation in the roots starts some four months after planting, that leaf weight shows a seasonal

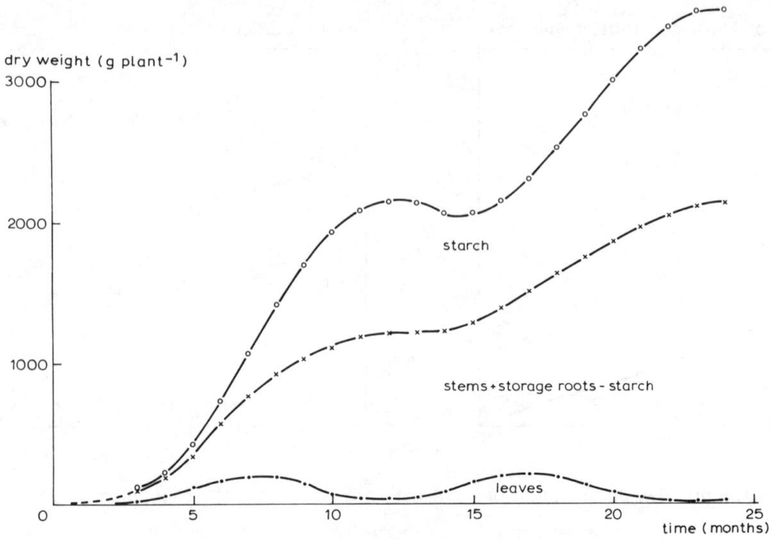

Figure 15. Dry weight of various plant parts of cassava in the course of time. (Source: Cours, 1948)

pattern of increase and decrease, and that this pattern is reflected in the dry matter accumulation in the other plant parts.

Boerboom's (1978) analysis of cassava production shows that the distribution of dry matter over the storage roots and stem or shoot is invariable with

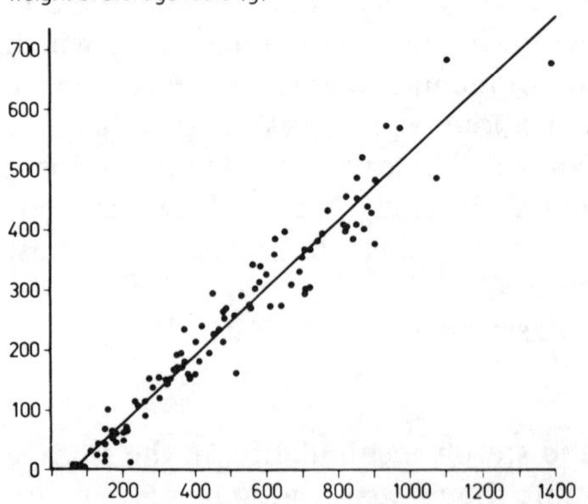

Figure 16. Relationship between dry weight of whole plant and dry weight of storage roots for cassava plants. (Source: Boerboom, 1978)

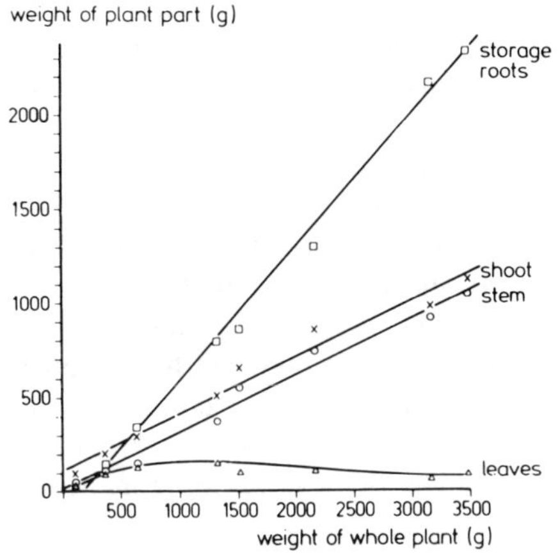

Figure 17. Relationships between dry weight of whole plant and dry weight of plant parts for cassava cv. SPP. (Source: Boerboom, 1978)

time for an extended period (Figures 16 and 17). The main parameters describing the distribution pattern over the different plant parts are: the total plant weight at which storage root production starts, represented by the intercept with the x axis and the efficiency of storage – root production, represented by the slope of the regression lines in Figures 16 and 17.

Experimental data were used to determine the partitioning factors in the same way as for rice (Figure 18). For the present procedure, the growth period

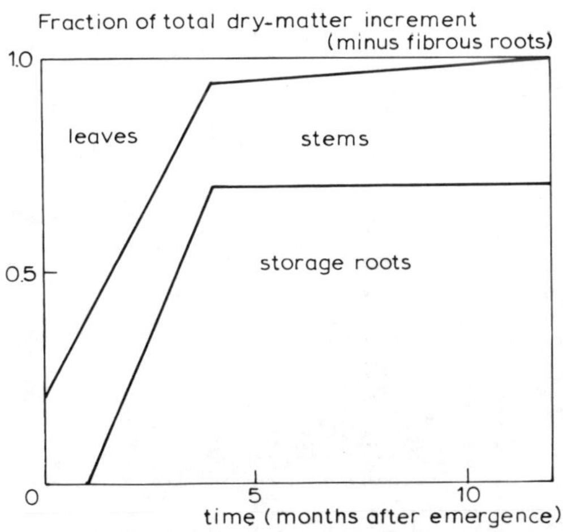

Figure 18. Partitioning factors for plant parts of cassava in the course of time.

is restricted to one year, because it is impossible to determine the partitioning factors for longer growth periods, as foliage formation in the second year mainly occurs at the expense of existing reserves and not from current dry matter production (Figure 15).

The independent variable in Figure 18 is not the development stage, as for rice and maize, but time after emergence. The experiments to which this graph refers were carried out at Buitenzorg, (Java, Indonesia) having a photoperiod of about 12 hours, i.e. under optimum conditions of day length. This description was adopted because it appeared impossible to determine temperature requirements for reaching the various development stages, or to establish development rates from existing experimental data. To achieve consistency in data handling, a constant development rate over the year could be assumed, giving an x axis with a scale from 0 to 1, i.e. adopting a development rate of 1/365, or 0.00274 d^{-1}.

2.3 A simple model of potential crop production

H. van Keulen

2.3.1 Introduction

In Section 2.1, potential crop production was defined as the total dry matter production of a green crop surface that, during its entire growth period, is optimally supplied with water and all essential nutrient elements, and grows without interference from weeds, pests and diseases. From this concept, a step may be made to the estimation of potential yield, i.e. the production of economically useful plant parts, by taking into account the phenological development of a particular crop species or cultivar, and the associated partitioning of dry matter over various organs of the plant, as outlined in Section 2.2. In this Section a scheme is presented to calculate both total dry matter production and economic yield for a number of crops, based on radiation and temperature regime, only.

The principle of the procedure is that repetitive calculations are performed, starting at some point in time at which the state of the crop can be described in quantitative terms, either determined from experimental data or estimated from other known relations. For most crops a suitable point in time is emergence, which is defined as the moment of transition from growth of the seedling from the reserves in the seeds to growth originating from carbohydrates formed in the process of assimilation. Transplanted rice is a special case, because the seedlings, growing on a nursery bed, are uprooted after some time and replanted on the site where they will eventually mature. The moment of transplanting is then a better starting point.

The state of the crop at the start of the calculations is characterized by measurable quantities, e.g. the weight of the aerial plant parts, the weight of the roots and the green leaf area, active in the assimilation process. From this state and the environmental conditions in the following period the rates of the relevant processes, such as assimilation and respiration, are calculated. These basic processes govern the rates of change of the various quantities that can thus be calculated. Realization of these rates over the relevant time interval and addition to the quantities present at the beginning of the period yield the magnitude of the quantities at the end of the period. Or, in mathematical notation:

$$Q_{t+\Delta t} = Q_t + R_q \cdot \Delta t \qquad (7)$$

where

$Q_{t+\Delta t}$ is quantity at time $t + \Delta t$
Q_t is quantity at time t
R_q is the rate of change of quantity Q during time interval Δt
Δt is time interval between the beginning of the period and the end of the period.

Exercise 10
If the unit of Q is kg ha^{-1}, and the unit of Δt is days, what is the unit of R_q? Suppose $Q_t = 200$, $R_q = 15$ and $\Delta t = 10$, what will $Q_{t+\Delta t}$ be?

The calculations are then repeated for the next time interval, and so on, until the end of the growth period of the crop. In this way the growth curve, i.e. the cumulative dry matter production (Section 2.1) is obtained. By partitioning the dry matter produced during each time interval according to the coefficients given in Section 2.2, the weight of the various organs can be calculated. The partitioning coefficients are a function of the development stage of the vegetation and that 'quantity' must therefore also be calculated. This may be done in the way suggested in Section 2.2, by adding the average air temperatures in the course of the growing period and dividing the accumulated temperature sum at any moment by the sum required for the completion of a certain phenological phase. The ratio obtained is the required quantity, which is defined as the development stage.

The approach followed assumes that the rates of change calculated at the beginning of a time interval do not change during that interval. This assumption puts a restriction on the length of the time interval applied. In theory, infinitely small time intervals would have to be applied, because realization of a rate of change over even a small interval results in different values for the quantities and this would thus lead to a different rate of change for the next small time interval. That would, however, hardly be possible from a practical point of view. Moreover the deviations are often within reasonable limits, even if the time interval has a finite size. In our approach we have chosen a period of ten days which, on the one hand, permits calculations for an entire growth period of some hundreds of days to be performed in a reasonable time on a pocket calculator, and, on the other hand, yields acceptable results for the purpose pursued here.

The principles of the calculation procedure outlined sofar are those underlying the state variable approach in systems analysis and modelling. This approach will not be further elaborated upon in this volume; for descriptions of this approach see de Wit & Goudriaan (1978) and Penning de Vries & van Laar (1982).

Exercise 11
In the biological sciences, often the growth rate – i.e. the rate of increase of a quantity – is proportional to the quantity present, thus:

$$R_q = a \cdot Q_t$$

Calculate the time course of the quantity Q for a thirty day period when Q_0 = 5 kg ha^{-1} (Q_0 is the quantity at time zero) and the value of the proportionality factor a = 0.1 d^{-1}; use for Δt a value of 5 days.
Repeat the calculation for a value of Δt = 3 days. Compare the results. What do you notice? Explain the difference.

2.3.2 *An actual example*

This example concerns an experiment with the rice variety IR8, one of the so-called high yielding varieties (HYV) developed at the International Rice Research Institute. The experiment was carried out in Paramaribo, Suriname, South America (5°49' N, 55°09' W). The rice was transplanted on 10 November 1972 (Van Slobbe, 1973). The air temperatures used in the calculations were obtained from reported ten-day averages for the experimental period. Radiation was calculated from monthly averages of sunshine duration reported (Section 3.1). These data were used to calculate potential gross CO_2 assimilation (Section 2.1), which is given in Table 9.

Table 9. Potential daily gross assimilation expressed in CH_2O for Paramaribo, Suriname.

Date		F_{gs} (kg ha^{-1} d^{-1})
Jan.	15	332
Feb.	15	344
March	15	368
April	15	364
May	15	354
June	15	378
July	15	417
Aug.	15	454
Sept.	15	
Oct.	15	
Nov.	15	336
Dec.	15	283

Table 10. Calculation scheme for potential production.

Column no.	1	2	3	4	5	6	7	8	9	10	11	12
Period	T_a	TSUM	DVS	GASS	MRES	ASAG	DMI	FR	IWRT	WRT	FL	IWLV
										40		
1 Nov. 2	27.2	272	0.18	60.5	2.1	58.4	40.9	0.35	14.3	183	0.395	16.2
2 Nov. 3	26.3	535	0.36	127.2	8.2	119.0	83.3	0.165	13.7	320	0.445	37.1
3 Dec. 1	25.8	793	0.53	216.0	20.7	195.3	136.7	0.075	10.3	423	0.48	65.6
4 Dec. 2	26.4	1057	0.70	260.4	41.2	219.2	153.4	0.07	10.7	530	0.40	61.4
5 Dec. 3	26.3	1320	0.88	295.0	64.2	230.8	161.6	0.07	11.3	643	0.265	42.8
6a Jan. 1-7	26.0	1502	1.0	316.0	88.5	227.5	159.3	0.025	4.0	671	0.06	9.6
6b Jan. 8-10	26.0	1580	1.10	316.0	71.2	244.8	195.8	0	0	671	0	−48.0
7 Jan. 2	26.0	1840	1.43	332.0	75.7	256.3	205.0	0	0	671	0	−45.1
8 Jan. 3	26.0	2100	1.75	336.0	91.7	244.3	195.4	0	0	671	0	−36.1
9 Feb. 1-8	26.0	2308	2.0	319.6	107.6	212.0	169.6	0	0	671	0	−28.8

Table 10. (continued)

Column no. Period	13 WLV	14 FS	15 IWST	16 WST	17 FG	18 IWGR	19 WGR	20 LAI	21 TADW	22 TDW	23 TDWL	24 Line no.
1 Nov. 2	100			0				0.25	100	140	140	a
2 Nov. 3	262	0.255	10.4	104	0	0	0	0.65	366	549	549	b
3 Dec. 1	633	0.39	32.5	429	0	0	0	1.6	1062	1382	1382	c
4 Dec. 2	1289	0.445	60.8	1037	0	0	0	3.2	2326	2749	2749	d
5 Dec. 3	1903	0.53	81.3	1850	0	0	0	4.75	3753	4283	4283	e
6a Jan. 1-7	2331	0.665	107.6	2925	0	0	0	5.8	5256	5899	5899	f
6b Jan. 8-10	2398	0.225	35.8	3176	0.69	125.6	879	6.0	6453	7124	7124	g1
7 Jan. 2	2254	0	0	3176	1.0	195.8	1466	5.6	7040	7711	7567	g2
8 Jan. 3	1803	0	0	3176	1.0	205.0	3516	4.5	9090	9761	9166	h
9 Feb. 1-8	1442	0	0	3176	1.0	195.4	5470	3.6	11044	11715	10757	i
	1212	0	0	3176	1.0	169.6	6827	3.0	12402	13073	11886	j

The calculation procedure is illustrated in Table 10; the letters in the following text refer to the lines in Table 10, which are indicated in the last column (No. 24) of the table.

Line a specifies the situation at time zero. Because it concerns transplanted rice, the day of transplanting is chosen. Root dry matter after transplanting equals 40 kg ha^{-1} (Column 10); above ground 100 kg ha^{-1} is present, which consists entirely of leaf blades (Column 13). The green area, intercepting solar energy for the assimilation process, is calculated from the weight of the leaf blades, assuming a constant ratio between the area and the weight of leaf blades. This ratio is called the specific leaf area, expressed in square meters of green area per kg of dry matter of leaf blades. For rice, its value is 25, thus the area is 2500 m^2 ha^{-1}. From this, the leaf area index (LAI), i.e. the ratio of leaf area to soil surface area, is calculated. Since one hectare is 10 000 m^2, the leaf area index at transplanting time equals 0.25 (Column 20).

Line b describes the first full ten days of the growing period. The average daily air temperature during that period is 27.2 °C (Column 1), which when integrated over the period yields a temperature sum of 272 d °C (Column 2). As explained in Section 2.2, the accumulated temperature sum is a measure of the phenological development stage of the crop. For the variety IR8 the required temperature sum for anthesis is 1500 d °C, assuming a base temperature of 0 °C. The development stage (Column 3) is calculated as the ratio of the temperature sum accumulated and the value of 1500 d °C, hence 272/1500 = 0.18.

In Table 9, potential daily gross assimilation, F_{gs}, expressed in CH_2O is given for the middle of each month of the year in kg ha^{-1} d^{-1}. The value for any ten – day period is obtained by interpolation between the values given in Table 9. For the second ten – day period of November that value is found directly from the table: 336 kg $ha^{-1}d^{-1}$. This represents potential gross assimilation, i.e. that realized by a closed green canopy, which intercepts all incoming energy. For a leaf area index of 0.25, only part of the solar energy is

Table 11. Reduction factor for gross assimilation due to incomplete light interception.

LAI	Reduction factor
0.	0.
0.25	0.18
0.5	0.33
1.0	0.55
1.5	0.70
2.0	0.80
2.5	0.86
3.0	0.91
3.5	0.94
4.0	0.96
4.5	0.98
5.0	1.0

intercepted (Section 2.1), hence the potential as dictated by the environment is not realized. The reduction factor for various values of LAI is given in Table 11, calculated from Equation 5, in Section 2.1. For LAI = 0.25, the reduction factor equals 0.18. In Column 4, the rate of gross assimilation is introduced

$$\text{GASS} = 336 \times 0.18 = 60.5 \text{ kg ha}^{-1} \text{ d}^{-1}$$

As explained in Section 2.1, part of the energy fixed in the assimilatory process is respired by the crop to maintain the existing structures. For the vegetative material of a rice plant, the relative maintenance respiration rate is assumed to be 0.015 kg CH_2O per kg dry matter per day during the pre–anthesis phase when, especially in the potential production situation, the nitrogen content of the material is relatively high. Hence, the rate of maintenance respiration expressed in CH_2O is obtained by multiplying the total live dry matter present (Column 23) by the relative maintenance respiration rate, R_m:

$$\text{MRES} = 140 \times 0.015 = 2.1 \text{ kg ha}^{-1} \text{ d}^{-1} \text{ (Column 5)}$$

The amount of assimilation products available for increase in dry weight of the crop equals the difference between gross assimilation and maintenance respiration. Thus:

$$\text{ASAG} = 60.5 - 2.1 = 58.4 \text{ kg ha}^{-1} \text{ d}^{-1} \text{ (Column 6)}$$

The conversion of primary assimilation products into structural plant material again entails loss of energy. In the present approach, this growth respiration is represented by its complement, the conversion efficiency, E_g, (Section 2.1). This means that the dry weight increment is equal to the conversion efficiency times the available assimilation products. For vegetative material of average composition E_g equals 0.7. Thus:

$$\text{DMI} = 0.7 \times 58.4 = 40.9 \text{ kg ha}^{-1} \text{ d}^{-1} \text{ (Column 7)}$$

The total increase in dry matter is utilized concurrently for the growth of various plant parts. In the early stages there is growth of roots, leaf blades and leaf sheaths and stems. The fraction of the increment partitioned to each of the organs is, under potential growth conditions, primarily determined by the phenological state of the crop (Section 2.2). In Table 12, the fractions allocated to each of the organs are given as a function of development stage.

The instantaneous values of the partitioning factors for roots, leaf blades and stems plus leaf sheaths are read from this table through interpolation. The independent variable, i.e. the development stage, is taken as the value halfway between the beginning and the end of the ten–day period. For this period therefore, (0 + 0.18)/2 = 0.09.

Table 12. Partitioning factors for dry matter to various plant organs as a function of development stage

Development stage	f_r	f_l	f_s	f_g
0	0.6	0.375	0.025	0
0.1	0.325	0.40	0.275	0
0.2	0.225	0.425	0.35	0
0.3	0.14	0.46	0.4	0
0.4	0.075	0.485	0.44	0
0.5	0.075	0.475	0.45	0
0.6	0.07	0.42	0.51	0
0.7	0.07	0.32	0.61	0
0.8	0.055	0.21	0.735	0
0.9	0.04	0.1	0.36	0.5
1.0	0.	0.	0.	1.0
2.0	0.	0.	0.	1.0

In Column 8, the fraction allocated to the root is introduced, which is equal to 0.35. Thus the rate of increase in root dry weight is:

$$\text{IWRT} = 0.35 \times 40.9 = 14.3 \text{ kg ha}^{-1} \text{d}^{-1} \text{ (Column 9)}$$

The weight of the root system at the end of the period is obtained by adding the rate of increase from Column 9 multiplied by the length of the time interval to the weight at the end of the preceding ten – day period (Line a, Column 10). Thus:

$$\text{WRT} = 40 + 14.3 \times 10 = 183 \text{ kg ha}^{-1} \text{ (Column 10)}$$

The fraction of the dry weight increment allocated to the leaf blades is again obtained from Table 12 at development stage 0.09, which equals 0.395. The rate of increase in dry weight of the leaf blades is calculated as:

$$\text{IWLV} = 0.395 \times 40.9 = 16.2 \text{ kg ha}^{-1} \text{d}^{-1} \text{ (Column 12)}$$

The weight of the leaf blades at the end of the ten – day period is obtained by adding the rate of increase times Δt to the value at the beginning of the ten – day period (Line a, Column 13). Hence,

$$\text{WLV} = 100 + 16.2 \times 10 = 262 \text{ kg ha}^{-1} \text{ (Column 13)}$$

The remainder of the above ground vegetative material is designated 'stems' in the present approach. It consists not only of the true stems, but contains also the leaf sheaths and the ear structures other than the seed. For the present ten – day period the fraction of the increment allocated to the stem is obtained from Table 12 at development stage 0.09, which equals 0.255. The rate of increase in stem dry weight is equal to that fraction multiplied by the rate of total dry matter increase:

$$\text{IWST} = 0.255 \times 40.9 = 10.4 \text{ kg ha}^{-1} \text{ d}^{-1} \text{ (Column 15)}$$

The weight of the stem at the end of the ten – day period follows from the addition of the rate of increase times the length of the time interval to the weight at the end of the preceding ten – day period (Line a, Column 16):

$$\text{WST} = 0 + 10.4 \times 10 = 104 \text{ kg ha}^{-1} \text{ (Column 16)}$$

During this ten – day period, the crop is still in its vegetative phase, hence the fraction allocated to the grain is zero (Column 17). Therefore the values in Columns 18 and 19 also remain zero.

The leaf area index at the end of the ten – day period is obtained from the dry weight of the leaf blades (Column 13) by multiplying with the specific leaf area of 25 and taking into account the surface area:

$$\text{LAI} = 262 \times 25 \times 10^{-4} = 0.65 \text{ m}^2 \text{ m}^{-2} \text{ (Column 20)}$$

At the end of all calculations for the ten – day period, three auxiliary variables are calculated that are helpful for comparison with measured data. In Column 21, total above ground dry weight is introduced, which is the sum of the weight of leaf blades (Column 13), stems (Column 16) and grains (Column 19). Thus:

$$\text{TADW} = 262 + 104 + 0 = 366 \text{ kg ha}^{-1}$$

The total dry weight of the vegetation (Column 22) is equal to the above – ground dry weight, plus the weight of the root system (Column 10):

$$\text{TDW} = 262 + 104 + 0 + 183 = 549 \text{ kg ha}^{-1}$$

The total dry weight of live material (TDWL, Column 23) that is subject to maintenance respiration is equal to the total dry weight, because no dead material is present as yet. With this calculation, the treatment of the first ten – day period is finalized and the calculations can be repeated for the next ten – day period.

The conditions are not basically different for that period (Line c) from those in the previous one, therefore the line will be described in less detail.

Line c refers to the last ten – day period of November. The average air temperature then is 26.3 °C (Column 1). The accumulated temperature sum for the crop at the end of the period is obtained by adding the 263 d °C for this ten – day period to the value accumulated up till the beginning of the period (Column 2, Line b). Therefore the value in Column 2, Line c, equals:

$$TSUM = 272 + 263 = 535 \text{ d °C}$$

The corresponding value of the development stage is found by dividing the value in Column 2 by 1500 d °C, the required temperature sum for anthesis of this variety:

$$DVS = 535/1500 = 0.36 \text{ (Column 3)}$$

The value of potential gross assimilation for the last ten – day period of November is obtained by interpolation in Table 9:

$$GRA = 336 + 1/3 \times (283 - 336) = 318 \text{ kg ha}^{-1} \text{ d}^{-1}$$

The first and second value within the brackets are for the middle of December and the middle of November, respectively. To account for the influence of incomplete light interception, due to the low leaf area index, potential assimilation must be multiplied by the reduction factor from Table 11 for an LAI of 0.65:

$$0.33 + (0.15/0.5) \times (0.55 - 0.33) = 0.40$$

In Column 4, therefore, the rate of gross assimilation is introduced as:

$$GASS = 318 \times 0.40 = 127.2 \text{ kg ha}^{-1} \text{ d}^{-1}$$

Maintenance respiration for the period is calculated from the total live plant dry weight at the beginning (Column 23) and the relative maintenance respiration rate, 0.015 kg CH_2O per kg dry weight per day:

$$MRES = 549 \times 0.015 = 8.2 \text{ kg ha}^{-1} \text{ d}^{-1} \text{(Column 5)}$$

Carbohydrates available for increase in structural dry weight of the vegetation are equal to:

$$ASAG = 127.2 - 8.2 = 119.0 \text{ kg ha}^{-1} \text{ d}^{-1} \text{ (Column 6)}$$

From this, the total rate of increase in dry weight is calculated, taking into account the conversion efficiency:

$$DMI = 119.0 \times 0.7 = 83.3 \text{ kg ha}^{-1} \text{ d}^{-1} \text{ (Column 7)}$$

The partitioning factors for the various plant organs are obtained from Table 12, at the appropriate value of the development stage:

$$DVS = 0.18 + 0.5 \times (0.36 - 0.18) = 0.27$$

Thus:

FR = 0.165	(Column 8)
FL = 0.445	(Column 11)
FS = 0.39	(Column 14)
FG = 0.	(Column 17)

The rate of increase in root dry weight is calculated by multiplying the increase in total dry weight by the partitioning factor:

$$IWRT = 0.165 \times 83.3 = 13.7 \text{ kg ha}^{-1} \text{ d}^{-1} \text{ (Column 9)}$$

and the total root weight at the end of the ten – day period equals:

$$WRT = 183 + 13.7 \times 10 = 320 \text{ kg ha}^{-1} \text{ (Column 10)}$$

The rate of increase in leaf dry weight is obtained by multiplying FL and DMI, hence:

$$IWLV = 0.445 \times 83.3 = 37.1 \text{ kg ha}^{-1} \text{ d}^{-1} \text{ (Column 12)}$$

and the dry weight at the end of the period follows from addition of the increment to that present already:

$$WLV = 262 + 37.1 \times 10 = 633 \text{ kg ha}^{-1} \text{ (Column 13)}$$

The rate of increase in stem dry weight is calculated from total increase in dry weight and the fraction partitioned to the stem:

$$IWST = 0.39 \times 83.3 = 32.5 \text{ kg ha}^{-1} \text{ d}^{-1} \text{ (Column 15)}$$

Total stem weight at the end of the ten – day period equals:

$$WST = 104 + 32.5 \times 10 = 429 \text{ kg ha}^{-1} \text{ (Column 16)}$$

The leaf area index at the end of the ten – day period follows from the dry weight of the leaf blades:

$$LAI = 633 \times 25 \times 10^{-4} = 1.6 \text{ m}^2 \text{ m}^{-2} \text{ (Column 20)}$$

Total above – ground dry weight at the end of this ten – day period equals:

TADW = 633 + 429 + 0 = 1062 kg ha^{-1} (Column 21)

The total dry weight of the vegetation is:

TDW = 633 + 429 + 0 + 320 = 1382 kg ha^{-1} (Column 22)

which is all live material, hence:

TDWL = 1382 kg ha^{-1} (Column 23)

The calculations for the *Lines d, e* and *f* follow exactly the same pattern as the preceding ones; they are therefore not treated here.
Line g refers to the first ten – day period of January.

Column 1: average air temperature during the period is 26.0 °C
Column 2: accumulated temperature sum at the end of the period equals
1320 + 260 = 1589 d °C!

That is a point beyond anthesis. The temperature relations with respect to development are different for the period before anthesis and after anthesis, therefore these ten days cannot be treated in one line, but are split into two parts. The first part covers the period until the anthesis date, the length of that part is determined by the remainder of the required temperature sum, i.e.:

(1500 – 1320)/26.0 = 7 days

The sixth ten – day period can thus be depicted as:

Line g1, which refers to the first seven days of the period.
Column 1: average air temperature during the period is 26.0 °C
Column 2: accumulated temperature sum at the end of the period equals:
1320 + 7 x 26.0 °C = 1502 d °C
Column 3: corresponding development stage is calculated as:
1502/1500 = 1.0 (i.e. anthesis)

Column 4: potential gross assimilation during the period follows from Table 9:
283 + 2/3 x (332 – 283) = 316 kg ha^{-1} d^{-1};
The light interception factor equals 1 (Table 11), thus the gross assimilation rate equals:
316 kg ha^{-1} d^{-1}
Column 5: maintenance respiration is calculated from total canopy dry weight:

$5899 \times 0.015 = 88.5 \text{ kg ha}^{-1} \text{ d}^{-1}$

Column 6: assimilate availability for increase in structural material equals:
$316.0 - 88.5 = 227.5 \text{ kg ha}^{-1} \text{ d}^{-1}$

Column 7: the rate of increase in dry weight of structural material equals:
$227.5 \times 0.7 = 159.3 \text{ kg ha}^{-1} \text{ d}^{-1}$

Column 8: the fraction of dry weight allocated to the roots is obtained from Table 12 at development stage 0.94 and equals 0.025

Column 9: the rate of increase in root dry weight equals:
$0.025 \times 159.3 = 4 \text{ kg ha}^{-1} \text{ d}^{-1}$

Column 10: total dry weight of the roots at the end of the period equals:
$643 + 4 \times 7 = 671 \text{ kg ha}^{-1}$

Column 11: the fraction allocated to the leaf blades equals: 0.06

Column 12: the rate of increase in dry weight of the leaf blades equals:
$0.06 \times 159.3 = 9.6 \text{ kg ha}^{-1} \text{ d}^{-1}$

Column 13: total dry weight of the leaf blades at the end of the period is:
$2331 + 9.6 \times 7 = 2398 \text{ kg ha}^{-1}$

Column 14: the fraction of dry weight allocated to the stems equals: 0.225

Column 15: the rate of increase in dry weight of the stems equals:
$0.225 \times 159.3 = 35.8 \text{ kg ha}^{-1} \text{ d}^{-1}$

Column 16: total dry weight of the stems at the end of the period equals:
$2925 + 35.8 \times 7 = 3176 \text{ kg ha}^{-1}$

Column 17: the fraction of dry weight allocated to the grains, derived from Table 12 equals 0.69. This period is before anthesis. However, a substantial part of the assimilates produced in this period is temporarily stored in the stems, and later translocated to the growing grain (Yoshida, 1980). In the present approach, these assimilates are directly added, therefore, to grain dry weight

Column 18: the rate of increase in grain dry weight equals:
$159.3 \times 0.69 \times 0.8/0.7 = 125.6 \text{ kg ha}^{-1} \text{ d}^{-1}$
The ratio 0.8/0.7 is introduced to account for the fact that the efficiency of conversion of primary photosynthates into grain structural dry matter is higher than in vegetative dry matter.

Column 19: grain dry weight at the end of the period equals:
$125.6 \times 7 = 879 \text{ kg ha}^{-1}$

Column 20: the leaf area index at the end of the period is calculated from leaf dry weight:
$2398 \times 25 \times 10^{-4} = 6.0 \text{ m}^2 \text{ m}^{-2}$

Column 21: total above ground dry weight of the vegetation equals:
$2398 + 3176 + 879 = 6453 \text{ kg ha}^{-1}$

Column 22: total dry weight of the vegetation equals:
$2398 + 3176 + 879 + 671 = 7124 \text{ kg ha}^{-1}$

Column 23: there is still no dead material present, hence, DWLV equals 7124 kg ha^{-1}

Line g2 refers to the remainder of the first ten–day period of January, a period of three days.

Column 1: average temperature during the period is 26.0 °C

Column 2: the temperature sum till the end of the period equals:
$1502 + 3 \times 26.0 = 1580$

Column 3: for the post–anthesis phase, the development scale runs from the value 1 at anthesis to the value 2 at maturity. During the post–anthesis phase a required temperature sum of 800 d °C for this (and most other) varieties has been established. The development stage is calculated as the total temperature sum above 1500 d °C divided by 800, added to the value of 1 at anthesis. Thus:
$(1580 - 1500)/800 + 1 = 1.10$

Column 4 : potential gross assimilation rate is equal to that in the previous period at 316 kg ha^{-1} d^{-1}

Column 5: maintenance respiration is calculated from the total dry weight. However, the relative maintenance respiration rate is changed after anthesis from 0.015 to 0.01 kg CH_2O per kg dry matter per day. This change is mainly related to the fact that the nitrogen content of the vegetative material decreases after anthesis, even when the nitrogen supply to the vegetation is abundant (Section 4.1). Thus:
$7124 \times 0.01 = 71.2$ kg ha^{-1} d^{-1}

Column 6: assimilate availability for increase in dry weight is equal to:
$316 - 71.2 = 244.8$ kg ha^{-1} d^{-1}

Column 7: the rate of dry weight increase during the period, taking into account conversion efficiency equals:
$244.8 \times 0.8 = 195.8$ kg ha^{-1} d^{-1}

Columns 8,
9 and 10: after anthesis the sink strength of the above–ground organs and notably that of the growing grains is such that no more assimilates are diverted to the root system.

Column 11: leaf growth also ceases after anthesis, hence FL = 0

Column 12: leaves have a limited life span, thus after a certain time period, they senesce and stop functioning. This process is accelerated after anthesis, when translocation of essential substances to the developing grains takes place; therefore the weight of active leaf blades declines after anthesis; it is assumed that the relative rate of decline is constant, i.e. each day a constant fraction of the leaves senesces. The value of this constant equals 0.02 kg leaf blades per kg leaf blades per day. The rate of decline in dry weight of the leaf blades equals WLV x 0.02, thus:
$2398 \times 0.02 = 48$ kg ha^{-1} d^{-1}

Column 13: the dry weight of the living leaf blades at the end of the period equals:

$2398 - 48 \times 3 = 2254$ kg ha^{-1}

Columns 14 and 15: after anthesis there is no increase in stem dry weight, hence both values are zero.

Column 16: stem dry weight remains constant at 3176 kg ha^{-1}

Column 17: after anthesis all available assimilates are monopolized by the growing grain, thus that fraction equals 1

Column 18: the rate of increase in grain dry weight over the period equals:

$1 \times 195.8 = 195.8$ kg ha^{-1} d^{-1}

Column 19: grain dry weight at the end of the period equals:

$879 + 195.8 \times 3 = 1466$ kg ha^{-1}

Column 20: leaf area index follows from leaf dry weight:

$2254 \times 25 \times 10^{-4} = 5.6$ m^2 m^{-2}

Column 21: total above ground dry weight equals:

$2254 + 3176 + 1466 + 144 = 7040$ kg ha^{-1}

Column 22: total canopy dry weight equals:

$7040 + 671 = 7711$ kg ha^{-1}

Column 23: live tissue, subject to maintenance respiration equals:

$7711 - 144 = 7567$ kg ha^{-1}

Line h refers to the second ten – day period of January.

Column 1: average air temperature during the period is 26.0 °C

Column 2: the temperature sum at the end of the period equals:

$1580 + 260 = 1840$ d °C

Column 3: the corresponding development stage equals:

$(1840 - 1500)/800 + 1 = 1.43$

Column 4: potential gross assimilation rate equals 332 (Table 9) and the leaf area index permits full light interception, therefore gross assimilation equals:

$332 \times 1 = 332$ kg ha^{-1} d^{-1}

Column 5: maintenance respiration is calculated from the live tissue weight:

$7567 \times 0.01 = 75.7$ kg ha^{-1} d^{-1}

Column 6: assimilate availability for growth equals:

$332 - 75.7 = 256.3$ kg ha^{-1} d^{-1}

Column 7: rate of increase in structural dry weight, taking into account conversion efficiency, is:

$256.3 \times 0.8 = 205.0$ kg ha^{-1} d^{-1}

Columns 8 and 9: values remain zero

Column 10: weight of the root system remains constant: 671 kg ha^{-1}

Column 11: fraction to the leaves is zero

Column 12: the rate of decrease in weight of live leaves is approximated by:
$2254 \times 0.02 = 45.1$ kg ha^{-1} d^{-1}

Column 13: live leaf blade dry matter at the end of the period equals:
$2254 - 45.1 \times 10 = 1803$ kg ha^{-1}

Columns 14 and 15: values are zero.

Column 16: weight of stems is constant at 3176 kg ha^{-1}

Column 17: fraction allocated to the grain is 1.0

Column 18: the rate of increase in dry weight of the grain equals that in total dry weight: 205.0 kg ha^{-1} d^{-1}

Column 19: total grain dry weight at the end of the period is:
$1466 + 205.0 \times 10 = 3516$ kg ha^{-1}

Column 20: leaf area index at the end of the period equals:
$1803 \times 25 \times 10^{-4} = 4.5$ m^2 m^{-2}

Column 21: total above ground dry weight of the vegetation equals:
$1803 + 3176 + 3516 + 595 = 9090$ kg ha^{-1}

Column 22: total dry weight of the vegetation equals:
$9090 + 671 = 9761$ kg ha^{-1}

Column 23: total live dry weight equals 9166 kg ha^{-1}

The calculations for *line (i)* follow exactly the pattern of the previous line and they are not treated here.

Line j refers to the first ten – day period of February.

Column 1: average air temperature during the period is 26.0 °C

Column 2: accumulated temperature sum till the end of the period equals:
$2100 + 260 = 2360$

Column 3: corresponding development stage is:
$(2360 - 1500)/800 + 1 = 2.075$
development stage 2 corresponds to maturity, therefore the duration of this period is only:
$(800 - (2100 - 1500))/26.0 = 8$
Hence, eight days till maturity.

Column 2: accumulated temperature sum till the end of the period equals:
$2100 + 8 \times 26.0 = 2308$

Column 3: corresponding development stage at the end of the period is:
$1 + (2308 - 1500)/800 = 2.0$

Column 4: gross assimilation rate for the first ten – day period of February equals:
$332 + 2/3 \times (344 - 332) = 340$ kg ha^{-1} d^{-1}
The reduction factor for light interception equals 0.94, gross assimilation rate for the period thus equals:
$340 \times 0.94 = 319.6$ kg ha^{-1} d^{-1}

Column 5: maintenance respiration is calculated from total live dry weight:
$10757 \times 0.01 = 107.6 \text{ kg ha}^{-1} \text{ d}^{-1}$

Column 6: assimilate availability for increase in dry weight equals:
$319.6 - 107.6 = 212.0 \text{ kg ha}^{-1} \text{ d}^{-1}$

Column 7: rate of increase in total dry weight taking into account conversion efficiency equals:
$212.0 \times 0.8 = 169.6 \text{ kg ha}^{-1} \text{ d}^{-1}$

Column 8 and 9: fraction to roots and root growth is zero.

Column 10: root dry weight remains at 671 kg ha^{-1}

Column 11: fraction to leaf blades is zero.

Column 12: rate of decrease in leaf dry weight is
$1442 \times 0.02 = 28.8 \text{ kg ha}^{-1} \text{ d}^{-1}$

Column 13: live leaf dry weight at maturity equals:
$1442 - 28.8 \times 8 = 1212 \text{ kg ha}^{-1}$

Columns 14 and 15: fraction to the stem and increase in stem dry weight is zero.

Column 16: stem dry weight is constant at 3176 kg ha^{-1}

Column 17: fraction allocated to the grain equals one.

Column 18: rate of increase in grain dry weight equals:
$69.6 \text{ kg ha}^{-1} \text{ d}^{-1}$

Column 19: total grain dry weight at maturity equals:
$5470 + 169.6 \times 8 = 6827 \text{ kg ha}^{-1}$

Column 20: green leaf area index at maturity is calculated from leaf blade dry weight:
$1212 \times 25 \times 10^{-4} = 3.0$

Column 21: total above ground dry matter equals:
$1212 + 3176 + 6827 + 1187 = 12402 \text{ kg ha}^{-1}$

Column 22: total plant dry weight at harvest equals:
$12402 + 671 = 13073 \text{ kg ha}^{-1}$

Column 23: total live weight at harvest equals:
11886 kg ha^{-1}

2.3.3 Comparison with measurements

In Figure 19, the calculated time course of dry matter production is compared to the measured data and a very satisfactory agreement between both is evident. The measured grain yield (at 12% moisture content) was 7.5 t ha^{-1}, which again is close enough to the calculated value of 7.7 t ha^{-1} (6827×1.12).

The calculation procedure, outlined in the preceding Subsection was also applied to a set of data from IRRI, Los Baños. In a maximum annual production trial, rice was grown year — round, three different cultivars being used (Yoshida et al., 1972). The first one was IR8, for which parameters identical

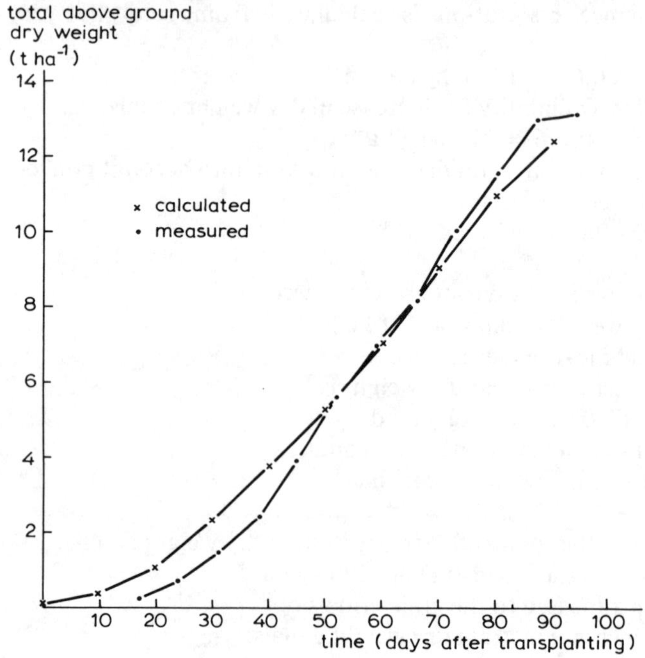

Figure 19. Comparison of measured and calculated above-ground dry-matter accumulation for bunded rice grown in Paramaribo, Suriname.

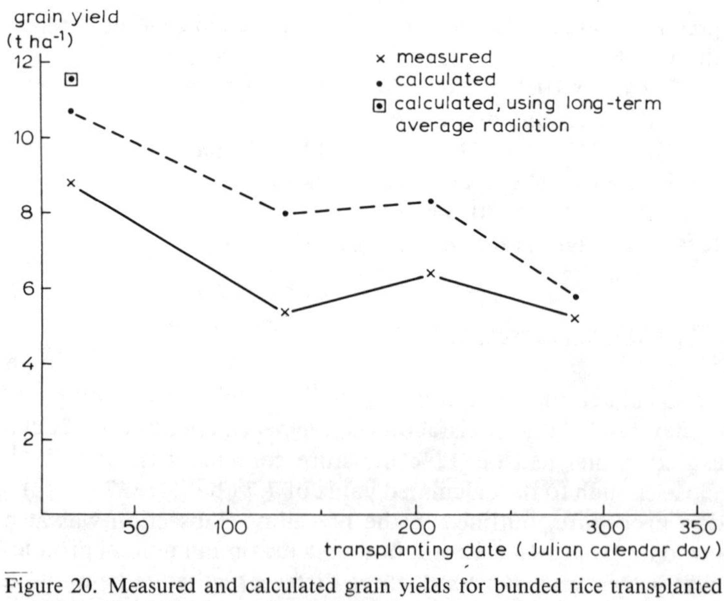

Figure 20. Measured and calculated grain yields for bunded rice transplanted on different dates at IRRI, Los Banos, the Philippines.

to those of the preceding section were used. The other two were the early naturing cultivars, IR—747—B2 and IR—667—98, respectively. For these varieties, the required temperature sum from transplanting to anthesis was set at 1100 d °C, i.e. a development rate of 0.0182 d^{-1} at a temperature of 20 °C.

The results of the calculations are presented in Fugure 20, along with the measured data. The figure shows that the pattern of grain yield with time of transplanting is identical for the measured and the calculated data, but that the calculations are consistently of a higher level. It would seem, therefore, that in the experiments the potential, dictated by weather conditions was not fully reached. Reasons for the discrepancy can only be speculated upon, but nitrogen application of 125 — 150 kg N ha^{-1}, more than two — thirds of which applied as a basal dressing seems hardly sufficient for yields of over 6000 kg ha^{-1} (Section 4.1). It would seem, therefore, that the conclusion reached by the authors that a maximum annual yield of over 28000 kg ha^{-1} is possible, is valid. The more so, if it is considered that the year 1971 was unfavourable in terms of radiation as is shown in Figure 20 by the result calculated with long — term average radiation data.

In Figure 21, the measured and calculated growth curves are shown for a

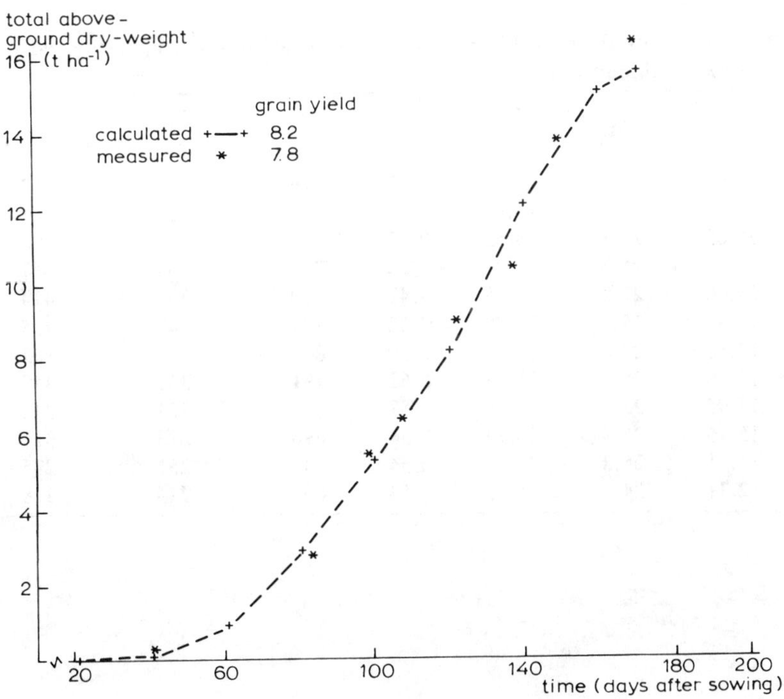

Figure 21. Measured and calculated above-ground dry-matter accumulation for spring wheat, grown in Israel.

spring wheat crop grown in the Central Negev Desert in Israel under irrigation (Hochman, 1982). The variety Lachish used in this experiment requires a temperature sum of 1500 d °C from emergence to anthesis and 850 d °C from anthesis to maturity, both at a base temperature of 0 °C. The partitioning functions used in the model are given in Section 3.4, where the same experiment is used to illustrate the effect of water shortage on production.

These examples show that potential yield and production may be estimated with reasonable accuracy on the basis of crop characteristics and weather conditions.

Exercise 12

Calculate the grain yield for the rice variety IR8, transplanted in Los Baños on January 20. Use the basic data given in Table 13. Assume for each month three ten – day periods as in Table 10. The values for F_{gs} are averages for the month and are not, as in Table 9, applicable to the middle of the month.

Table 13. Basic data for Exercise 12, F_{cl} and F_{ov} expressed in CO_2, F_{gs} expressed in CH_2O, Los Banos, Philippines, 14° N.

Month	H_a ($MJ\ m^{-2}d^{-1}$)	T_a (°C)	H_g ($MJ\ m^{-2}d^{-1}$)	f_o –	F_{cl} ($kg\ ha^{-1}d^{-1}$)	F_{ov} ($kg\ ha^{-1}d^{-1}$)	F_{gs} ($kg\ ha^{-1}d^{-1}$)
J	14.07	23.4	22.60	0.47	616	252	303
F	18.31	24.5	25.11	0.34	674	281	368
M	20.03	24.8	28.55	0.37	734	311	393
A	23.81	25.3	30.82	0.28	781	334	447
M	21.00	27.0	31.55	0.42	800	342	414
J	18.48	26.0	31.58	0.52	804	342	385
J	16.80	26.0	31.60	0.59	803	343	363
A	15.70	25.3	31.28	0.62	791	339	348
S	15.96	25.9	29.69	0.58	757	323	344
O	15.16	25.9	26.73	0.54	698	293	327
N	13.31	25.3	23.46	0.54	633	261	295
D	12.31	24.2	21.78	0.54	600	244	278

3 CROP PRODUCTION AS DETERMINED BY MOISTURE AVAILABILITY

3.1 Potential evapotranspiration

J.A.A. Berkhout and H. van Keulen

3.1.1 Introduction

The atmosphere is the main source of carbon dioxide, which is needed by plants in the assimilation process. The rate of supply of carbon dioxide depends both on the concentration difference between the atmosphere and the active sites in the plant, and on the resistance to carbon dioxide transport, part of which is in the stomata in the leaves of the plant (Section 3.3).

The water status of a plant influences the rate of CO_2 supply, because stomatal opening, and hence the resistance, is affected by the water potential in the plant. The stomata of a plant start to close at some critical lower water content of the plant, resulting from a negative balance between the uptake of water through its roots and the amount of water lost through the stomata in the process of transpiration. This closure of the stomata continues until the balance of uptake and transpiration is restored, but now at a lower level. Hence, a direct but complicated interrelation exists between transpiration, assimilation and the water supply in the soil (Section 3.2).

In this section the transpiration process is examined under conditions of non-limiting supply of water to the plant, which is defined as potential transpiration. The analysis of Penman (1956) is followed. In that approach the radiation sources that provide the energy to evaporate the water are considered in combination with the turbulence of the air, to remove the water vapour. This procedure enables the estimation of potential evapotranspiration, using data obtained from standard meteorological stations.

3.1.2 Radiation

The sun, which has a temperature of 6000 K, emits energy like any black body with a temperature above absolute zero. The solar constant, i.e. the mean energy received at the earth's mean distance from the sun, outside the earth's atmosphere, on a surface normal to the incident radiation, has a value of approximately 1.4 kJ m^{-2} s^{-1}.

Exercise 13
In the old system of units the solar constant was expressed in cal cm^{-2} min^{-1}.
What is the value of the solar constant in these units?

In passing through the atmosphere, a variable part of the radiant energy is absorbed, scattered and reflected, due to the presence of ozone, water vapour, clouds and dust in the atmosphere. Total global radiation at the surface of the earth is measured at a limited number of meteorological stations. For other stations it may be possible to estimate the average total global radiation by means of an empirical relation, using the measured duration of bright sunshine (Black et al., 1954). The form of this relationship — the Angström formula – is:

$$R_I = R_A (a_A + b_A nN^{-1}) \tag{8}$$

where

R_I is the radiation actually received ($J\ m^{-2}\ d^{-1}$).

R_A is Angot's value, or the theoretical amount of radiation that would reach the earth's surface in the absence of an atmosphere ($J\ m^{-2}$) d^{-1}. Values of R_A are tabulated as a function of day of the year and latitude in Table 14.

nN^{-1} is the ratio of actual duration of bright sunshine (n) and the maximum possible length on a cloudless day (N), both in h. The actual duration of sunshine is recorded in most stations, whereas the maximum duration during the day is again tabulated as a function of date and latitude in Table15.

a_A and b_A are empirical constants.

The numerical value of these constants depends on the location or latitude (Glover & McCulloch, 1958) and the climate (Rietveld, 1978). Indicative values used by F.A.O. (Frere & Popov, 1979) are:

for cold and temperate zones	$a_A = 0.18$ and	$b_A = 0.55$
for dry tropical zones	0.25	0.45
for humid tropical zones	0.29	0.42

The Angström formula gives fair results for weekly or monthly averages.

Table 14. Angot's values (10^7 J $m^{-2} d^{-1}$) as a function of latitude and day of the year.

Date for northern latitudes

Northern Latitude	13 Jan.	4 Feb.	26 Feb.	21 Mar.	13 Apr.	6 May	29 May	22 June	15 July	8 Aug.	31 Aug.	23 Sep.	16 Oct.	8 Nov.	30 Nov.	22 Dec.
90	—	—	—	—	1.77	3.23	4.18	4.51	4.16	3.20	1.75	—	—	—	—	—
80	—	—	0.03	0.65	1.77	3.18	4.12	4.44	4.10	3.16	1.75	0.64	0.03	—	—	—
70	—	0.10	0.55	1.28	2.20	3.13	3.93	4.24	3.91	3.11	2.17	1.27	0.54	0.10	—	—
60	0.30	0.61	1.15	1.87	2.66	3.39	3.91	4.10	3.89	3.35	2.63	1.85	1.14	0.61	0.30	0.21
50	0.86	1.21	1.75	2.41	3.06	3.63	4.01	4.14	3.99	3.60	3.03	2.38	1.73	1.20	0.85	0.74
40	1.46	1.82	2.31	2.87	3.38	3.81	4.07	4.15	4.05	3.77	3.34	2.83	2.28	1.80	1.46	1.33
30	2.07	2.38	2.80	3.24	3.62	3.89	4.05	4.08	4.02	3.85	3.58	3.20	2.77	2.36	2.06	1.95
20	2.64	2.89	3.22	3.52	3.74	3.86	3.91	3.91	3.89	3.83	3.70	3.48	3.18	2.87	2.62	2.53
10	3.15	3.33	3.54	3.69	3.75	3.74	3.69	3.65	3.67	3.71	3.71	3.65	3.49	3.30	3.13	3.07
0	3.58	3.67	3.75	3.75	3.65	3.50	3.37	3.31	3.35	3.47	3.61	3.70	3.71	3.64	3.56	3.53
−10	3.92	3.92	3.85	3.69	3.45	3.18	2.96	2.88	2.95	3.15	3.41	3.65	3.81	3.88	3.90	3.91
−20	4.16	4.05	3.84	3.52	3.14	2.76	2.48	2.37	2.47	2.74	3.10	3.48	3.80	4.01	4.14	4.18
−30	4.29	4.07	3.72	3.24	2.74	2.27	1.95	1.82	1.94	2.25	2.70	3.20	3.67	4.03	4.27	4.36
−40	4.32	3.99	3.46	2.87	2.25	1.73	1.38	1.24	1.37	1.71	2.23	2.83	3.43	3.95	4.30	4.43
−50	4.26	3.80	3.15	2.41	1.71	1.16	0.81	0.69	0.80	1.15	1.69	2.38	3.11	3.77	4.24	4.42
−60	4.15	3.54	2.73	1.87	1.13	0.59	0.28	0.20	0.28	0.58	1.11	1.85	2.70	3.52	4.13	4.38
−70	4.18	3.28	2.25	1.28	0.53	0.10	—	—	—	0.10	0.53	1.27	2.23	3.26	4.16	4.52
−80	4.38	3.33	1.82	0.65	0.03	—	—	—	—	—	0.03	0.64	1.80	3.31	4.36	4.74
−90	4.45	3.39	1.82	—	—	—	—	—	—	—	—	—	1.80	3.35	4.42	4.81

Table 15. Maximum duration of bright sunshine (hours) as a function of latitude and day of the year.

Date for northern/southern latitudes

Latitude (°N or °S)	15 Jan. July	15 Feb. Aug.	15 Mar. Sep.	15 Apr. Oct.	15 May Nov.	15 June Dec.	15 July Jan.	15 Aug. Feb.	15 Sept. Mar.	15 Oct. Apr.	15 Nov. May	15 Dec. June
0	12.1	12.1	12.1	12.1	12.1	12.1	12.1	12.1	12.1	12.1	12.1	12.1
10	11.7	11.8	12.0	12.4	12.5	12.7	12.7	12.5	12.2	11.9	11.7	11.5
20	11.1	11.5	12.0	12.6	13.1	13.3	13.2	12.9	12.3	11.7	11.3	11.0
30	10.4	11.2	11.9	12.9	13.6	14.0	13.9	13.3	12.4	11.5	10.8	10.3
40	9.9	10.7	11.8	13.3	14.3	14.9	14.7	13.9	12.5	11.2	10.1	9.5
50	9.2	10.2	11.6	13.8	15.3	16.2	15.9	14.6	12.6	10.9	9.3	8.3
60	7.5	9.2	11.6	14.6	17.0	18.5	18.0	15.9	12.9	10.3	7.2	6.4
70	3.1	7.4	11.4	16.2	18.5			18.7	13.4	9.2	4.7	
80		1.9	10.6						14.8	6.0		

Exercise 14
Meteorological station Don Muang, Thailand (humid tropical zone)
Position: 13°55′ N, 100°36′ E

Mean monthly values of measured daily hours of sunshine (n):

J	F	M	A	M	J	J	A	S	O	N	D
7.1	5.9	5.9	5.1	4.4	4.0	3.7	2.9	3.3	3.9	4.9	5.9

Calculate R_I. Use Equation 8, and Tables 14 and 15.

Part of the solar radiation reaching the earth's surface is reflected. The term 'albedo' denotes that reflection fraction; its value is determined by a combination of surface properties (open water, dry soil, moist to wet soil, vegetation, etc.) and the inclination of the sun. Indicative mean albedo values are 0.05 for a water surface and 0.25 for a green crop surface completely covering the ground (Penman, 1963).
The atmosphere (including the clouds) and the earth's surface, which both have temperatures above the absolute zero, emit thermal radiation. The magnitude of this radiation is proportional to the temperature. As the atmosphere is partly transparent and the temperature of the earth's surface is higher than that of the atmosphere, the outgoing long-wave radiation component will be dominant. Empirical relations have been formulated to calculate this net outgoing long-wave radiation on the basis of air temperature, humidity and cloudiness. Penman (1956) used the following expression, which is derived from the Brunt-formula (Brunt, 1932):

$$R_B = \sigma (T_a + 273)^4 \cdot (0.56 - 0.079\, e_a^{0.5}) \cdot (0.1 + 0.9\, nN^{-1}) \qquad (9)$$

where

- R_B is net outgoing long-wave radiation ($J\ m^{-2}\ d^{-1}$)
- σ is the Stefan Boltzman constant ($4.9 \times 10^{-3}\ J\ m^{-2}\ d^{-1}\ K^{-4}$)
- T_a is mean air temperature at screen height in °C. The factor 273 is added to convert to absolute temperature.
- e_a is actual vapour pressure of the air at screen height (mbar).
- nN^{-1} is the ratio of actual and maximum hours of sunshine (Equation 8)

Again, Equation 9 is not intended for periods shorter than a week.

Exercise 15

e_a is a measure of the amount of water vapour in the air. The unit mbar expresses this in the partial pressure of the water vapour. Conversion of this dimension into alternative ones may be obtained from the following expressions:

(a) 1 mbar = 100 Pa
where Pa is the abbreviation for Pascal, which is the SI unit for pressure.

(b) 1mbar = $T_k / 0.217 \times W_w$
where W_w is the concentration of water vapour in the air in kg m^{-3}, and T_k stands for the absolute temperature in K.

(c) 1 mbar = 0.75 mm Hg
where mm Hg stands for millimeters of mercury, which was used previously to express vapour pressure, among others in the original work of Penman (1948, 1956).

In Equation 9 the multiplication factor for e_a equals 0.079 for e_a expressed in mbar. What would be its value for e_a expressed in kg m^{-3} (20 °C), or in mm Hg?

Exercise 16

Meteorological station Don Muang mean monthly value of measured daily:

	Temperature (°C)	Vapour pressure (mbar)
J	26.0	22.5
F	27.4	25.6
M	28.9	27.9
A	29.8	29.8
M	29.3	30.2
J	28.7	29.2
J	28.2	28.7
A	28.0	30.2
S	28.2	29.8
O	28.1	29.3
N	27.4	27.0
D	25.6	22.7

Calculate the mean monthly outgoing longwave radiation (R_B)

Net radiation is given by:

$$R_N = R_I (1 - r_a) - R_B \quad (10)$$

where

R_N	is net radiation ($J\ m^{-2}\ d^{-1}$)
R_I	is incident shortwave radiation ($J\ m^{-2}\ d^{-1}$)
r_a	is the albedo (dimensionless)
R_B	is net outgoing longwave radiation ($J\ m^{-2}\ d^{-1}$)

Exercise 17
Calculate the mean monthly net radiation (R_N) for the Don Muang meteorological station for an open water surface.

3.1.3 Evaporation

Now the heat balance of a thin, extended water layer, that covers a black isolated floor is considered. The net radiation that is absorbed can be calculated from meteorological data (Equation 10) for an assumed albedo of 0.05. This net radiation (R_N) heats the water until the sensible heat loss to the surrounding air plus the heat loss due to the evaporation of water equals this net radiation, or:

$$R_N = H + LE \quad (11)$$

where

R_N	is net radiation ($J\ m^{-2}\ d^{-1}$)
H	is sensible heat loss ($J\ m^{-2}\ d^{-1}$)
E	is the rate of water loss from the surface ($kg\ m^{-2}\ d^{-1}$)
L	is latent heat of vaporization of water ($2450 \times 10^3\ J\ kg^{-1}$)
LE	is the evaporative heat loss ($J\ m^{-2}\ d^{-1}$)

The loss of sensible heat of a surface to its surroundings is proportional to the temperature difference, according to:

$$H = h_u (T_s - T_a) \quad (12)$$

where

H is sensible heat loss ($J\ m^{-2}\ d^{-1}$)
h_u is the sensible heat transfer coefficient ($J\ m^{-2}\ d^{-1}\ °C^{-1}$)
T_s and
T_a are the temperature of the evaporating surface and the temperature at standard screen height, respectively (°C)

The value of the sensible heat transfer coefficient, h_u, depends on the atmospheric turbulence and may be expressed as an empirically determined function of mean wind velocity at a defined height (Penman, 1948):

$$h_u = a_u (1 + b_u \bar{u}) \tag{13}$$

where
$\bar{u}$ is mean wind velocity ($m\ s^{-1}$)
a_u and
b_u
are empirical constants.
For a smooth land surface and wind velocity measured at a standard height of 2 m, an indicative value for a_u is $6.4 \times 10^5\ J\ m^{-2}\ d^{-1}\ °C^{-1}$ and for b_u, $0.54\ s\ m^{-1}$ (Frere & Popov, 1979).

Exercise 18
Mean monthly value of measured daily wind velocity at Don Muang, measured at standard height (2 m) in $m\ s^{-1}$

J	F	M	A	M	J	J	A	S	O	N	D
2.0	2.5	2.8	2.7	2.5	2.5	2.4	2.5	2.3	2.1	2.1	1.4

Calculate the mean monthly h_u values.

Analogous to the sensible heat, a water surface looses water vapour, in proportion to the vapour pressure difference between the surface and the surrounding air:

$$E = k_u (e_s - e_a) \tag{14}$$

where

E is the rate of water vapour loss in ($H_2O\ m^{-2}\ d^{-1}$)
k_u is the vapour transfer coefficient ($kg\ m^{-2}\ d^{-1}\ mbar^{-1}$)
e_s and
e_a are the vapour pressure at the surface and at standard screen height,

respectively (mbar)

The air at the water surface is water vapour saturated. This saturated vapour pressure is related to the surface temperature T_s, a relation that may be approximated (Goudriaan, 1977) by:

$$e_s = 6.11 \times e^{(17.4T_s/(T_s + 239))} \qquad (15)$$

where

e_s is the saturated vapour pressure (mbar)
T_s is the surface temperature (°C)

The exchange of sensible heat and that of evaporative heat are governed by the same physical processes of turbulence and diffusion. Therefore a relationship exists between the sensible heat transfer coefficient and the vapour transfer coefficient (Bowen's ratio), or

$$\gamma = h_u k_u^{-1} L^{-1} \qquad (16)$$

where

γ is the psychrometer constant. Its value is about 0.66

Exercise 19
What is a psychrometer?
What is the dimension of γ?

Substituting Equations 12, 14 and 15 in the balance Equation 11, yields:

$$R_N = h_u(T_s - T_a) + (h_u/\gamma) \cdot (e_s - e_a) \qquad (17)$$

This equation contains two unknowns, the temperature of the water surface, T_s, and the saturated vapour pressure in the air at this surface, e_s. Both variables are related in the manner described in Equation 15. Hence, there are two equations with two unknowns, so that the surface temperature and the vapour pressure at the surface can be solved. This enables calculation of the evaporation of the water surface from Equation 14.

Exercise 20

- Calculate the evaporation rate at Don Muang in January for a thin extended water surface with an albedo of 0.05, using the data of Exercises 14, 17, and 18.
- Calculate for this purpose the sensible and evaporative heat loss for assumed water surface temperatures of 20, 21, 22, 23, and 24 °C. Plot the sum of both against these temperatures and draw in the same graph the net radiation. Read from the graph the equilibrium temperature, where the water surface neither gains nor looses heat. Calculate then the evaporation rate.

The iterative procedure illustrated in Exercise 20 is rather cumbersome and can be avoided by linearizing the relation between temperature and saturated vapour pressure (Equation 15). This was first done by Penman (1948), who used the approximate relation:

$$(e_s - e_a) = \Delta(T_s - T_d) \qquad (18)$$

where

T_d is the dewpoint of the air, i.e. the temperature at which the vapour in the air would start to condense or, in other words, the temperature at which the actual vapour pressure in the air would be the saturated vapour pressure. The relation expressed in Equation 18 is depicted in Figure 22.

Δ is the slope of the saturation vapour pressure curve between the average air temperature and dewpoint.

Exercise 21

What is the dewpoint of air with a vapour pressure of 10, 15, 50 mbar?

Exercise 22

What was the value of Δ in Don Muang in January?
Calculate the value of Δ at temperatures of 2, 6, 10 .....30, 34 and 38 °C. Plot the values of Δ versus temperature.

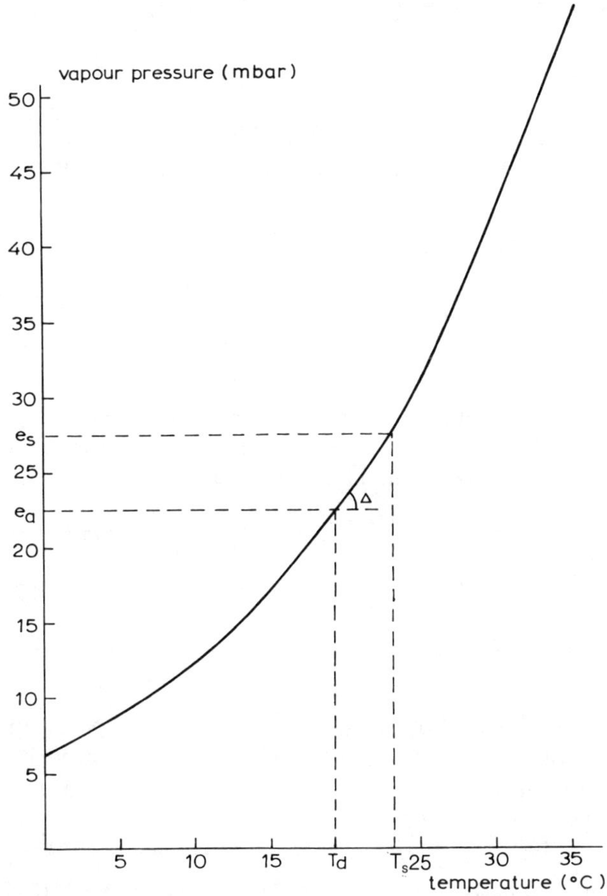

Figure 22. The relation between temperature and saturated vapour pressure. T_a and T_d represent air temperature and dewpoint, respectively; e_s and e_a are the saturated and actual vapour pressure in the atmosphere, respectively.

Substituting Equation 18 into Equation 17 gives:

$$R_N = h_u (T_s - T_a) + (h_u \Delta/\gamma)(T_s - T_d) \tag{19}$$

In this equation the surface temperature, T_s, is the only unknown and can be made explicit:

$$T_s = T_a + \frac{\gamma}{\Delta + \gamma} R_N/h_u - \frac{\Delta}{\Delta + \gamma}(T_a - T_d) \tag{20}$$

Exercise 23
Verify that the psychrometric constant, γ, and the slope of the saturated vapour pressure curve, Δ, have the same dimension and that R_N/h_u has the dimension of °C.

Exercise 24
Under what conditions is the temperature of the surface equal to the temperature of the air? When is it certainly higher and when is it certainly lower?
Derive an equation for the relation between the wet bulb temperature and dewpoint for a psychrometer that is well – shielded from radiation.

Combining Equations 11, 12 and 20 results in:

$$LE = \frac{\Delta}{\Delta + \gamma} (R_N + h_u (T_a - T_d)) \qquad (21)$$

This is the well – known Penman equation for the calculation of evaporation from a free water surface. $(T_a - T_d)$ in this equation may be replaced by $(e_d - e_a)/\Delta$ in which e_d is the saturation vapour pressure at air temperature. Then the Penman equation can be written as:

$$LE = \frac{1}{\Delta + \gamma} (\Delta R_N + h_u(e_d - e_a)) \qquad (22)$$

Exercise 25
– Calculate the evaporation rate from a free water surface at Don Muang for the months January, March, May, July, September and November.

3.1.4. Potential evapotranspiration

Plant leaves are protected against water loss by a cuticular layer that is almost impermeable to water. However, there are numerous small stomata (see Section 3.3) in these leaves to enable the entrance of carbon dioxide. The walls of the substomatal cavities are wet, therefore the air inside the stomata is saturated with water vapour. In comparison to evaporation, this water vapour has to overcome an additional resistance to move out of the cavity to the surface of the leaves. Therefore the transpiration of an extended single leaf surface is smaller than that of an extended surface of water. On the other hand, the exchange surface is larger for a canopy with an LAI higher than one. Taking these effects into account, Monteith (1965) has modified the original equation for the evaporation of a free water surface for the description of canopy transpiration. It is beyond the scope of this book to go into the details of his theory, but the result is an equation that very closely resembles Equation 22:

$$LE = \frac{1}{((\Delta + \gamma) \cdot (1/h_u + 1/C_s)/1/h_u)} \cdot (\Delta R_N + h_u(e_d - e_a)) \qquad (23)$$

where

C_s is conductance for water vapour, expressed in the same units as h_u.

In the case of a crop canopy, C_s represents the conductance of a large number of leaves placed in parallel and is referred to as the surface conductance. For a well – watered crop, i.e. under conditions of potential transpiration, C_s appears much larger than h_u. The influence of the correction factor in Equation 23 is therefore rather small. For practical purposes, therefore, the Penman equation in its original form appears to be a useful estimate of transpiration losses by crops. In the case of water shortage, the closure of stomata is reflected in a decrease of the canopy conductance C_s.

However, the difference in albedo between a water surface and a green crop surface, which is 0.05 for the first and roughly 0.25 for the latter, has to be taken into account. The water loss that is calculated in this way is referred to as the potential evapo – transpiration of a closed, short green crop surface well supplied with water. The prefix 'evapo' is used, because no distinction is made between water loss by transpiration from the leaves and that by evaporation from the wet soil surface under the crop.

Exercise 26

– Calculate the evapo-transpiration at Don Muang for the months January, March, May, July, September and November.

3.2 The water balance of the soil

P.M. Driessen

As discussed in Section 2.1, plant production involves intake of atmospheric CO_2 through stomatal openings in the epidermis. Most of the water that plants take up from the soil is again lost to the atmosphere by transpiration through the same openings (Section 3.1). The daily turnover of water can be considerable: transpiration of 0.4 cm of water from a crop surface on a clear sunny day corresponds to a water loss from the root zone of no less than 40 000 kg ha^{-1} d^{-1}. If soil moisture taken up by the roots is not replenished, the soil will dry out to such an extent that the plants wilt and — ultimately — die. The tenacity with which the soil retains its water is equalled by the suction that the roots must exert to be able to take up soil moisture. This suction, known as the soil moisture potential or 'matric suction', can be measured. In hydrology, potential (ψ) is usually expressed as energy per unit weight of soil water, with the dimension of length (van Bakel, 1981). An optimum ψ range exists within which a given plant takes up water freely. Above or below this range the plant senses stress. It reacts by actively curbing its daily water consumption through partial or complete closure of its stomata. The consequence is evident: this stomatal closure interferes with CO_2 intake and reduces assimilation and, consequently, dry-matter production.

Any model of the production capacity of (dryland) crops must therefore keep track of the soil moisture potential to determine when and to what degree a crop is exposed to water stress. This is commonly done with the aid of a water balance equation, which compares, for a given period of time, incoming water in the rooted surface soil with outgoing water and quantifies the difference between the two as a change in the amount of soil moisture stored. This rooted surface soil, or 'root zone', is a continuous soil layer with an upper boundary (the soil surface), and a lower boundary at a depth of RD cm (the rooting depth). Water enters and leaves the root zone via these two boundaries, but water is also removed directly from the interior parts of the root zone, viz. the water taken up by plant roots. This uptake is almost entirely discharged as transpiration.

The rate of change in soil moisture content of the root zone can thus be described with a water balance equation of the following nature:

$$RSM = (IM + (CR - D) - T)/RD \qquad (24)$$

where

RSM	is the rate of change in moisture content of the root zone (cm^3 cm^{-3} d^{-1})
IM	is the rate of net influx through the upper root zone boundary ($cm\ d^{-1}$)
(CR – D)	is the rate of net influx through the lower root zone boundary ($cm\ d^{-1}$)
T	is the rate of crop transpiration ($cm\ d^{-1}$)
RD	is the depth of the root zone (cm)

Figure 23 presents the various incoming and outgoing fluxes of water in a crop-soil-atmosphere system. This figure illustrates that supply of water at the upper root zone boundary is composed of precipitation (at a rate P, in cm d^{-1}), irrigation (at a rate I_e, in cm d^{-1}) and, possibly, water that was stored on top of the soil surface (at a rate DS, in cm d^{-1}). There is also loss of water from the soil surface, viz. as evaporation (at a rate E_a, in cm d^{-1}). Thus the net rate of water supply at the soil surface is ($P + I_e + DS - E_a$) cm d^{-1}. This net surface supply infiltrates into the soil at a rate of IM cm d^{-1}. During some time intervals, the net surface supply rate may exceed the maximum rate at which water can infiltrate in that particular soil (IM_{max}). In first instance, the excess supply is stored on the surface. The maximum surface storage capacity (SS_{max}, in cm) depends on the surface properties and the slope angle of the land. If excess supply exceeds the surplus storage capacity $(SS_{max} - SS_t)/\Delta t$, the remaining water leaves the system as surface runoff (SR, in cm d^{-1}).

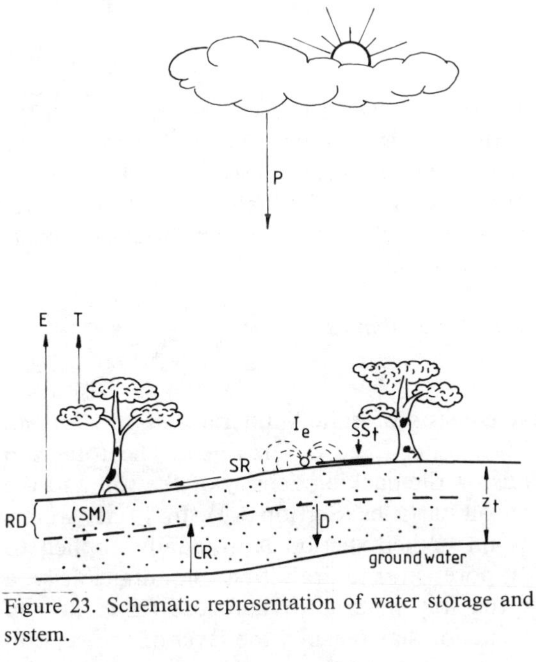

Figure 23. Schematic representation of water storage and flow in a plant-soil-atmosphere system.

In accord with the above, the actual rate of infiltration of water through the upper root zone boundary can be described as:

$$IM = P + I_e - E_a + DS - SR \tag{25}$$

where

IM is the actual infiltration rate (cm d^{-1})
P is the actual precipitation rate (cm d^{-1})
I_e is the effective irrigation rate (cm d^{-1})
E_a is the actual evaporation rate (cm d^{-1})
DS is the rate of decline of surface storage (cm d^{-1}) (DS is defined as positive if surface storage decreases and negative if surface storage increases)
SR is the rate of surface run-off (cm d^{-1})

At the lower boundary of the root zone (i.e. at depth RD), vertical flow of water between root zone and groundwater may take place. Water flow from the groundwater into the root zone is called capillary rise (CR), water loss from the root zone to the groundwater is called downward percolation (D). There is, of course, no percolation ($D = 0$) if $CR > 0$, and vice versa, but an equilibrium is possible in which both CR and D are zero. The rate of water flow through the lower root zone boundary can thus be described as:

$$(CR - D) = CR - D \tag{26}$$

The water balance equation (24) can be solved for each time interval in the growth cycle of a crop. The individual terms are quantified on the basis of the state of the system at the beginning of the time interval (Section 2.3).

Before embarking on a discussion of the individual terms of the water balance equation, it seems useful to go briefly into some important concepts used in the description of soil-moisture dynamics.

3.2.1 Basic principles of soil-moisture dynamics

Soil moisture storage

Soil is a porous medium that consists of solid (mineral and/or organic) particles and pores that are either water-filled or air-filled. The total soil porosity, SM_0 ($cm^3\ cm^{-3}$) depends on the packing density of the soil and the specific density of the individual soil particles (Section 5.3). In a fully water-saturated soil, matric suction is nil. When suction is gradually applied to saturated soil material, the widest pores start to drain first. Simultaneously, a matric suction, ψ, builds up in the soil material. Water loss from the soil material ceases when the matric suction has reached the level of the suction externally applied by the extraction apparatus. In other words, if the matric suction, ψ, is nil, then the corresponding soil moisture content, SM_ψ, equals SM_0; if $\psi > 0$ cm suction, then $SM_\psi < SM_0$. The decrease in SM_ψ is not necessarily proportional to the increase in matric suction ψ. A coarse, sandy soil has many wide pores and loses already much of its moisture when exposed to a low or moderate suction. A clay soil however consists of fine particles and is finely porous. These narrow pores retain water with a higher (capillary) force than wide pores do. They release this water only after the high capillary forces are compensated by an equally high matric suction. It follows that the soil moisture content, SM_ψ, is determined by the matric suction, ψ, and also by the volume and size distribution of all soil pores. The latter characteristics are largely a function of the particle-size distribution and packing density of the soil material.

Figure 24 illustrates this point by presenting typical $SM_\psi - \psi$ relations for a

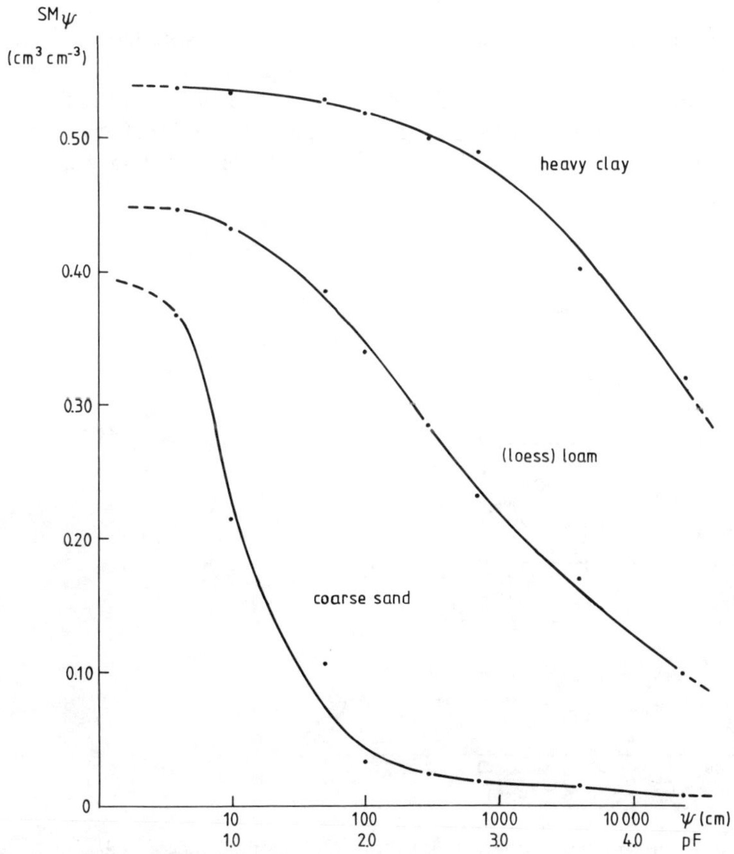

Figure 24. Some characteristic $SM\psi - \psi$ relations of soils of different texture.

fine (clay) soil, a medium-fine (loam) soil and a coarse (sand) soil. The graph shows that at a given matric suction, a finely grained soil retains more water than a coarser one. (In the schematized presentation used here, effects of hysteresis, causing different ψ – SM_ψ relations upon wetting and upon drying, are neglected.) Matric suction values observed in soils range between 0 and > 20 000 cm. To facilitate graphical presentation of the SM_ψ – ψ relation over the entire ψ range, soil scientists commonly relate SM_ψ to log ψ; the SM_ψ – log ψ curve is known as the soil's 'pF curve'. A matric suction of, for instance, 16 000 cm corresponds with a pF value of log(16 000) = 4.2. This notation will not be used in this monograph, but is important for those who search for information in literature to be aware of.

The particle-size distribution can be established for any soil material. To this end, the size of the individual matrix particles is determined in the laboratory and attributed to one of the following three particle size classes: 'sand'(0.05 – 2.0 mm in diameter), 'silt' (0.002 – 0.05 mm) and 'clay' (< 0.002 mm).

The relative proportions of sand, silt and clay particles in the soil determine to which 'texture'class the soil material belongs. Figure 25 is a graphical representation of the sand-silt-clay composition of each texture class. As the figure shows, a soil material consisting of 0.35 g g^{-1} clay particles and 0.30 g g^{-1} silt, would be classified as a 'clayloam'. It contains of course also 0.35 g g^{-1} of sand grains.

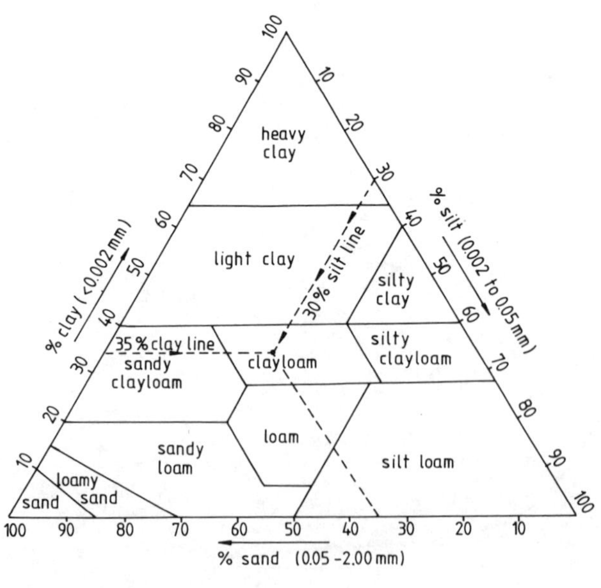

Figure 25. Graphical representation of the sand-silt-clay composition of mineral soil material in the different soil texture classes.

In 'normal' soils, which are neither extremely loose nor extremely compact, the total pore volume and pore-size distribution are reasonably well correlated with the texture class. This correlation can be put to use: as shown in Figure 24, an increase in matric suction is associated with a decrease in soil-moisture content, but the actual relationship is co-determined by pore characteristics. In a schematized set-up, the latter are conveniently represented by a texture-specific constant, γ.
Analysis of $SM_\psi - \psi$ relationships measured in the Netherlands, suggests that SM_ψ can satisfactorily be described by:

$$SM_\psi = SM_0 \cdot e^{-\gamma \cdot (\ln\psi)^2} \quad (27)$$

Indicative values of SM_0 and γ for the various soil texture classes are presented in Table 16. Equation 27 has a limited applicability in the case of certain tropical soils. This will be discussed in more detail in Section 5.3.

Exercise 27
A soil material consists of 0.20 g g^{-1} clay and of 0.40 g g^{-1} silt. Determine its texture class (use Figure 25). Determine the soil moisture content, SM_ψ, if the soil's matric suction, ψ, is 1000 cm (Use Equation 27 and Table 16).

Table 16. Indicative values of SM_0 and γ for the various soil texture classes.

Texture class*	SM_0 ($cm^3\,cm^{-3}$)	γ (cm^{-2})
coarse sand	0.395	0.1000
fine sand	0.364	0.0288
loamy sand	0.439	0.0330
fine sandy loam	0.504	0.0207
silt loam	0.509	0.0185
loam	0.503	0.0180
loess loam	0.455	0.0169
sandy clay loam	0.432	0.0096
silty clay loam	0.475	0.0105
clay loam	0.445	0.0058
light clay	0.453	0.0085
silty clay	0.507	0.0065
heavy clay	0.540	0.0042
peat**	0.863	0.0112

* Texture class criteria are depicted in Figure 25.
** Highly variable class.

Soil-moisture flow

Moisture stored in the soil is not entirely static; it can move through the pores at a rate that is proportional to the 'hydraulic gradient' and inversely proportional to the resistance in the soil pores. The driving force is called the hydraulic head, which is composed of matric forces and gravity forces. In the case of a water-saturated soil, water flow is driven solely by gravity because there is no matric suction. The ratio of hydraulic head, H_n, and flow distance, L_n, is the hydraulic gradient. The reciprocal value of pore resistance is known as the hydraulic conductivity. The hydraulic conductivity depends on the volume and properties of all pores that can be used for water transport; it is thus also a function of the soil-moisture content or suction. The commonly used symbol for the hydraulic conductivity is k_ψ of the soil-moisture content or suction. The commonly used symbol for the hydraulic conductivity is k_ψ. Under conditions of water saturation, k_ψ equals k_0, the 'saturated conductivity' (cm d^{-1}), and vertical saturated water flow between two arbitrary points n_1 and n_2 follows Darcy's law (an analogue of Ohm's law):

$$F = K_0 \cdot H_n/L_n \tag{28}$$

where

- F is the flow rate between points n_1 and n_2 (cm d^{-1})
- k_0 is the saturated hydraulic conductivity (cm d^{-1})
- H_n is the total difference in hydraulic head between points n_1 and n_2 (cm)
- L_n is the distance between points n_1 and n_2 (cm)

If $\psi > 0$ cm, the total hydraulic head is composed of both gravity forces and matric suction:

$$H_n = \psi + g_n \tag{29}$$

where

- H_n is the total hydraulic head at point n (cm)
- ψ is the matric suction at point n (cm)
- g_n is the gravity head at point n, equal to the vertical distance between point n and the groundwater level (cm)

The gravity head, g_n, is defined as negative in downward direction.

Exercise 28

A Dutch silty clay soil with a groundwater table at 1 m below soil surface

contains 0.3 cm^3 cm^{-3} moisture at a point 10 cm below soil surface. Calculate the matric suction at that point (use Table 16 and isolate ψ from Equation 27). Calculate the gravity head at that point.
What would the total hydraulic head be at that point, if the soil were a loam soil instead of a silty clay soil?

If $\psi > 0$ cm, the hydraulic conductivity, k_ψ, is lower than k_0 because some of the pores have emptied and are no longer functioning as transport channels for water ('unsaturated flow'). The remaining water-filled pores are on average narrower, and present, therefore, a higher resistance to flow, which implies that k_ψ decreases at an increasing rate with increasing suction. The relation between k_ψ and ψ is not simple. If ψ is very low, i.e. below 10 cm, most soils loose no water at all because they lack pores that are wide enough to drain at such low levels of suction. It follows that $k_\psi = k_0$ at matric suctions below this so-called 'air entry point'. This effect is shown in Figure 26 but it is disregarded in the model. If ψ rises and pores drain, k_ψ decreases as a function of its momentary value and of the compounded pore characteristics, expressed in an empirical texture-specific constant α:

$$dk_\psi/d\psi = k_\psi \cdot \psi$$

Integration of this expression yields:

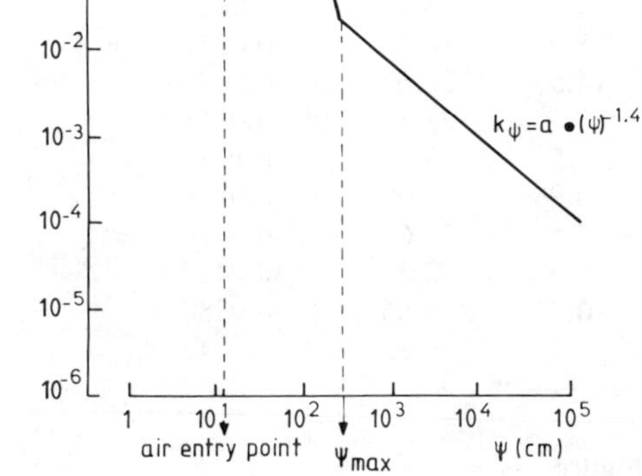

Figure 26. Generalized $k\psi - \psi$ relations, as described by Rijtema (1965).

$$k_\psi = k_0 \cdot e^{-\alpha \cdot \psi} \tag{30}$$

where

k_ψ is the hydraulic conductivity at matric suction ψ (cm d^{-1})
k_0 is the texture – specific saturated hydraulic conductivity (cm d^{-1})
α is a texture-specific empirical constant (cm^{-1})

Indicative values of k_0 and α for the various texture classes are presented in Table 17. Measurements have shown that Equation 30 does not hold over the entire range of ψ values occurring in soils. If ψ exceeds a texture-specific suction limit, ψ_{max}, an empirical equation has to be used to describe the relation between k_ψ and ψ:

$$k_\psi = a \cdot \psi^{-1.4}, \qquad \text{if } \psi > \psi_{max} \tag{31}$$

where

a is a texture-specific empirical constant ($cm^{2.4}$ d^{-1})
ψ_{max} is a texture-specific suction limit (cm)

Table 17. Indicative values of suction limit, ψ_{max}, saturated hydraulic conductivity, k_0, and constants a and α for the various texture classes.

Texture class*	ψ_{max} (cm)	k_0 (cm d^{-1})	α ($cm^{2.4}$ d^{-1})	α (cm^{-1})
coarse sand	70	1120	0.080	0.224
fine sand	175	50	10.9	0.0500
loamy sand	200	26.5	16.4	0.0398
fine sandy loam	290	12.0	26.5	0.0248
silt loam	300	6.5	47.3	0.0200
loam	300	5.0	14.4	0.0231
loess loam	130	14.5	22.6	0.0490
sandy clay loam	200	23.5	33.6	0.0353
silty clay loam	170	1.5	36.0	0.0237
clay loam	300	0.98	1.69	0.0248
light clay	300	3.5	55.6	0.0174
silty clay	50	1.3	28.2	0.0480
heavy clay	80	0.22	4.86	0.0380
peat**	50	5.3	6.82	0.1045

(Source: Rijtema, 1969, adapted).
* Texture class criteria are depicted in Figure 25.
** Highly variable class.

Indicative values of a and ψ_{max} for the various texture classes are presented in Table 17.

Observing the suction limit, ψ_{max}, a dynamic version of Equation 28 can be used to describe water flow in all situations:

$$F = k_{\psi} \cdot \delta H / \delta L \qquad (32)$$

where

- F is the flow rate (cm d^{-1})
- δH is the difference in hydraulic head (cm)
- δL is the distance of flow (cm)

Exercise 29
Calculate the indicative hydraulic conductivity of a loam soil at a matric suction of 100 cm. Use Table 17. Do the same for a loam soil at ψ = 1000 cm; use Table 17.

3.2.2 *The individual terms of the water balance*

Water transport through the upper root – zone boundary
The actual infiltration rate, IM, is determined by the total water supply at the soil surface, but also by the maximum possible infiltration rate of water into the soil, IM_{max}. This implies that IM can only be determined if IM_{max} and all components of surface water supply are known (Equation 25). Therefore, these variables will be discussed first.

The maximum rate of infiltration (IM_{max})
Infiltration of water into the soil results in wetting of the surface layer. Within the wetted surface soil three zones may be distinguished:

a. the uppermost soil layer; this zone is completely saturated with water.
b. the 'wetting front'; this is a relatively thin zone marking the transition between the wetted and the unaltered soil matrix.
c. the 'transmission zone' between a and b, which has a uniform soil moisture content throughout.

The maximum infiltration rate is determined by matrix forces and gravity forces. The influence of matrix forces is expressed in the 'sorptivity' (S, in cm $d^{-0.5}$) of the material. The sorptivity value indicates the rate at which a soil absorbs water if the matric suction is the only driving force. Sorptivity varies with the particle size distribution of the matrix material and the initial moisture content of the matrix.

Table 18 specifies 'standard sorptivity values', S_0, for each of the standard soil texture classes. Standard sorptivity represents the sorptivity of a completely dry matrix, i.e. a soil material with a moisture content, SM_ψ, of 0 $cm^3\ cm^{-3}$. Stroosnijder (1976) demonstrates that sorptivity decreases linearly with increasing moisture content of the soil. The following approximate relation is therefore suggested:

$$S = S_0 \cdot (1 - SM_\psi / SM_0) \qquad (33)$$

where

S	is the actual sorptivity ($cm\ d^{-0.5}$)
S_0	is the standard sorptivity ($cm\ d^{-0.5}$)
SM_ψ	is the soil moisture content ($cm^3\ cm^{-3}$)
SM_0	is the total pore space ($cm^3\ cm^{-3}$)

At prolonged infiltration, the importance of sorptivity decreases relative to the importance of gravity forces. The maximum infiltration rate is ultimately

Table 18. Indicative values of standard sorptivity (S_0)* and transmission zone permeability (A) for the various soil texture classes.

Texture class**	S_0 ($cm\ d^{-1/2}$)	A ($cm\ d^{-1}$)
coarse sand	50.16	119.23
fine sand	21.44	30.33
loamy sand	19.20	17.80
fine sandy loam	17.57	9.36
silt loam	14.46	5.32
loam	11.73	3.97
loess loam	13.05	8.88
sandy clay loam	19.05	16.51
silty clay loam	6.15	1.18
clay loam	4.70	0.76
light clay	10.74	2.94
silty clay	3.98	0.80
heavy clay	1.93	0.15
peat	7.44	1.86

* Based on values reported by Stroosnijder (1976).
** Texture class criteria are depicted in Figure 25.

limited to a value dictated by the hydraulic conductivity of the transmission zone (A, expressed in cm d^{-1}). The value of A depends on the moisture content of the transmission zone and on the geometry of the matrix material. Because matrix geometry and soil texture are normally related, indicative A – values have been postulated for each soil texture class as presented in Table 18.

In reality, the maximum infiltration rate (i.e. IM_{max}, expressed in cm d^{-1}) is determined by the combined effects of matrix forces and gravity forces:

$$IM_{max} = S_0 \cdot (1 - SM_{\psi}/SM_0) \cdot (\Delta t)^{-0.5} + A \qquad (34)$$

Exercise 30

In 3 hours of continuous rain, a total of 1.5 cm of precipitation is measured in a rain gauge. If this rain falls on a flat and level tract of moist ($SM_{\psi} = 0.35$ cm^3 cm^{-3}) loam soil, will the land flood? Suppose it rains for 6 hours at a stretch on the same land and with the same intensity. Does that make any difference? Explain your answer.

The precipitation rate (P)

Not all of the gauged precipitation reaches the soil surface and is thus potentially available for infiltration. A small part is intercepted by the crop and evaporates directly from the vegetation. The interception during a given period of time is strongly influenced by rainfall distribution. This becomes evident as one realizes that the storage capacity of the vegetation is limited; once this storage capacity is fully used, any additional precipitation will reach the soil surface. The quantitative effect of interception on the soil moisture balance is difficult to analyze. On the one hand evaporation of intercepted water from the leaf surfaces reduces the efficiency of precipitation as a source of available soil moisture, while, on the other hand, 'steered drip' from the vegetation can increase the availability of water to a crop. Interception will conceivably reduce transpiration to some extent, depending on the roughness of the crop surface, wind velocity and the apparent diffusion resistance of the crop. This would counteract the effect of interception on the soil – moisture balance.

In view of the uncertain effect of interception and the low confidence level of interception estimates in situations where weather records are less than perfect, gauged precipitation is introduced in the water balance.

The effective irrigation rate (I_e)
Effective irrigation represents the net input of irrigation water into the root zone. It may be measured directly in the field, e.g. with rain gauges in the case of sprinkler irrigation. More often, only the gross amount of water released at the project headworks is measured and then effective irrigation water inputs are approximated by multiplying the rate of water release at the headworks with an overall efficiency factor:

$$I_e = I \cdot E_p \qquad (35)$$

where

I_e	is the effective irrigation rate (cm d^{-1})
I	is the rate of water release at headworks (cm d^{-1})
E_p	is the overall efficiency factor

The overall efficiency factor, E_p, is composed of three partial reduction factors which express losses incurred during conveyance of the water from headworks to field canals, during field canal flow, and at application of the irrigation water:

$$E_p = E_d \cdot E_f \cdot E_i \qquad (36)$$

where

E_d	is the field application efficiency factor
E_f	is the field canal efficiency factor
E_i	is the conveyance efficiency factor

Table 19 presents indicative efficiency factors as suggested by Doorenbos & Pruitt (1977). These values illustrate that the gross quantity of water to be supplied at the project headworks under unfavourable conditions (e.g. small parcelling, unlined canals and/or poor management), greatly exceeds the net quantity that is available to the crop.

Exercise 31
Determine, using Table 19, the overall efficiency (E_p) of irrigation in a small (< 1000 ha), ineffectively managed scheme, where maize on a loam soil receives rotational surface irrigation according to a pre – determined schedule. The scheme consists of small (< 20 ha) blocks served from unlined canals.

Table 19. Conveyance (E_i), field canal (E_f), distribution ($E_i \times E_f$), and field application (E_d) efficiencies.

Conveyance efficiency (E_i)	
Continuous supply with no substantial change in flow	0.9
Rotational supply in projects of 3000-7000 ha and rotation areas of 70-300 ha, effective management	0.8
Rotational supply in large schemes (>10000 ha) and small schemes (<1000 ha) with resp. problematic communication and less effective management:	
based on predetermined schedule	0.7
based on advance request	0.65
Field canal efficiency (E_f)	
Blocks larger than 20 ha:	
— unlined	0.8
— lined or piped	0.9
Blocks up to 20 ha:	
— unlined	0.7
— lined or piped	0.8
Distribution efficiency ($E_i \times E_f$)	
Average for rotational supply with management and communication:	
— adequate	0.65
— sufficient	0.55
— insufficient	0.4
— poor	0.3
Field application efficiency (E_d)	
Surface methods:	
— sandy soils	0.55
— loamy soils	0.7
— clayey soils up to	0.6
Subsurface methods	
— up to	0.8
Sprinkler:	
— hot dry climate	0.6
— moderate climate	0.7
— humid and cool climate	0.8
Rice	0.32

(Source: Doorenbos & Pruitt, 1977.)

The actual evaporation rate (E_a)

Actual evaporation from the soil surface depends on:

- the potential evaporation rate, an expression of the evaporative demand of the atmosphere (E_0, expressed in cm d^{-1}; Section 3.1)

– the hydraulic conductivity of the surface soil
– the shading effect of the canopy (= f (LAI))

In the absence of a crop, the maximum rate of evaporation, E_m, is assumed to be equal to the potential evaporation rate, E_0, if sufficient water can reach the soil surface. Only in exceptional cases, e.g. in permeable strongly capillary soils with a shallow water table, or when there is water on the soil surface, could such a situation persist for a prolonged period of time. This would not be without danger. Particularly in semi–arid areas, where evaporative demand is high, accumulation of considerable quantities of soluble salts at or near the soil surface would be the undesirable result. In the presence of a crop, E_m is always lower than E_0, even if the rate of water supply from below is not limiting the rate of evaporation. Ritchie (1972) suggested the following relation to account for the reduction in the rate of evaporation due to shading of the soil surface under a crop canopy:

$$E_m = E_0 \bullet e^{-0.4 \times LAI} \qquad (37)$$

where

LAI is the leaf area index ($m^2\,m^{-2}$)

In all cases where there is water on the soil surface (i.e. $SS_t > 0$) or where the rate of water loss through evaporation is lower than the rate at which capillary rise from the groundwater can supply water to the soil surface, the actual rate of evaporation, E_a, equals the maximum rate, E_m:

$$E_a = E_0 \bullet e^{-0.4 \times LAI}, \quad \text{if} \quad SS_t > 0 \quad \text{or} \quad E_m < CR \qquad (38)$$

with CR the rate of capillary rise ($cm\,d^{-1}$)

Normally, cropping requires that the groundwater table is (kept) at some depth and capillary rise cannot cover evaporation losses. When this occurs, a very high matric suction builds up in the upper few centimetres of the soil and a thin, air dry ‘mulch layer’ forms. It is very difficult to quantify the reduction in evaporation due to the presence of such a mulch layer. In the present model the following assumptions are made:

– the moisture content of air dry soil material (SM_a) is set at an arbitrary value of 0.33 x SM_{16000}
– evaporation from the soil is at a maximum when the soil is water–saturated
– if E_m exceeds CR, the ratio between actual and maximum evaporation from the soil surface is equal to the ratio between the actual and maximum

quantities of water that could ultimately evaporate from the rooted soil. It follows that:

$$E_a = E_m \cdot (SM_\psi - SM_a)/(SM_0 - SM_a)\text{ , if } E_m > CR \qquad (39)$$

If water is present on the soil surface, i.e. $SS_t > 0$, E_a assumes the value E_m. The following example illustrates how the actual rate of evaporation is approximated.

Assume a loam soil with a matric suction, ψ, of 500 cm throughout the root zone. The groundwater depth, z_t, is 200 cm. The soil is cropped with maize with a LAI of 4.0; the potential evaporation rate, E_0, is 0.5 cm d^{-1}

The maximum evaporation rate, E_m, is calculated using Equation 37:

$$E_m = 0.5 \cdot e^{-0.4 \times 4.0} = 0.10 \text{ cm } d^{-1}$$

The next step is to verify whether the calculated maximum evaporation rate, E_m, exceeds the rate of capillary rise in this situation. The theory behind the quantification of capillary rise will be discussed later in this section; for now it suffices to know that Tables 21 – 34 specify for each soil texture class the relation between flow distance, matric suction and rate of capillary rise. The use of these tables is simple: the matric suction of the root zone and the flow distance to the surface are known (in this case 500 cm and 200 cm, respectively). From the proper table (Table 26 for loam soils) the CR value that corresponds with the known combination of matric suction and distance, is read off: CR = 0.06 cm d^{-1}.
Hence, E_m is slightly higher than CR, so a mulch layer forms. The actual rate of evaporation must therefore be calculated using Equations 37 and 39:

$$E_a = 0.10 \times (0.25 - 0.03)/(0.50 - 0.03) = 0.047 \text{ cm } d^{-1}$$

Exercise 32
It is a common notion among soil scientists that medium – textured soils do not easily develop a mulch layer; coarse – textured sandy soils and fine – textured clay soils develop a mulch layer more rapidly. Consider the following field situation: ψ = 1000 cm, z_t = 100 cm and E_0 = 0.5 cm d^{-1}. The land is cropped; LAI = 3.0. Check for each of the 14 standard soil texture classes whether or not a mulch layer would form under these conditions.
(Compare E_m with CR; use Tables 21 – 34).

The surface storage capacity (SS_{max})
In a number of situations, agricultural land may be flooded. The obvious

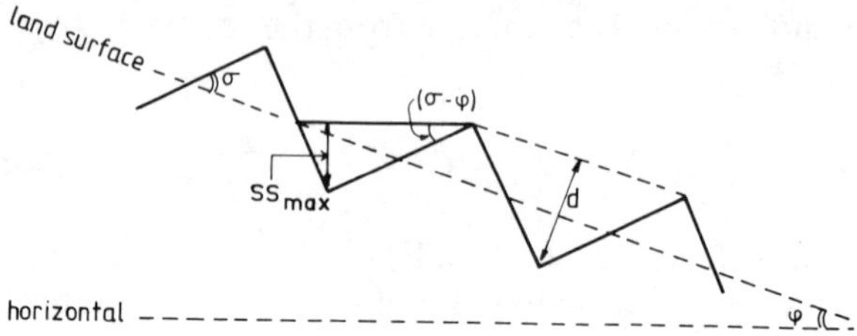

Figure 27. Schematic representation of the surface storage capacity, SS_{max}, of land. σ = clod/furrow angle (degrees). ϕ = slope of the land (degrees). d = surface roughness (cm).

example is a rice field where semi–permanent flooding is a cultivation measure. Temporary flooding may also occur where upland crops are grown, e.g. during and directly after a heavy rain shower or as a consequence of irrigation. The quantity of water that can potentially be stored on top of the surface (SS_{max}, expressed in cm) is determined by the surface properties and the slope angle of the land (Figure 27).

The surface storage capacity is mathematically described as:

$$SS_{max} = 0.5 \cdot \frac{\sin^2(\sigma-\phi)}{\sin\sigma} \cdot \frac{\cot(\sigma+\phi) + \cot(\sigma-\phi)}{2 \cdot \cos\sigma \cdot \cos\phi} \quad (40)$$

where

SS_{max}	is the surface storage capacity (cm)
d	is the surface roughness (cm)
σ	is the clod angle/furrow angle (°)
ϕ	is the slope angle of the land (°)

In most cases the clod/furrow angle, σ, lies between 30 and 45°; the field slope, ϕ, of land which can still be used for agriculture is commonly less than 13.5° (= 30 percent). The surface roughness, d, is of the order of 20 cm for contour–ploughed land, 6 to 8 cm for land tilled with light equipment and 1 to 2 cm for untilled land. Figure 28 presents a nomogram based on Equation 40, giving SS_{max} as a function of surface roughness, d, and field slope, ϕ; the clod/furrow angle, σ, is fixed at 30°.

Equation 40 does not apply in the case of bunded rice fields, where surface storage capacity is an exogenous variable, determined by plot dimensions, slope angle and effective bund height. In all other situations, the quantity of water actually stored on the surface at the beginning of a time interval (SS_t) can never exceed SS_{max}.

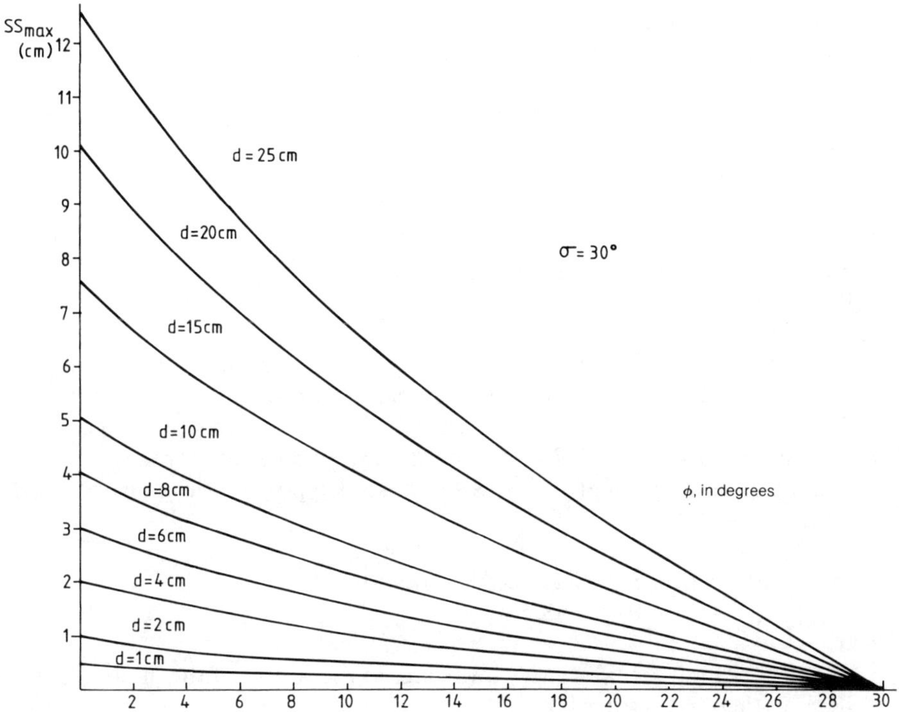

Figure 28. Nomogram of SS_{max} as a function of surface roughness, d, and field slope, ϕ, at a fixed clod/furrow angle (σ) of 30°.

Exercise 33

A tract of land, with a slope of 10°, is ploughed along the contours. The furrow depth, d, is 20 cm; $\sigma = 30°$.

Calculate the surface storage capacity, SS_{max}, using Equation 40. Verify the result with the aid of Figure 28.

The actual surface storage (SS_t) and run off (SR)

As discussed earlier in this section, the quantity of water stored on the surface can remain the same, or increase or decrease in the course of a time interval. Whether a change in surface storage takes place and, if so, in which direction, depends on the infiltration capacity of the soil, IM_{max}, and the sum of rainfall and irrigation, diminished by losses due to evaporation.

Quantification of DS, the change in surface storage and of SR, the surface run off, is now a matter of simple book keeping:

– first, the equilibrium situation is considered in which the supply equals the soil's infiltration capacity:
if

$$P + I_e - E_a = IM_{max}$$

then

$$DS = 0 \text{ (there is no change in surface storage)}$$
$$SR = 0 \text{ (there is no surface run off)} \qquad (41)$$

– if supply is lower than the infiltration capacity of the soil, i.e. if $P + I_e - E_a < IM_{max}$, then all surface water supplied can infiltrate and there is still some infiltration capacity left. This will cause a decline in surface storage ($DS > 0$); there will be no run off ($SR = 0$). The decline in surface storage can never exceed the quantity present ($DS \leqslant SS_t/\Delta t$), so two possibilities must be considered:

1. excess infiltration capacity exceeds actual storage:

$$IM_{max} - (P + I_e - E_a) \geq SS_t/\Delta t,$$

then

$$DS = SS_t/\Delta t \text{ (all actual storage infiltrates in the course of the interval } \Delta t)$$
$$SR = 0 \text{ (there is no run off)} \qquad (42a)$$

2. excess infiltration capacity is lower than actual storage:

$$IM_{max} - (P + I_e - E_a) < SS_t/\Delta t,$$

then

$$DS = IM_{max} - (P + I_e - E_a) \text{ (the decline in surface storage is limited by the available infiltration capacity)}$$
$$SR = 0 \text{ (there is no run off)} \qquad (42b)$$

– if surface supply exceeds the infiltration capacity of the soil, i.e. if $P + I_e - E_a > IM_{max}$, then the excess supply is, in the first instance, stored on the surface. The storage capacity available, amounts to $(SS_{max} - SS_t)$. If excess supply exceeds the available storage capacity, then $(SS_{max} - SS_t)$ is filled first and the rest is lost as surface run off, SR. Consequently, there are again two possibilities:

1. excess supply exceeds available storage capacity:

$$P + I_e - E_a - IM_{max} > (SS_{max} - SS_t)/\Delta t,$$

then

$$DS = -(SS_{max} - SS_t)/\Delta t \text{ (all available surface storage capacity is used up)}$$
$$SR = (P + I_e - E_a) - IM_{max} - DS \qquad (43a)$$

2. excess supply is equal to or lower than the available storage capacity:

$(P + I_e - E_a) - IM_{max} \leqslant (SS_{max} - SS_t)/\Delta t$,

then

$DS = IM_{max} - (P + I_e - E_a)$ (all excess supply can be stored in available storage capacity)

$SR = 0$ (there is no surface run off) (43b)

In Equation 25, the rate of infiltration of water at the upper root zone boundary was described as:

$$IM = P + I_e - E_a + DS - SR$$

The actual infiltration rate cannot exceed a maximum, IM_{max}, that has been quantified, as have all other terms in Equation 25. Therefore, the actual infiltration rate IM can now be quantified. A generally applicable procedure for the quantification of surface influx, IM, is as follows:

1st step: Determine SM_{ψ} using Equation 27 and Table 16. Determine the maximum infiltration rate, IM_{max}, using Equation 34, and Table 18.

2nd step: Determine the surface storage capacity, SS_{max}, using Equation 40, or the nomogram presented in Figure 28.

3rd step: Determine from your data set the precipitation rate, P.

4th step: Calculate the effective irrigation rate, I_e, using Equations 35 and 36 and Table 19.

5th step: Determine SS_t and determine the rate of capillary rise, CR, using the appropriate table from Tables 21 - 34. Determine E_m using Equation 37.
if $SS_t > 0$, or $E_m \leqslant CR$, then $E_a = E_m$
if $E_m > CR$ then calculate E_a using Equation 39.

6th step: Calculate the difference between supply and maximum infiltration rate: $P + I_e - E_a - IM_{max} = Q$.
if $Q = 0 \rightarrow$ step 7
if $Q < 0 \rightarrow$ step 8
if $Q > 0 \rightarrow$ step 9

7th step: If $Q = 0$, then $IM = P + I_e - E_a$ (Combination of Equations 25 and 41)
$DS = 0$ and $SR = 0$.

8th step: If $Q < 0$, determine whether $-Q \geq SS_t/\Delta t$ or $-Q < SS_t/\Delta t$.
if $-Q \geq SS_t/\Delta t$, then $IM = P + I_e - E_a + SS_t/\Delta t$ (Combination of Equations 25 and 42a)
$DS = SS_t/\Delta t$ and $SR = 0$
if $-Q < SS_t/\Delta t$, then $IM = IM_{max}$ (Combination of Equations 25 and 42b).
$DS = -Q$ and $SR = 0$.

9th step: If $Q > 0$, then calculate $(SS_{max} - SS_t)/\Delta t$.
if $Q > (SS_{max} - SS_t)/\Delta t$, then $IM = IM_{max}$ (Combination of Equations 25 and 43a).
$DS = (SS_{max} - SS_t)/\Delta t$, and $SR = Q - (SS_{max} - SS_t)/\Delta t$.
if $Q \leqslant (SS_{max} - SS_t)/\Delta t$, then $IM = IM_{max}$ (Combination of Equations 25 and 43b).
$DS = - Q$ and $SR = 0$.

Exercise 34

During a ten-day time interval, a gauged 0.5 cm d^{-1} of rain falls on a tract of clay loam soils ($\psi = 623$ cm) under maize. The crop has a leaf area index (LAI) of 4.0. The land is dry at the surface ($SS_t = 0$). The slope, ϕ, is 10°, the surface roughness, d, is 2 cm and the clod angle, σ, 30°. The potential evaporation rate, E_0, amounts to 0.4 cm d^{-1}. The groundwater table is at a depth of 150 cm below soil surface. No irrigation is applied.
Calculate the actual infiltration rate, IM, during the time interval.
Calculate the rate of change in surface storage – if any – during the time interval.
Calculate the rate of surface run off – if any – during the time interval.
Note: Carefully follow the nine-step calculation procedure presented in this paragraph. Retain the calculation results for later use.

Water flow through the lower root zone boundary (CR – D)

The infiltration rate, IM, describes the net influx of water into the soil via the upper boundary of the rooted soil volume. It is, of course, also possible that water enters or leaves the root zone via its lower boundary (at a depth of RD cm; Figure 23).
As explained in Subsection 3.2.1, water flow in soils is driven by the total hydraulic head, H_n, which consists of a matric suction, ψ, and a gravity head, g_n:

$$H_n = \psi + g_n$$

The gravity head, g_n, was defined as negative in downward direction. This means that the total hydraulic head, H_n, at a given point above the groundwater is positive if $\psi > g_n$, i.e. if the matric suction ψ (cm) exceeds the vertical distance (cm) between that point and the groundwater. A positive hydraulic head allows positive (in other words, upward) water flow from the groundwater to the point with matric suction ψ. This is called capillary rise (CR, expressed in in cm d^{-1}). If $\psi < g_n$, the resulting total hydraulic head is negative and drives downward water flow from the rooted surface soil to the groundwater at depth z_t. This downward flow is called percolation (D, expressed in cm d^{-1}). As water flows only in one direction at a time, CR = 0 if D > 0, and D = 0 if CR > 0. For the sake of convenience, the rate of influx into the root zone at depth RD is represented in the water-balance equation (Equation 24) by only one term with notation (CR – D), expressed in cm d^{-1}. The two components of this term, CR and D, will be separately discussed in the following.

The rate of capillary rise (CR)

The steady state solution of the universal flow Equation 32 makes it possible to calculate capillary rise:

$$CR = k_\psi \cdot \left(\frac{d\psi}{dz} - 1\right), \quad \text{for } z_t \geq z \geq RD \tag{44}$$

As k_ψ is described by different equations for low-suction conditions (Equation 30) and high-suction conditions (Equation 31), the integration of Equation 44 has to be done for two ψ ranges.

For the low-suction range ($\psi \leqslant \psi_{max}$), the relation between CR, ψ and the flow distance (z_t – RD) is rather straightforward; this relation is elaborated by Rijtema (1965). In the present approach, Rijtema's equation is used in a slightly adapted form because the ψ range below air entry point is not separately considered. The adapted equation reads:

$$CR = \frac{k_0 \cdot (e^{-\alpha \cdot \psi} - e^{-\alpha \cdot (z_t - RD)})}{e^{-\alpha \cdot (z_t - RD)} - 1}, \quad \text{if } \psi \leqslant \psi_{max} \tag{45}$$

For the high-suction range, the relation between CR, ψ and (z_t – RD) has to be calculated by numerical integration. This technique does not consider water flow over the distance (z_t – RD) in one step, but divides this distance in small increments, $\Delta(z_t - RD)$, using the expression:

$$CR = k_{\bar{\psi}} \cdot \left(\frac{\bar{\psi}}{\Delta(z_t - RD)} - 1\right) \tag{46}$$

where

$\bar{\psi}$ is the mean suction in the increment $\Delta(z_t - RD)$
$k_{\bar{\psi}}$ is $a \cdot \bar{\psi}^{-1.4}$ (see Equation 31)

Numerical integration techniques are very time-consuming without computers. The reason for this is that the accuracy of the ultimately calculated CR-value is best if the individual $\Delta(z_t - RD)$ increments are very small. The number of calculations required to cover the entire distance (z_t − RD) becomes very large then and this prohibits the use of this technique in the approach presented here. This problem is avoided by the use of tables relating the rate of capillary rise over the distance (z_t − RD) to any combination of ψ and (z_t − RD). Such tables have been prepared by Rijtema (1969) for the various soil texture classes; they are presented as Tables 21-34.

Table 21. The vertical distance of capillary flow, cm, in relation to flow rate and matrix potential for the soil texture class coarse sand.

ψ (cm)	CR ($cm\ d^{-1}$)							
	0.5	0.4	0.3	0.2	0.15	0.1	0.06	0.02
20	20.0	20.0	20.0	20.0	20.0	20.0	20.0	20.0
50	43.4	44.1	45.0	46.1	46.8	47.6	48.4	49.4
100	44.4	45.4	46.7	48.5	49.8	51.6	53.9	58.8
250	44.5	45.5	46.7	48.6	49.9	51.7	54.0	59.2
500	44.5	45.5	46.8	48.6	49.9	51.8	54.1	59.5
1000	44.5	45.5	46.8	48.6	49.9	51.8	54.2	59.7
2500	44.5	45.5	46.8	48.6	49.9	51.8	54.2	59.8
5000	44.5	45.5	46.8	48.6	50.0	51.8	54.3	59.9
10000	44.5	45.5	46.8	48.6	50.0	51.9	54.3	60.0
16000	44.5	45.5	46.8	48.6	50.0	51.9	54.3	60.0

Table 22. The vertical distance of capillary flow, cm, in relation to flow rate and matrix potential for the soil texture class loamy sand.

ψ (cm)	CR ($cm\ d^{-1}$)							
	0.5	0.4	0.3	0.2	0.15	0.1	0.06	0.02
20	3.7	4.3	5.0	6.0	6.6	7.5	8.3	9.4
50	11.6	13.6	16.5	20.9	24.2	28.9	34.4	43.1
100	13.9	16.4	20.1	26.2	31.0	38.3	48.1	68.8
250	14.5	17.2	21.1	27.6	32.2	41.1	52.7	81.5
500	14.7	17.5	21.5	28.2	33.7	42.4	54.8	87.6
1000	14.9	17.7	21.8	28.7	34.3	43.3	56.4	92.3
2500	15.1	17.9	22.1	29.1	34.9	44.1	57.7	96.4
5000	15.2	18.0	22.3	29.4	35.2	44.6	58.5	98.7
10000	15.3	18.1	22.4	29.5	35.5	45.0	59.1	100.6
16000	15.3	18.2	22.5	29.6	35.6	45.2	59.5	101.6

Table 23. The vertical distance of capillary flow, cm, in relation to flow rate and matrix potential for the soil texture class loamy fine sand.

ψ (cm)	CR (cm d-1)							
	0.5	0.4	0.3	0.2	0.15	0.1	0.06	0.02
20	19.4	19.6	19.7	19.8	19.8	19.9	19.9	20.0
50	47.2	47.8	48.3	48.8	49.1	49.4	49.6	49.9
100	83.2	85.7	88.6	91.8	93.6	95.6	97.2	99.0
250	101.2	106.8	114.1	124.4	131.8	142.3	155.8	185.1
500	103.2	109.2	117.3	129.3	138.2	151.8	171.1	225.9
1000	104.6	111.1	119.8	133.0	143.1	159.2	183.2	260.5
2500	105.9	112.6	121.9	136.1	147.3	165.5	193.7	291.4
5000	106.6	113.6	123.1	138.0	150.0	169.1	199.8	309.7
10000	107.2	114.4	124.1	139.4	151.8	172.1	204.8	324.6
16000	107.6	114.7	124.7	140.2	152.8	173.7	207.4	332.5

Table 24. The vertical distance of capillary flow, cm, in relation to flow rate and matrix potential for the soil texture class fine sandy loam.

ψ (cm)	CR (cm d-1)							
	0.5	0.4	0.3	0.2	0.15	0.1	0.06	0.02
20	19.1	19.3	19.5	19.7	19.7	19.8	19.9	20.0
50	47.0	47.5	48.1	48.7	49.0	49.4	49.6	49.9
100	88.8	90.1	92.3	94.7	95.9	97.2	98.3	99.4
250	137.0	145.1	155.5	169.8	179.5	192.3	206.9	230.0
500	140.9	150.0	162.0	179.4	192.0	210.6	235.5	297.1
1000	143.3	153.1	166.0	185.3	199.4	222.4	254.7	350.6
2500	145.4	155.6	169.4	190.4	206.7	232.5	271.6	399.9
5000	146.6	157.1	171.4	193.4	210.7	238.4	281.5	429.3
10000	147.5	158.3	173.0	195.8	213.9	243.2	289.5	453.2
16000	148.0	158.9	173.9	197.1	215.6	245.8	293.7	465.9

Table 25. The vertical distance of capillary flow, cm, in relation to flow rate and matrix potential for the soil texture class silt loam.

ψ (cm)	CR (cm d-1)							
	0.5	0.4	0.3	0.2	0.15	0.1	0.06	0.02
20	18.3	18.6	18.9	19.3	19.4	19.6	19.8	19.9

Table 25. (continued)

ψ (cm)	CR (cm d^{-1})							
	0.5	0.4	0.3	0.2	0.15	0.1	0.06	0.02
50	44.2	45.3	46.3	47.5	48.1	48.7	49.2	49.7
100	81.2	84.2	87.6	91.3	93.3	95.4	97.2	99.0
250	127.7	137.2	149.2	165.7	176.8	191.3	207.3	231.3
500	134.4	145.4	160.0	181.4	197.2	220.5	252.1	326.9
1000	138.6	150.7	167.0	191.9	211.1	241.0	285.4	414.7
2500	142.3	155.3	173.0	201.0	223.2	259.1	315.2	500.3
5000	144.4	157.9	176.6	206.3	230.2	269.6	322.8	552.2
10000	146.1	160.1	179.4	210.6	236.0	278.2	347.0	594.7
16000	147.0	161.2	180.9	212.8	239.0	282.8	354.6	617.4

Table 26. The vertical distance of capillary flow, cm, in relation to flow rate and matrix potential for the soil texture class loam.

ψ (cm)	CR (cm d^{-1})							
	0.5	0.4	0.3	0.2	0.15	0.1	0.06	0.02
20	17.7	18.2	18.6	19.0	19.3	19.5	19.7	19.9
50	42.2	43.5	45.0	46.5	47.3	48.2	48.9	49.6
100	74.0	77.7	82.1	87.0	89.8	92.9	95.6	98.5
250	102.6	111.0	122.1	137.8	148.8	164.0	182.0	214.3
500	104.6	113.7	125.7	143.1	155.7	174.2	198.4	256.5
1000	105.9	115.3	127.8	146.3	160.1	180.6	209.1	287.2
2000	107.0	116.7	129.7	149.1	163.8	186.2	218.3	314.4
5000	107.7	117.5	130.7	150.7	165.9	189.4	223.7	330.5
10000	108.2	118.2	131.6	152.0	167.7	192.0	228.0	343.5
16000	108.5	118.5	132.1	152.7	168.6	193.4	230.3	350.5

Table 27. The vertical distance of capillary flow, cm, in relation to flow rate and matrix potential for the soil texture class loess loam.

ψ (cm)	CR (cm d^{-1})							
	0.5	0.4	0.3	0.2	0.15	0.1	0.06	0.02
20	18.9	19.1	19.3	19.5	19.7	19.8	19.9	20.0
50	43.8	44.9	46.0	47.3	47.9	48.6	49.1	49.7
100	65.4	69.0	73.3	78.9	82.4	86.8	91.1	96.6
250	71.5	76.4	83.0	92.7	100.1	111.2	126.6	165.5

Table 27. (continued)

ψ (cm)	CR (cm d⁻¹) 0.5	0.4	0.3	0.2	0.15	0.1	0.06	0.02
500	74.2	79.8	87.4	99.3	108.8	124.0	147.3	218.3
1000	76.2	83.3	90.8	104.4	115.6	134.1	163.8	264.7
2000	78.0	84.5	93.7	108.8	121.3	142.7	178.2	307.0
5000	79.0	85.8	95.4	111.3	124.7	147.8	186.7	332.1
10000	79.8	86.8	96.8	113.4	127.5	151.9	193.5	352.5
16000	80.3	87.3	97.5	114.4	128.9	154.1	197.1	363.4

Table 28. The vertical distance of capillary flow, cm, in relation to flow rate and matrix potential for the soil texture class sandy clay loam.

ψ (cm)	CR (cm d⁻¹) 0.5	0.4	0.3	0.2	0.15	0.1	0.06	0.02
20	19.4	19.5	19.6	19.7	19.8	19.9	19.9	20.0
50	47.3	47.8	48.3	48.9	49.1	49.4	49.7	49.9
100	85.1	87.5	90.1	93.0	94.6	96.3	97.7	99.2
250	110.0	116.3	124.4	136.0	144.3	156.0	170.9	201.7
500	113.9	121.2	130.9	145.7	157.0	174.6	200.3	272.7
1000	117.0	125.0	136.0	153.2	167.0	189.4	224.5	338.6
2500	119.6	128.2	140.3	159.6	175.6	202.3	245.8	400.5
5000	121.1	130.1	142.8	163.4	180.6	209.8	258.3	437.6
10000	122.3	131.6	144.8	166.5	184.6	215.9	268.4	467.9
16000	123.0	132.4	145.9	168.1	186.8	219.1	273.8	484.0

Table 29. The vertical distance of capillary flow, cm, in relation to flow rate and matrix potential for the soil texture class silty clay loam.

ψ (cm)	CR (cm d⁻¹) 0.5	0.4	0.3	0.2	0.15	0.1	0.06	0.02
20	14.0	14.9	15.9	17.1	17.7	18.4	19.0	19.7
50	31.0	33.5	36.5	40.0	42.1	44.4	46.4	48.8
100	48.1	53.1	59.4	67.9	73.3	80.0	86.6	94.9
200	58.2	65.3	75.0	89.5	100.1	115.3	134.7	175.0
500	61.4	69.4	80.5	97.4	110.6	130.8	159.4	235.7
1000	64.7	73.5	85.9	105.6	121.3	146.7	185.2	305.6

Table 29. (continued)

ψ (cm)	CR (cm d⁻¹)							
	0.5	0.4	0.3	0.2	0.15	0.1	0.06	0.02
2500	67.5	76.9	90.5	112.5	130.5	160.4	208.0	371.7
5000	69.1	79.0	93.2	116.5	135.9	168.5	221.4	411.5
10000	70.4	80.6	95.3	119.8	140.2	175.0	232.2	443.9
16000	71.1	81.5	96.5	121.5	142.6	178.5	238.0	461.1

Table 30. The vertical distance of capillary flow, cm, in relation to flow rate and matrix potential for the soil texture class clay loam.

ψ (cm)	CR ($cm\ d^{-1}$)							
	0.5	0.4	0.3	0.2	0.15	0.1	0.06	0.02
20	12.1	13.1	14.3	15.8	16.7	17.7	18.5	19.5
50	25.6	28.3	31.7	36.0	38.6	41.7	44.7	48.1
100	37.6	42.4	48.8	57.7	63.8	71.8	80.3	92.0
250	43.6	49.7	58.2	71.2	80.9	95.2	113.7	153.9
500	43.9	50.1	58.7	71.8	81.8	96.4	115.8	160.2
1000	44.0	50.3	58.9	72.2	82.3	97.2	117.1	164.0
2500	44.1	50.4	59.1	72.5	82.7	97.9	118.2	167.2
5000	44.2	50.5	59.3	72.7	83.0	98.2	118.8	169.1
10000	44.3	50.6	59.4	72.9	83.2	98.5	119.4	170.7
16000	44.3	50.6	59.4	73.0	83.3	98.7	119.6	171.5

Table 31. The vertical distance of capillary flow, cm, in relation to flow rate and matrix potential for the soil texture class light clay.

ψ (cm)	CR ($cm\ d^{-1}$)							
	0.5	0.4	0.3	0.2	0.15	0.1	0.06	0.02
20	17.1	17.6	18.1	18.7	19.0	19.3	19.6	19.9
50	40.8	42.4	44.0	45.8	46.8	47.8	48.6	49.5
100	73.4	77.4	81.9	87.0	89.9	93.0	95.6	98.5
250	114.5	124.7	137.9	156.0	168.3	184.5	202.4	229.3
500	122.0	134.0	150.1	173.8	191.4	217.3	252.0	332.9
1000	127.1	140.3	158.3	186.1	207.7	241.3	290.7	432.8
2500	131.3	145.6	165.4	196.7	221.8	262.4	325.5	532.3
5000	133.8	148.7	169.6	203.0	230.1	274.9	346.1	593.1

Table 31. (continued)

ψ (cm)	CR (cm d^{-1})							
	0.5	0.4	0.3	0.2	0.15	0.1	0.06	0.02
10000	135.8	151.2	173.0	208.0	236.8	284.9	362.9	642.9
16000	136.9	152.6	174.7	210.7	240.4	290.3	371.8	669.6

Table 32. The vertical distance of capillary flow, cm, in relation to flow rate and matrix potential for the soil texture class silty clay.

ψ (cm)	CR (cm d^{-1})							
	0.5	0.4	0.3	0.2	0.15	0.1	0.06	0.02
20	12.3	13.3	14.5	15.9	16.8	17.2	18.6	19.5
50	22.3	24.8	28.0	32.3	35.2	38.8	42.4	47.1
100	28.0	31.7	36.7	44.4	50.0	58.1	67.9	84.8
250	33.8	38.9	46.2	58.1	67.7	83.0	105.0	158.0
500	37.1	43.1	51.7	66.3	78.5	98.8	130.2	220.6
1000	39.7	46.2	56.0	72.6	86.9	111.3	150.7	277.1
2500	41.9	49.0	59.6	78.0	94.1	122.1	168.6	329.4
5000	43.1	50.5	61.7	81.2	98.4	128.4	179.1	360.7
10000	44.2	51.8	63.4	83.8	101.8	133.5	187.6	386.1
16000	44.7	52.5	64.3	85.1	103.6	136.3	192.1	399.7

Table 33. The vertical distance of capillary flow, cm, in relation to flow rate and matrix potential for the soil texture class heavy clay.

ψ (cm)	CR (cm d^{-1})							
	0.5	0.4	0.3	0.2	0.15	0.1	0.06	0.02
20	4.7	5.5	6.7	8.6	10.0	12.0	14.3	17.6
50	7.9	9.5	11.7	15.5	18.5	23.1	29.0	39.8
100	9.4	11.3	14.1	19.0	23.0	29.5	38.7	60.1
250	10.4	12.6	15.8	21.6	26.5	34.6	47.0	82.2
500	11.0	13.3	16.8	23.0	28.4	37.5	51.7	95.9
1000	11.4	13.8	17.6	24.1	29.9	39.7	55.4	106.8
2500	11.8	14.3	18.2	25.1	31.1	41.5	58.5	116.1
5000	12.0	14.6	18.5	25.6	31.8	42.6	60.3	121.5
10000	12.2	14.8	18.8	26.0	32.4	43.5	61.8	125.9
16000	12.3	14.9	19.0	26.3	32.7	44.0	62.6	128.3

Table 34. The vertical distance of capillary flow, cm, in relation to flow rate and matrix potential for the soil texture class peat.

ψ (cm)	CR (cm d^{-1})							
	0.5	0.4	0.3	0.2	0.15	0.1	0.06	0.02
20	15.4	16.1	16.9	17.8	18.3	18.8	19.3	19.7
50	22.9	24.8	27.1	30.4	32.7	35.8	39.3	45.0
100	24.4	26.6	29.6	34.0	37.4	42.5	49.5	66.5
250	25.9	28.5	32.0	37.6	42.2	49.5	60.8	95.6
500	26.7	29.5	33.4	39.7	44.9	53.5	67.4	114.4
1000	27.3	30.3	34.4	41.2	46.9	56.6	72.5	129.5
2500	27.9	30.9	35.3	42.5	48.7	59.2	76.9	142.5
5000	28.2	31.3	35.8	43.3	49.7	60.8	79.5	150.1
10000	28.4	31.6	36.2	43.9	50.5	62.0	81.5	156.3
16000	28.5	31.8	36.3	44.3	51.0	62.7	82.6	159.6

Exercise 35
A loam soil cropped with maize (RD = 100 cm) has a matric suction, ψ, in the root zone of 300 cm. The water table depth, z_t, is 250 cm below soil surface. Determine whether there is capillary rise of water from the groundwater at depth z_t to the lower boundary of the root zone at depth RD.
Calculate CR with Equation 45 (Remember: $\psi \leqslant \psi_{max}$). Use Table 17 to identify texture-specific k_0 and α.
Now determine CR with the aid of Table 26.
What is the value of (CR − D) in this case?

The rate of percolation (D)

When the value of the matric suction is lower than the value of the gravity head, the total hydraulic head is negative and water movement will be downwards. In that situation, the rooted surface soil looses water to the subsoil and eventually to the groundwater through percolation. The rate of percolation, D (cm d^{-1}), can in principle be quantified with Equation 32. In the present approach, percolation is treated as infiltration of soil moisture from the lower boundary of the root zone (i.e. depth RD) into the subsoil. It is unlikely that the percolation rate will fluctuate very suddenly because the percolation water comes from the rooted surface soil, which buffers the rate of water release at depth RD. The percolation rate will be largely dictated by the gravity forces. When discussing the maximum infiltration rate, IM_{max}, a matric suction of the transmission zone of 10 cm was postulated, which results in a hydraulic conductivity of that zone of $k_0 \cdot e^{-\alpha \cdot 10} = A$ in cm d^{-1}. In the case of percolation, the role of the transmission zone is taken over by the root zone with matric suction ψ, so that percolation would proceed at a rate:

$$D = k_\psi \qquad (47)$$

Exercise 36

A loam soil cropped with maize (RD = 100 cm) has a matric suction, ψ, in the root zone of 100 cm. The water table depth, z_t, is 250 cm below soil surface.
Determine whether there is percolation of water from the lower boundary of the root zone (RD) to the groundwater at depth z_t.
Calculate the percolation rate, D.
What is the value of the term (CR – D) in this case?

It is clear that there is neither capillary rise nor percolation if the matric suction ψ is compensated by an equally high (but negative) gravity head g_n (see Equation 29). In that case the total hydraulic head is nil, there is no driving force and consequently no flow: (CR – D) = 0.

A generally applicable procedure for the calculation of (CR – D) is as follows:

1st step: Establish the total hydraulic head at the lower root zone boundary, H_{RD}, by subtracting from the matric suction, ψ, the distance between lower root zone boundary and water table:
$H_{RD} = \psi - (z_t - RD)$
2nd step: If $H_{RD} = 0$, then (CR – D) = 0
3rd step: If $H_{RD} < 0$, then (CR – D) = $-k_\psi$. Use Table 17 and Equation 30 (if $\psi \leqslant \psi_{max}$), or Equation 31 (if $\psi > \psi_{max}$)
4th step: If $H_{RD} > 0$, then (CR – D) = CR. The appropriate value of CR can be established using Tables 21-34.

Exercise 37

In Exercise 34, a situation was described with maize growing on a clay loam soil. Assume that the rooting depth of the crop, RD, is 75 cm.
Calculate, for the situation described in Exercise 34, the rate of water flow through the lower root-zone boundary, (CR – D).
Note: Retain the calculation result for later use.

Direct water loss from the interior root zone (T)

In addition to water movement via the upper and lower root zone boundaries, there is also water loss from the interior root zone through uptake by plant roots. According to the continuity principle, the quantity of water taken up from the root zone must be equal to the sum of crop transpiration and the increment in weight of all water contained in the crop. This latter portion is negligible compared with the quantity lost through crop transpiration. Therefore, the present model defines water loss from the interior root zone as equal to crop transpiration.

A crop growing under conditions of optimum water supply transpires at a maximum rate, T_m (expressed in cm d^{-1}). If water availability to the roots is less than optimum (i.e. if the matric suction, ψ, is higher or lower than optimum), plants actively curb their water consumption, which causes the actual transpiration rate, T, to be lower than T_m.
Consequently, it is necessary to:

- quantify the maximum transpiration rate, T_m, of each crop at a specific site at each moment
- define at which matric suction the crop senses water stress and reduces its transpiration rate
- relate the degree of reduction in transpiration (T/T_m) to the soil moisture content, SM_ψ.

The maximum transpiration rate (T_m)

The maximum transpiration rate under conditions of optimum water supply is a function of the total surface area of all transpiring leaves and the potential transpiration rate, T_0. The latter represents transpiration, under the prevailing environmental conditions, of a well watered, closed, short, green, standard crop as defined by Penman (Section 3.1). Assuming proportionality between light interception by the leaf surface and transpiration, the maximum transpiration rate is approximated by:

$$T_m = (1 - e^{-0.8 \times LAI}) \cdot T_0 \qquad (48)$$

The potential transpiration rate, T_0, is found by subtracting soil evaporation under 'Penman conditions' from the potential evapotranspiration rate, ET_0. (see Section 3.1). Assuming the leaf area index of a closed standard crop to be 5 to 6, the potential transpiration rate can be written as:

$$T_0 = ET_0 - E_0 \cdot e^{-0.4 \times (5 \text{ to } 6)}$$

or

$$T_0 = ET_0 - 0.1 \cdot E_0 \qquad (49)$$

so that the approximated maximum transpiration rate, T_m, becomes:

$$T_m = (1 - e^{-0.8 \times LAI})(ET_0 - 0.1 \times E_0) \quad (50)$$

The concept of water stress

Plants experience water stress, if the matric suction, ψ, is either too low or too high. Both situations will be discussed.

– ψ lower than optimum:

On empirical grounds, a soil-moisture content SM_{100} has been chosen as the highest moisture content at which water supply to the plant is optimal. Matric suction values lower than 100 cm are associated with some degree of oxygen deficiency in the root zone. This statement is not true for all soil texture classes: coarse sandy soils and peat soils contain ample air, even if their moisture content is somewhat in excess of SM_{100}; heavy clays may be oxygen deficient at matric suctions over 100 cm. It would be better to assume oxygen deficiency when the soil-air content runs below a fixed value of, say, 10 percent. This has not been done in the present model because a number of concepts and data borrowed from literature are based on SM_{100} as a discriminating soil-moisture value. Oxygen deficiency interferes with water uptake, presumably because maintenance of root activity requires energy. Under conditions of oxygen deficiency, therefore, stomata close to reduce transpiration. In the present model, transpiration is arbitrarily assumed to be nil when the root zone contains less than 0.05 $cm^3\ cm^{-3}$ of air. Crops exposed to this condition for a continuous period of more than 20 days perish. This set of rules does not apply to crops with air ducts ('aerenchym') in their roots, such as rice.

– ψ higher than optimum:

By convention, the quantity of water retained by the soil at a matric suction ψ of 100 cm ('field capacity'), diminished with the quantity of water retained at a suction of 16 000 cm ('permanent wilting point'), is considered the quantity that is ultimately available for uptake by plants:

$$W_a = RD \cdot (SM_{100} - SM_{16000}) \quad (51)$$

where

W_a is the quantity of water ultimately available (cm)

Plants experience water stress at a matric suction well below 16000 cm. They actively curb their water use when the matric suction becomes higher than a system-specific, 'critical' suction (ψ_{cr}). The corresponding critical moisture content, SM_{cr}, lies somewhere between SM_{100} and permanent wilting point and depends on plant characteristics, soil characteristics and evaporative demand. Irrigation engineers have established for a large number of crops, which fraction (p between 0 and 1.0) of W_a can be freely taken up by a specific crop at a given maximum transpiration rate, T_m. When the quantity of stored soil moisture decreases below $((1 - p) \cdot W_a + RD \cdot SM_{16000})$ cm, water uptake is impaired and the crop begins to close its stomata. The corresponding critical soil moisture content, SM_{cr}, can thus be found with:

$$SM_{cr} = (1 - p)(SM_{100} - SM_{16000}) + SM_{16000} \tag{52}$$

Table 20 presents indicative p – values for the most important field crops at different values of T_m (Doorenbos et al., 1978).

Reduction in transpiration under stress (T/T_m)
As explained in the foregoing, the actual rate of crop transpiration, T, equals the maximum transpiration rate, T_m, only if the soil-moisture content of the

Table 20. Soil water depletion fraction (p) as a function of maximum transpiration rate, T_m, for different crop groups.

Crop group*	T_m in cm d^{-1}								
	0.2	0.3	0.4	0.5	0.6	0.7	0.8	0.9	1.0
1	0.45	0.38	0.30	0.25	0.23	0.20	0.18	0.16	0.15
2	0.60	0.50	0.43	0.35	0.30	0.28	0.25	0.23	0.20
3	0.75	0.65	0.55	0.45	0.40	0.38	0.33	0.30	0.25
4	0.85	0.75	0.65	0.55	0.50	0.48	0.43	0.38	0.35
5	0.92	0.85	0.75	0.65	0.60	0.55	0.50	0.48	0.45

*

Crop group	Crops
1	leaf vegetables, strawberry
1-2	cabbage, onion
2	clover, carrot, early tobacco
2-3	banana, pepper
3	grape, pea, potato
3-4	bean, sunflower, tomato, water melon, grass
4	citrus, groundnut, pineapple
4-5	alfalfa, cotton, tobacco, cassava, sweet potato, most grains
5	olive, safflower, sorghum, soybean, sugarcane

(Source: Doorenbos et al., 1978)

root zone is in the optimum range between SM_{100} and SM_{cr}. Actual transpiration is lower than maximum if $SM_\psi > SM_{100}$ or $SM_\psi < SM_{cr}$

In the low suction range ($SM_0 - 0.05 \geq SM_\psi > SM_{100}$), a linear relation is assumed between uptake of water for transpiration and soil air content:

$$T/T_m = ((SM_0 - 0.05) - SM_\psi)/((SM_0 - 0.05) - SM_{100}) \tag{53}$$

In the high suction range ($SM_{cr} \geq SM_\psi > SM_{16000}$), a linear relation is assumed between soil moisture availability and transpiration:

$$T/T_m = (SM_\psi - SM_{16000})/(SM_{cr} - SM_{16000}) \tag{54}$$

Summarizing, the actual transpiration rate, T, can be approximated for any value of SM_ψ with the aid of the following set of equations.

If $SM_\psi > SM_0 - 0.05$:

$$T = 0 \tag{55a}$$

If $(SM_0 - 0.05) \geq SM_\psi \geq SM_{100}$:

$$T = ((SM_0 - 0.05) - SM_\psi)/((SM_0 - 0.05) - SM_{100}) \cdot T_m \tag{55b}$$

If $SM_{100} > SM_\psi > SM_{cr}$:

$$T = T_m \tag{55c}$$

If $SM_{cr} \geq SM_\psi \geq SM_{16000}$:

$$T = (SM_\psi - SM_{16000})/(SM_{cr} - SM_{16000}) \cdot T_m \tag{55d}$$

If $SM_{16000} > SM_\psi$:

$$T = 0 \tag{55e}$$

A generally applicable procedure for the calculation of actual crop transpiration rate, T, is as follows:

1st step: Identify ET_0 and E_0 and approximate the potential transpiration rate, T_0, using Equation 49.
2nd step: Calculate the maximum transpiration rate, T_m, using Equation 48.
3rd step: Calculate the soil-moisture content of the root zone, SM_ψ, using Table 16 and Equation 27.
4th step: Calculate the critical moisture content of the root zone, SM_{cr}, using

Tables 16 and 20 and Equations 27 and 52. Retain the values of SM_{100} and SM_{16000}, which you must establish to solve Equation 52.

5th step: Calculate (SM_0 – 0.05) using Table 16.

6th step: Compare SM_ψ with (SM_0 – 0.05), SM_{100}, SM_{cr} and SM_{16000} and select the proper line from Equation 55 to calculate the actual transpiration rate, T.

Exercise 38

In Exercise 34, a situation was described where a maize crop with a leaf area index of 4.0 grows on a clay loam soil. Assume that the potential evapotranspiration rate, ET_0, averaged 0.35 cm d^{-1} over the ten-day time interval considered. Recall that the potential evaporation rate, E_0, was given in Exercise 35 as 0.4 cm d^{-1}; the matric suction, ψ, amounted to 623 cm.

Calculate the actual transpiration rate, T, of this crop. Use the 6 steps calculation procedure presented above.

Note: Retain the calculated value of T for later use.

3.2.3 *Solution of the water balance equation for all time intervals in a growing season*

Once the infiltration rate, IM, the subsurface flow rate, (CR – D), and crop transpiration rate, T, are quantified, the water balance in Equation 24 is simple to solve. The equation yields RSM, the rate of change in moisture content of the root zone.

The calculated rate of change in soil-moisture content can be used to establish the moisture content of the root zone at the end of the time interval. This is done by adding to the initial moisture content, SM_ψ, the calculated change, RSM x Δt. The result, SM_ψ + RSM x Δt, represents also the soil-moisture content at the beginning of the next time interval and therewith the value of SM_ψ for water-balance calculations over that interval. To keep variables belonging to different intervals apart, they are best labeled with a subscript:

$$SM_{\psi,(t+\Delta t)} = SM_{\psi,t} + RSM \cdot \Delta t \qquad (56)$$

where

$SM_{\psi,(t+\Delta t)}$ is soil-moisture content at the end of a time interval starting at day t and ending at day (t + Δt) ($cm^3\ cm^{-3}$)

$SM_{\psi,t}$ is soil-moisture content at the beginning of the time interval (i.e. at day t) ($cm^3\ cm^{-3}$)

RSM is rate of change in soil-moisture content over the time interval ($cm^3\ cm^{-3}\ d^{-1}$)

Evidently, the soil-moisture content is not the only variable that requires adjustment before the water-balance calculation of the next time interval can commence. A change in soil-moisture content is accompanied by changes in a number of physical system parameters (for instance the depth of the groundwater table, z_t). Other variables need to be adjusted to reflect crop development over the analyzed time interval, such as LAI and RD. There are also some variables that are entirely exogenous, such as P and I. Their values must be established for each individual time interval on the basis of information contained in the basic data set.

The updating of physical system characteristics will be discussed in Subsection 3.2.4. Changes in crop parameters will be discussed in Section 3.4, after the interrelations between transpiration and dry matter production have been treated.

It may be useful here to point out that the length of the time interval (Δt, in d) is not without significance for the accuracy of the generated calculation results. The rates of change in state variables of the system are calculated assuming that their values remain constant throughout the interval analyzed. In a dynamic process as crop growth, this cannot be entirely correct because in reality there is gradual change during a given time interval and system characteristics are bound to change (Section 2.3). Nonetheless, the use of state variables can give good results, if the ratio of their value to their rate of change is high compared to the time interval used. This means that the length of the interval Δt must be chosen commensurate with the dynamics of the process studied. In general, the accuracy of the results of the modelling exercise improves if the length of the individual time intervals is reduced. However, there are good arguments for some restraint in the reduction of Δt. Computational reasons aside, it is pointless to consider intervals of, say, 1 day ($\Delta t = 1$), if the resolution of the available basic data is low, e.g. when only monthly weather records are available. In practice, the choice of interval length is a compromise between the accuracy pursued, the dynamics of the process studied, the resolution of the available basic data and the computation capacity at hand. Experience has shown that for our purposes an interval length of 10 days is satisfactory in many cases. It should be realized, however, that there are also situations where shorter intervals are required for good results, e.g. in the case of a high total surface-water supply rate and/or a shallow groundwater table.

Exercise 39

In Exercise 34, a situation was described where a maize crop grows on a clay loam soil. In that exercise the rate of infiltration, IM, was calculated. In

Exercise 37, (CR − D) was calculated for the same situation, and in Exercise 38 the rate of crop transpiration, T, was calculated. Calculate the rate of change in moisture content of the root zone. Use the results of Exercises 34, 37 and 38 and the water balance Equation 24 (Remember: RD was 75 cm).
Calculate the soil-moisture content at the end of the time interval of 10 days. Use Equation 56.

3.2.4 *Variable adjustment following the water balance analysis for one time interval*

The soil-moisture content
Adjustment of the soil-moisture content – and consequently of the matric suction – poses no problem:

$$SM_{\psi,(t+\Delta t)} = SM_{\psi,t} + RSM \cdot \Delta t \qquad (56)$$

The groundwater depth
The net result of flow through the lower root zone boundary, (CR − D), induces a change in groundwater depth: Δz. Exact quantification of Δz is not easy. For practical purposes where groundwater is shallow and Δt short, it is assumed that the soil-moisture content of the subsoil – equal to SM_ψ at depth RD – increases linearly with depth to reach a value SM_0 at depth z_t. This situation is illustrated by Line A in Figure 29. In the case of capillary rise, water moves into the root zone, which causes the groundwater depth to drop over a distance Δz. Simultaneously, a new moisture profile will establish itself over the subsurface layer between RD and ($z_t + \Delta z$), (Figure 29, Line B). The flow through the lower root-zone boundary, (CR − D) x Δt cm, is equal to the surface area under line A, diminished by the quantity of water left in the layer between RD and ($z_t + \Delta z$), i.e. the surface area under line B.

Mathematical description of these surface areas and subsequent isolation of Δz yields:

$$\Delta z = 2 \cdot (CR - D) \cdot \Delta t/(SM_0 - SM_\psi) \qquad (57)$$

In many cultivated lands, the depth of the groundwater table is controlled by means of artificial drainage. This involves groundwater flow from the field to the drains under the influence of gravity. The drainage rate, D_{max} (cm d^{-1}), is the product of the hydraulic gradient (cm cm^{-1}) and the saturated conductivity, k_0, in accordance with the Universal Flow Equation. For a situation with steady downward flow, the gradient was described as the ratio of total head over flow distance (Equation 28). In the case of field drainage, the situation becomes more complicated because flow has not only a vertical but has also a

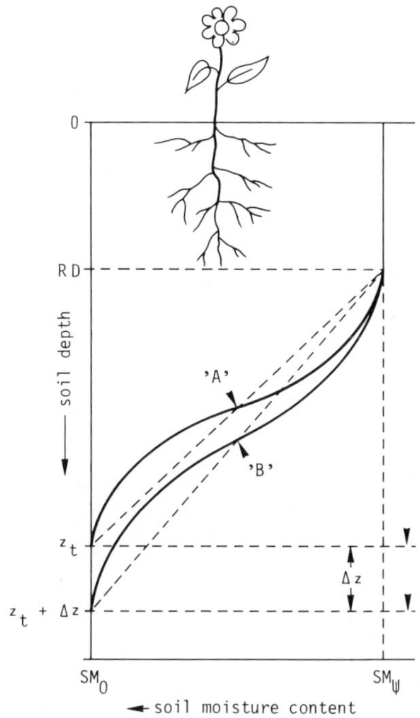

Figure 29. Schematic representation of the value of the matric suction, ψ, at various depths in the soil.

horizontal component, viz. flow from midway between drains to the drains, and a radial component for entry of water into the drains. In other words: in the case of drainage, the hydraulic gradient is composed of a vertical component, a lateral component and a radial component that depends on the wet perimeter of the drain.

To make drainage possible at all, there must be a gravity head. This means that the water table must be higher (= closer to soil surface), midway between drains than at the site of the drains; this difference in height, m_t, drives flow as schematically presented in Figure 30.

Clearly, the gradient of gravity head over flow distance is determined by the dimensions of the drain system. Kirkham (1958, 1961) has described this gradient as a function of m_t, drain spacing (L_d) and wet drain perimeter (πr_d), and has subsequently solved the Universal Flow Equation to obtain the drainage rate D_{max}. For drainage systems in deep and homogeneous soils, the resulting equation reads:

$$D_{max} = k_0 \cdot \frac{m_t}{m_t + \ln(L_d/\pi r_d) \cdot L_d/\pi} \qquad (58)$$

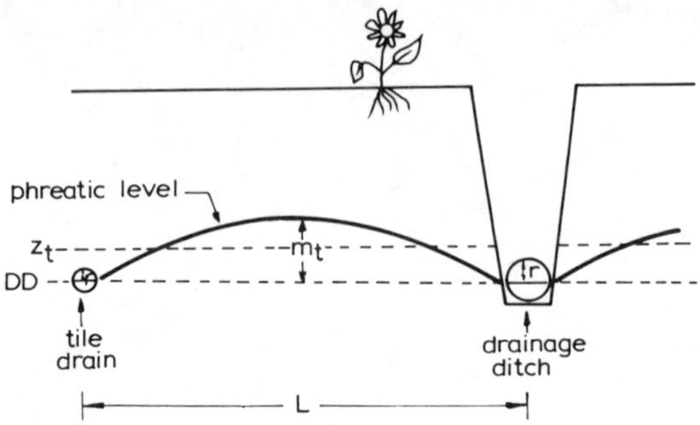

Figure 30. Schematic cross-section of a drained field.

where

D_{max} is drainage rate (cm d^{-1})
k_0 is saturated hydraulic conductivity (cm d^{-1})
m_t is hydraulic head midway between drains (cm)
L_d is drain spacing (cm)
r_d is drain radius (cm)

In the presence of an artificial drainage system, the change in groundwater depth, Δz, depends not only on unsaturated flow (CR – D), but also on the removal of groundwater by drainage (D_{max}). Consequently, Equation 57 must be extended to account for the loss of drainage water:

$$\Delta z = 2 \cdot \Delta t \cdot (D_{max} + (CR - D))/(SM_0 - SM_\psi) \quad (59)$$

The groundwater depth at the end of the time interval follows from:

$$z_{(t + \Delta t)} = z_t + \Delta z \quad (60)$$

where

$z_{(t+\Delta t)}$ is groundwater depth at the end of time interval (cm)
z_t is groundwater depth at the beginning of time interval (cm)

The hydraulic head midway between drains

In situations where artificial drainage is applied, a change in groundwater depth, Δz, affects the hydraulic head, m_t. Although the groundwater level between drains is slightly curved, the average groundwater depth is reasonably close to:

$$z_t = DD - 0.5 \cdot m_t \quad (61)$$

where

DD is drain depth (cm) (Figure 30)

It follows that the hydraulic head after adjustment of the groundwater depth, amounts to:

$$m_{(t+\Delta t)} = 2 \cdot (DD - z_{(t+\Delta t)}) \quad (62)$$

where

$m_{(t+\Delta t)}$ is hydraulic head midway between drains (cm) at the end of a time interval with a duration of Δt days.

Surface storage
Surface storage increases during a time interval if the supply of water exceeds the infiltration capacity, i.e. if $P + I_e - E_a > IM_{max}$. It decreases in the reverse case.

$$SS_{(t+\Delta t)} = SS_t - DS \cdot \Delta t \quad (63)$$

It follows from the discussion of DS in Subsection 3.2.2, that there are four possible situations:

– if $P + I_e - E_a - IM_{max} > (SS_{max} - SS_t)/\Delta t$,
then

$$SS_{(t+\Delta t)} = SS_{max} \quad (64a)$$

(Combination of Equations 63 and 43a).

– if $P + I_e - E_a - IM_{max} > 0$ and $\leqslant (SS_{max} - SS_t)/\Delta t$,
then

$$SS_{(t+\Delta t)} = SS_t + (P + I_e - E_a - IM_{max}) \cdot \Delta t \quad (64b)$$

(Combination of Equations 63 and 43b).

– if $P + I_e - E_a - IM_{max} \leqslant - SS_t/\Delta t$,
then

$$SS_{(t+\Delta t)} = 0 \quad (64c)$$

(Combination of Equations 63 and 42a).

– if $P + I_e - E_a - IM_{max} < 0$ and $> -SS_t/\Delta t$,
then

$$SS_{(t+\Delta t)} = SS_t + (P + I_e - E_a - IM_{max}) \cdot \Delta t \qquad (64d)$$

(Combination of Equations 63 and 42b).

Exercise 40
A cassava crop (RD = 50 cm) stands on an artificially drained tract of loam soil. The root zone has a matric suction of 100 cm. The dimensions of the drain system are: DD = 160 cm, L_d = 5000 cm, r_d = 8 cm, and the distance between groundwater level and drain depth, m_t = 60 cm. The interval length, Δt, is one day.
Calculate the initial groundwater depth, z_t, using Equation 61.
Calculate the drainage rate, D_{max}, using Equation 58 and Table 17.
Calculate (CR – D) using the calculation scheme given in Subsection 3.2.2.
Calculate the change in groundwater depth over the time interval, Δz, using Equation 59.
Calculate the groundwater depth at the end of the time interval, $z_{(t+\Delta t)}$, using Equation 60.
Calculate the hydraulic head midway between drains at the end of the time interval, $m_{(t+\Delta t)}$, using Equation 62.

3.3 The relation between water use and crop production

H. van Keulen and H.H. van Laar

3.3.1 *The concept of the transpiration coefficient*

To allow entrance of CO_2 into the plant, necessary for the assimilatory process, plants must have an open connection to the atmosphere. This connection is provided by small openings in the epidermis of the leaves, the stomata. In Figure 31, a schematic representation of a vertical section of a leaf is given, illustrating the position of the stomata. In this case stomata are only present on the lower side of the leaves, but in other species they may also be present on the upper side. Through these stomata, exchange of CO_2 takes places by diffusion.

The rate of diffusion can be described by a general Ohm's law equation:

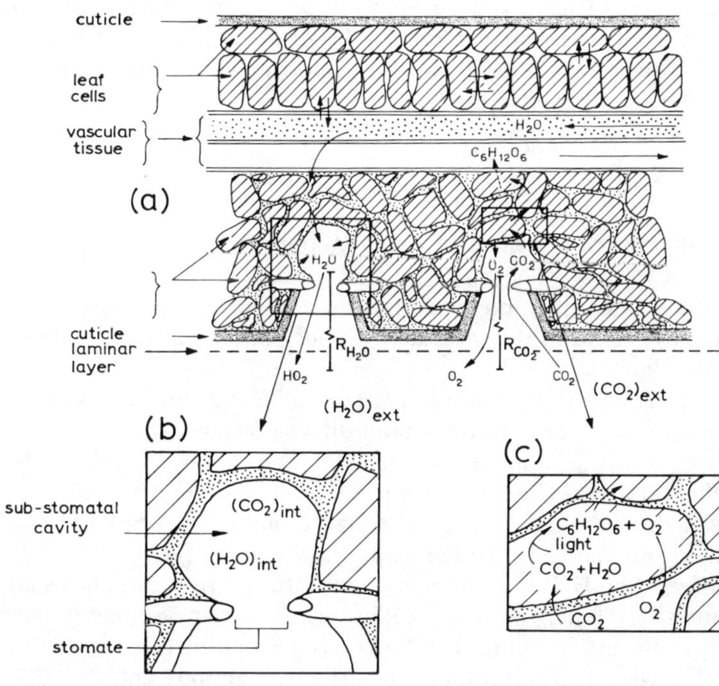

Figure 31. Schematic representation of a vertical section of a leaf (a). An enlargement showing a stomate (b) and an enlargement showing a photosynthetically active cell (c).

$$V_{CO_2} = \frac{(CO_2)_{ext} - (CO_2)_{int}}{R_{CO_2}} \quad (65)$$

where

V_{CO_2}	is the rate of CO_2 diffusion into the plant (kg $m^{-2}s^{-1}$)
$(CO_2)_{ext}$	is the concentration of CO_2 in the external air (kg m^{-3})
$(CO_2)_{int}$	is the concentration of CO_2 in the substomatal cavity (kg m^{-3})
R_{CO_2}	is the resistance to CO_2 diffusion (s m^{-1})

The total resistance to CO_2 diffusion is composed of two resistances in series: the resistance offered by a layer of still air, the boundary layer, just above the surface of the leaf; and the resistance offered by the stomata. In the total assimilation process a third resistance is of importance, that between the substomatal cavity and the chloroplasts, where the actual reduction of CO_2 takes place (Figure 31c). This resistance is referred to as the carboxylation resistance or the mesophyll resistance. In general, the latter resistance is small compared to the sum of the other two. The walls of the substomatal cavity (Figure 31b) are wet, hence the air inside the cavity is water-vapour saturated. When the stomata are open, this air is exposed to the atmosphere, which is normally not saturated with water vapour. The result is that water vapour diffuses from the substomatal cavity into the atmosphere. The rate of diffusion can be described analogous to Equation 65:

$$V_{H_2O} = \frac{(H_2O)_{int} - (H_2O)_{ext}}{R_{H_2O}} \quad (66)$$

V_{H_2O}	is the rate of diffusion of water vapour into the atmosphere (kg $m^{-2}s^{-1}$)
$(H_2O)_{int}$	is the concentration of water vapour inside the substomatal cavity, equal to the saturated vapour pressure at leaf temperature (kg m^{-3})
$(H_2O)_{ext}$	is the concentration of water vapour in the atmosphere (kg m^{-3}
R_{H_2O}	is the resistance to water vapour diffusion (s m^{-1})

Figure 31a illustrates that the resistance to diffusion of CO_2 from the substomatal cavity to the atmosphere, R_{CO_2}, is equal to that to diffusion of water vapour along the same pathway. Numerically the values differ, because CO_2 molecules are larger than H_2O molecules and the diffusion resistance is dependent on molecule size. Under the same conditions, R_{CO_2} is 1.66 times greater than R_{H_2O}. Taking this into account, it follows from Equations 65 and 66 that a relation exists between transpiration (water use) by a canopy and CO_2 assimilation. As explained in Section 2.1, the production of structural plant dry matter is directly proportional to CO_2 assimilation, adjusted for the respiratory losses. Because the latter depend also on the amount of dry matter present, it is not a fixed proportion of the gross CO_2 assimilation. Therefore, two different definitions are introduced to characterize the relation between water use and production: (i) the ratio of transpiration to assimilation (TAR), expressed in kg water transpired per kg of CO_2 fixed, (ii) the transpiration coefficient (TRC), expressed in kg water transpired per unit dry matter produced. The transpiration coefficient is not only dependent on the relation between assimilation and transpiration, but also on the quantity and composition of plant material present.

The relation between water use and dry-matter production was recognized already by early investigators in agricultural science and many experiments have been carried out to determine the exact dependencies between the two variables. Much of the earliest work was summarized by Briggs & Shantz (1913; 1914), who expressed the relation between water use and dry-matter production in the term 'water requirement', defined as the amount of water necessary to produce one unit of dry-matter weight (i.e. the equivalent of the transpiration coefficient). They established that differences exist between species, but also that for the same species different values were obtained in different years or in different environments. One of the main reasons for the latter variation may be deduced from a comparison of Equations 65 and 66. At a given value of R, that is at a fixed stomatal opening and a fixed thickness of the boundary layer, the rate of water loss will be higher when the difference in concentration of water vapour between the atmosphere and the substomatal cavity is larger. Thus, when the relative humidity in the atmosphere is lower, and the leaf temperature higher, more water will be lost. The gradient in CO_2 is hardly variable, because the CO_2 concentration in the atmosphere is fairly constant.

In a subsequent analysis, de Wit (1958) accounted for the influences of

Table 35. Transpiration coefficients for maize, variety North Western Dent measured in pots, in Akron Colorado in different years.

Year	Production (g)	Transpiration (kg)	Transpiration coefficient ($kg\ kg^{-1}$)
1911	160.7	58.8	365.9
1912	432.0	117.9	272.7
1913	356.3	141.1	396.0
1914	304.3	111.5	366.4
1915	112.0	28.4	253.6
1916	180.5	90.0	498.6
1917	336.8	120.5	357.8

(Source: Briggs & Shantz, 1913; 1914; Shantz & Piemeisel, 1927)

environmental conditions by relating production not directly to water use, but correcting the latter for the average evaporative demand (potential evapotranspiration, Section 3.1) during the growing period. This procedure is illustrated by the data in Table 35, giving values of the transpiration coefficient for maize grown in containers during various years. These values vary by a factor of almost two. In Figure 32 the production data are plotted against the ratio of total water loss and average evaporative demand during each growing period. The results illustrate that most of the variability in transpiration coefficient was indeed due to the varying evaporative demand during the various growing periods. This indicates that although dry-matter production and water use are closely associated, the actual quantitative relationships may be strongly modified by environmental conditions.

As mentioned before, also considerable differences were established between species growing under the same conditions. These differences mainly reflect different photosynthetic pathways, as explained in Section 2.1. The higher affinity to CO_2 of the carboxylating enzyme in C_4 species causes at a

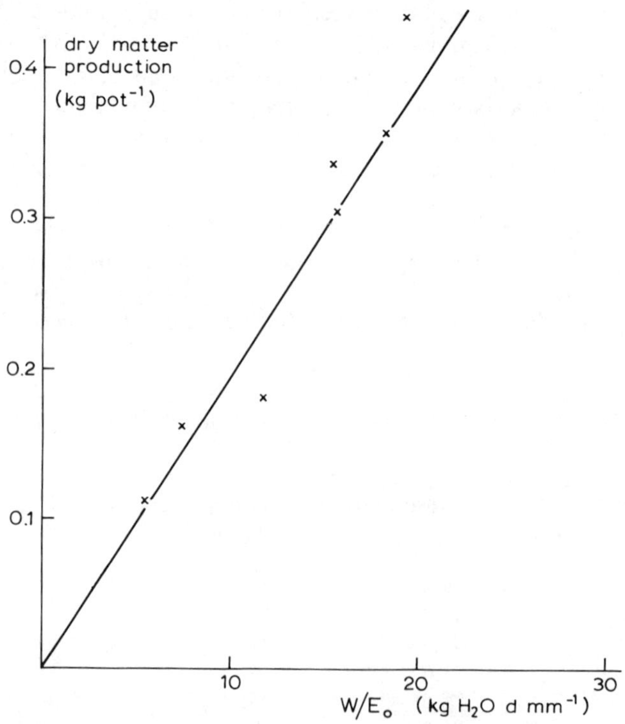

Figure 32. Relation between dry matter production and the ratio of transpiration to free water evaporation for maize cv. Northwestern Dent grown in pots. (Source: Briggs & Shantz, 1914)

given stomatal aperture, and hence a fixed transpiration rate, a higher rate of CO_2 assimilation in plants of the C_4 type than in plants of the C_3 type. The transpiration/assimilation ratio is therefore higher in species having the latter photosynthetic pathway.

Another source of variability in the transpiration coefficient is the type of stomatal control of the canopy. Under certain conditions, the assimilation process is controlled in such a way that the CO_2 concentration in the substomatal cavity ($(CO_2)_{int}$) is maintained at a constant level over a wide range of assimilation rates through adaptation of the stomatal opening (Goudriaan & van Laar 1978b). In terms of Equation 65 it means that a change in the gradient $(CO_2)_{ext} - (CO_2)_{int}$ is accompanied by a proportional change in the value of R_{CO_2}, so that V_{CO_2} remains constant. This is illustrated in Figure 33 where the relation between between V_{CO_2} and $1/R_{CO_2}$ is given for maize plants grown under controlled conditions. The result of this mechanism is that, for instance, under conditions of low light intensities, where CO_2 assimilation is determined by energy availability, plants having the regulatory mechanism

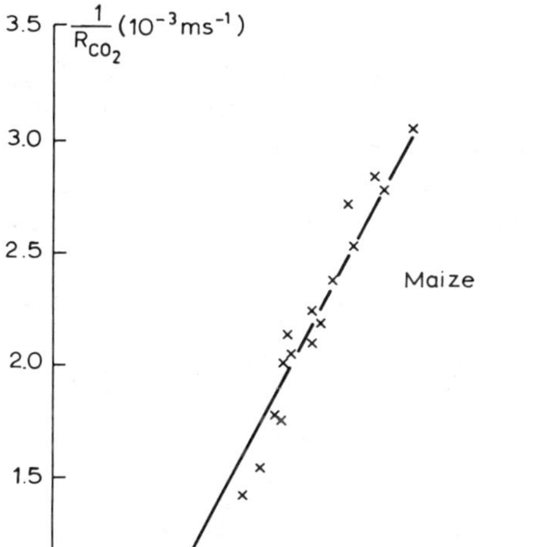

Figure 33. Relation between net CO_2 assimilation rate (v_{CO_2}) and total conductance for CO_2 for maize leaves.

will partially close their stomata, thus effectively reducing transpiration. Therefore plants having this regulatory mechanism exhibit a lower transpiration/assimilation ratio and hence a lower transpiration coefficient than plants in which the mechanism is absent. At high light intensities, the rate of CO_2 diffusion is determining assimilation. That rate is governed by the CO_2 gradient and, as the internal CO_2 concentration is higher in C_3 species than in C_4 species, again a difference in transpiration coefficient shows up at high light intensities.

Thus, in a schematized set-up, values of the transpiration/assimilation ratio in kg H_2O transpired per kg CO_2 fixed (or per kg CH_2O produced) can be determined for various conditions, i.e. for C_3 and C_4 species separately, for both groups further subdivided in regulating and non-regulating crops, and finally at different levels of vapour pressure deficit. An example of such data, calculated with a detailed simulation model of canopy assimilation and transpiration (de Wit et al., 1978), is given in Table 36.

3.3.2 Application of the transpiration coefficient

On the basis of crop characteristics and climatic conditions, potential daily gross CO_2 assimilation (F_{gc}) in kg ha^{-1} d^{-1} may be obtained for any arbitrary combination of time and place from Tables 1 and 2. By considering the average vapour pressure deficit during the period of interest, a value of the transpiration/assimilation ratio (TAR) may then be selected from Table 36. By multiplying the two values, the potential rate of transpiration is obtained in kg ha^{-1} d^{-1}.

Table 36. Transpiration/assimilation ratio (kg water per kg carbon dioxide) for various conditions.

C_3 species			
relative humidity (%)	25	50	75
R	84	70	56
NR	150	116	90
C_4 species			
R	57	45	35
NR	120	95	72

R denotes regulating stomata, NR non-regulating

Exercise 41
What is the conversion factor for the conversion of transpiration in kg ha^{-1} d^{-1} to the more common unit of mm d^{-1}.

In Section 3.1 potential evapotranspiration was calculated by Penman's combination method. In the case of a fully closed crop canopy, where soil evaporation is negligible, that value equals potential transpiration. In the preceding subsection it is argued that the value of the transpiration-assimilation ratio depends on plant type and stomatal behaviour. The considerable success of the Penman equation for the estimation of potential transpiration must be attributed to the fact that its value is very close to the value calculated for conditions of regulating stomata, which seems to be the most common field behaviour. The difference in transpiration-assimilation ratio between C_3 and C_4 species is mainly the result of differences in assimilation rate, transpiration being virtually identical, especially under high light conditions. In the situation where a crop has non-regulating stomata, the Penman equation underestimates potential crop transpiration. Correction factors for such conditions have been calculated, by comparing Penman values to transpiration values, calculated with a detailed physiologically based model of crop growth. The results are tabulated in Table 37 with a subdivision in C_3 and C_4 species and for overcast and clear conditions.

Under conditions of temporary water shortage in the soil, leading to partial stomatal closure, assimilation and transpiration are affected approximately to the same extent, hence the value of the transpiration coefficient remains constant. This characteristic permits an evaluation of the influence of moisture shortage on production. After determining potential transpiration, the amount of moisture available for plant uptake is determined from the soil-water balance, presented in Section 3.2. If that amount is lower than potential transpiration, actual transpiration falls short of the potential. Actual CO_2

Table 37. Correction factors for multiplying the Penman evapotranspiration to obtain potential crop transpiration in the absence of stomatal regulation.

Species	Sky condition	
	clear	overcast
C_3	1.47	2.1
C_4	1.6	3.0

assimilation (P_{gc}) is then directly determined from the amount of water transpired, by dividing the latter value by the earlier determined transpiration/assimilation ratio, hence $P_{gc} = T/TAR$, or
$P_{gc} = F_{gc} \times T/T_m$ (Section 3.2).

Exercise 42
Estimate the total dry-matter production for a hypothetical C_4 crop, exhibiting stomatal regulation, when 200 mm of moisture is available for transpiration during the growing period. Average relative humidity during that period is 75%. What would be the result if it was a C_3 crop lacking stomatal regulation?

3.3.3 The influence of nutritional status

A basic problem related to water-use efficiency is the influence of nutrient shortage on the relation between production and water use. The constancy of the transpiration coefficient, discussed in the preceding subsection, was restricted by de Wit (1958) in his analysis to situations where the 'nutrient status is not too low', and Viets (1962) concluded that 'all evidence indicates, that water use efficiency can be greatly increased, if fertilizers increase yield'.
The latter conclusion seems to be confirmed by experimental results obtained in natural herbaceous vegetation in the northern Negev desert of Israel (Figure 34), where the rate of water loss from the soil was practically identical under the nitrogen fertilized vegetation and the non-nitrogen fertilized one, although the former was growing at a substantially higher rate. Opposite results were, however, obtained in the Sahelian zone, where increased growth rates of natural pasture, resulting from fertilizer application, were accompanied by higher rates of transpiration (Stroosnijder & Koné, 1982).

For a more fundamental approach to the problem, the basic processes of CO_2 assimilation and transpiration must be considered. This is most conveniently done by measuring both exchange processes simultaneously under controlled conditions. The results of such an experiment are presented in Figure 35 for individual attached leaves of maize plants grown in the greenhouse, under both optimum and sub-optimum supply of nitrogen. The measured rate of net CO_2 assimilation is plotted versus the total conductance for water vapour exchange, that is the inverse of the sum of stomatal resistance and boundary layer resistance (R_{H_2O} in Equation 66), which is thus a direct measure of transpiration. The plants optimally supplied with nitrogen exhibit a markedly higher rate of net CO_2 assimilation than the plants under sub-optimum N supply, but they show a proportional increase in conductance for

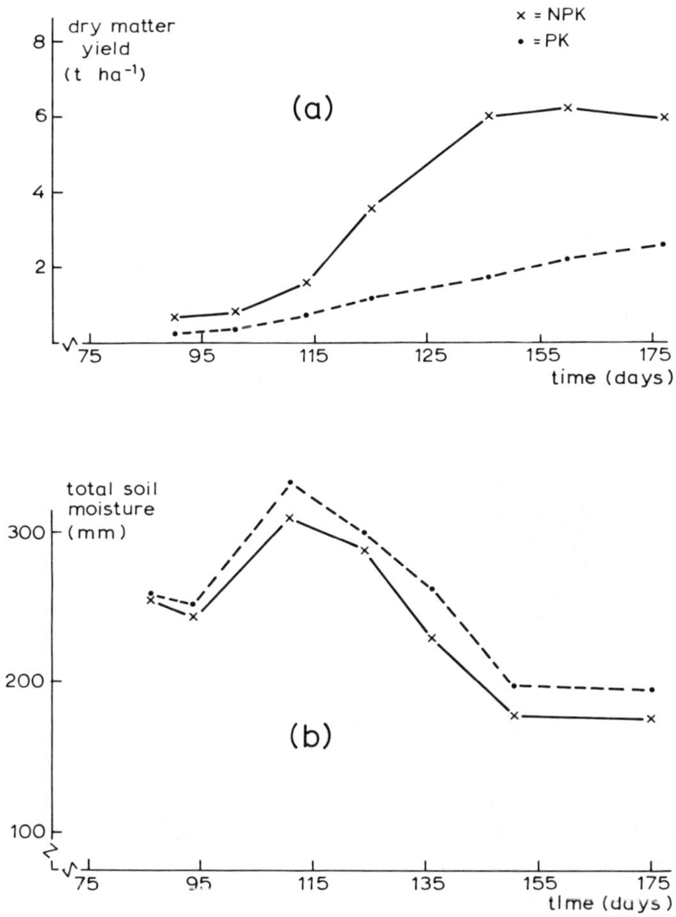

Figure 34. Cumulative dry matter accumulation of natural herbaceous vegetation with and without N fertilizer application (a) and time course of total soil moisture in the root zone (0–180 cm) of this vegetation (b). Data from Migda, Israel 1972/1973.

water vapour, hence a virtually constant assimilation/transpiration ratio. This behaviour is thus in accordance with that of the plants growing under the Sahelian conditions, and suggests, again, the existence of stomatal control by the internal CO_2 concentration, i.e. impaired assimilation due to nitrogen shortage is reflected in partial stomatal closure.

The slope of the line in Figure 35 represents the CO_2 gradient over the relevant resistance, i.e. the drop in CO_2 concentration from the atmosphere to the substomatal cavity. When the value of P_n at any point is divided by the relevant value of $1/R_{H_2O}$ (the latter multiplied by 1.66 to account for the

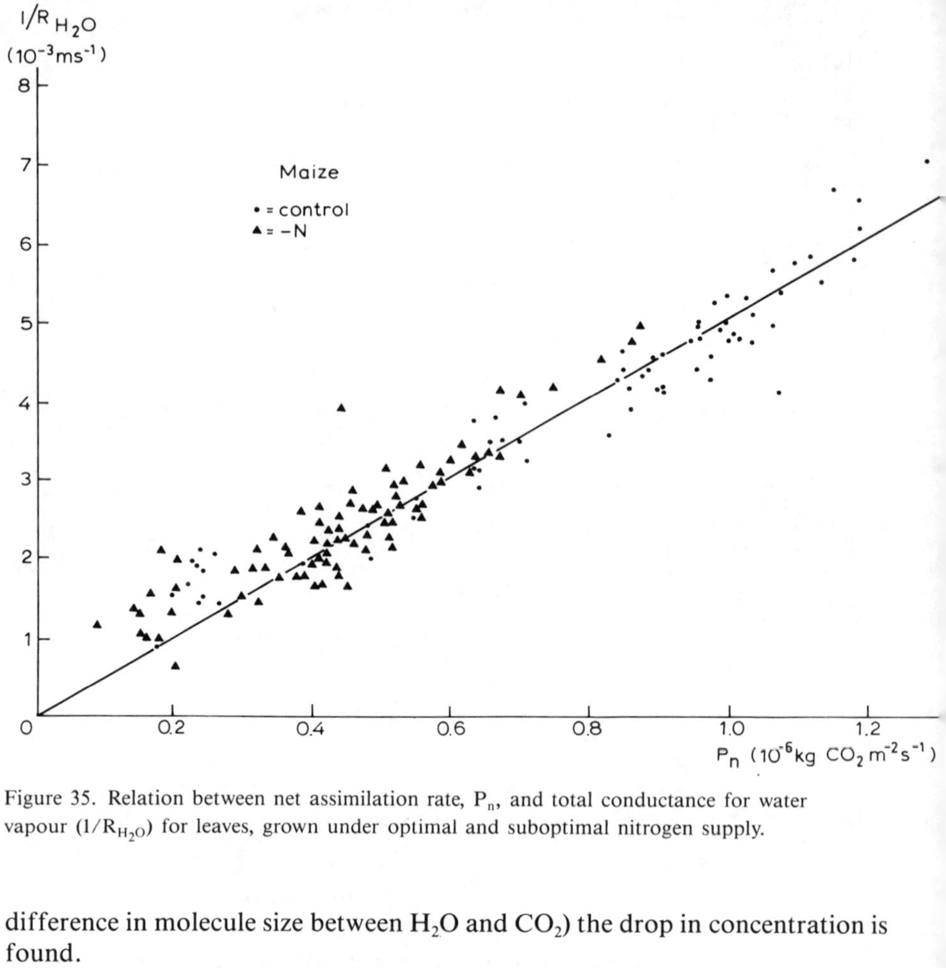

Figure 35. Relation between net assimilation rate, P_n, and total conductance for water vapour ($1/R_{H_2O}$) for leaves, grown under optimal and suboptimal nitrogen supply.

difference in molecule size between H_2O and CO_2) the drop in concentration is found.

Exercise 43
What is the concentration gradient for the maize leaves of Figure 33?
If the CO_2 concentration in the atmosphere is 330 vppm, what is the concentration in the substomatal cavity? Note that 1 mol of a gas at 20 °C has a volume of 24 dm^3.

In Figure 36, data similar to those for maize in Figure 35 are shown for Phalaris minor, a C_3 type grass species from the natural vegetation in the northern Negev. Here a tendency exists for a more favourable assimilation-

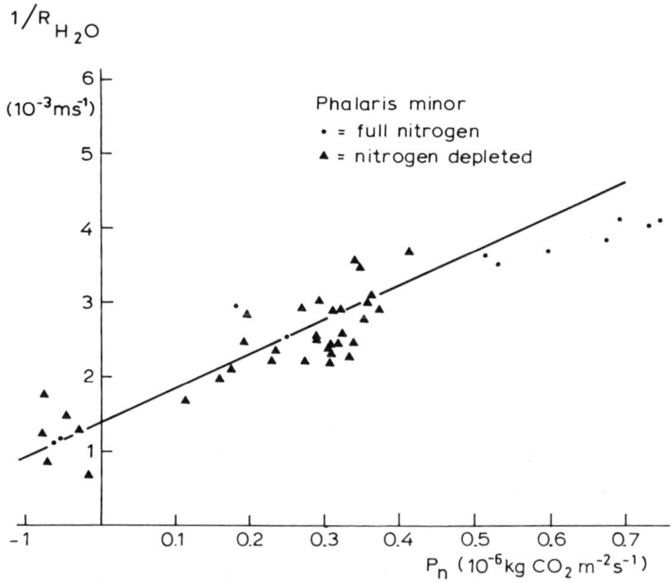

Figure 36. Relation between net assimilation rate P_n and total conductance for water vapour $(1/R_{H_2O})$ for leaves of *Phalaris minor*, grown with optimal and suboptimal nitrogen supply.

transpiration ratio in plants optimally supplied with nitrogen, although the evidence is not overwhelming. It could be, however, that at light intensities higher than those used in the measuring equipment, which are prevalent under field conditions, a more pronounced difference in assimilation-transpiration ratio would show up. In general, however, there are no strong indications for large differences in transpiration-assimilation ratio under different nutritional conditions. Therefore, in the present approach these differences are not taken into account.

3.3.4 Transpiration coefficient and water-use efficiency

Although the relation between transpiration and assimilation may not vary under different nutritional conditions, overall water-use efficiency, expressed as dry-matter production per unit of moisture applied, either because of rain or by irrigation, may well be affected. The vegetation growing at suboptimum nitrogen supply accumulates dry matter at a much slower rate (Figure 34) than the fertilized vegetation, which may lead to prolonged periods with incomplete soil cover. During those periods appreciable losses of moisture may occur by direct evaporation from the soil surface, especially when the surface remains wet by intermittent rain (Section 3.2). Such water losses do not contri-

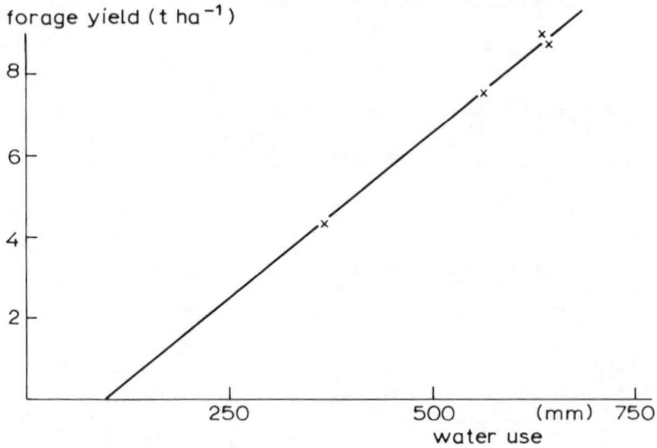

Figure 37. Relation between total forage yield and water use for alfalfa. (Source: Bauder & Bauer, 1978)

bute to production and must therefore be substracted before calculating the transpiration coefficient. This is illustrated in Table 38 and Figure 37, which refer to an irrigation experiment with alfalfa (Bauder & Bauer, 1978).
The tabulated data would suggest a decreasing transpiration coefficient with increasing application of irrigation. Figure 37 shows, however, that all points fit the same straight line, indicating a constant transpiration coefficient. The intercept with the x axis represents the amount of non-productive water loss, i.e. direct evaporation from the soil surface in this case, but it may also comprise deep drainage under other conditions. Such a graphical analysis is therefore a better basis for explanation.

Table 38. Forage yield, water use and transpiration coefficient (kg water per kg dry matter) for alfalfa.

Treatment	Forage yield (kg ha^{-1})	Water use (mm)	Water use efficiency (kg kg^{-1})
W1	4324	363	839
W2	7536	561	744
W3	8994	635	706
W4	8747	643	735

(Source: Bauder & Bauer, 1978).

Exercise 44
What is the transpiration coefficient for the alfalfa of Figure 37?

Especially under arid and semi-arid conditions where rainfall is low, the rainfall pattern is generally erratic, which means that the same amount of precipitation in different years may result from widely varying numbers of rainfall events. This erratic pattern has a direct influence on the proportion of precipitation lost by soil surface evaporation. A large number of relatively small showers will result in long periods in which the soil surface is wet, leading to prolonged high evaporation rates. If the same amount of precipitation results from a small number of relatively large showers, the soil surface will dry out some time after the shower, leading to the so-called self-mulching effect, i.e. the formation of a dry layer acting as an effective barrier to water transport. In that way the water in the deeper layers is protected from evaporation and is stored for transpiration by the plants. The magnitude of soil-evaporation losses is illustrated in Table 39, for one measured and two hypothetical rainfall distribution patterns (van Keulen, 1975). It shows that evaporation losses may be as high as 60% of the total moisture input, whereas under Sahelian conditions even higher values have been established (Stroosnijder & Koné, 1982).

It is important therefore, to distinguish between field water-use efficiency, which may include all sorts of losses and the true transpiration coefficient, which expresses the amount of water actually transpired per unit of dry matter produced. In the water-balance calculations both processes are treated therefore separately (Section 3.2).

Table 39. Partitioning of rainfall in evaporation and transpiration under different rainfall regimes.

Total rainfall (mm)	Number of showers	Evaporation (mm)	Transpiration (mm)
270	19	79	191
250	10	91	159
250	20	140	110

3.4 A simple model of water-limited production

H. van Keulen

3.4.1 Introduction

In the preceding sections the water balance of the soil, which determines the amount of water available for uptake by the plant, and the relation between water use and dry-matter production of crops have been treated. In this section that information will be used to present a calculation method for estimating crop yields under conditions where, at times during the growing season, water may be a limiting factor. The calculation procedure also yields information on the degree of water shortage for the crop during different periods. Such information gives an indication for the amount of irrigation water necessary to achieve potential production and for the timing of application of supplemental irrigation.

The calculation method is in principle identical to the one presented in Section 2.3, i.e. repetitive calculations are performed starting from a chosen point in time, where the state of the system can be described in quantitative terms. The state of the system, in this case, is not only defined by crop variables, but also by variables describing soil-moisture status. At the onset of the calculations this information must be available. The methodology is in first instance illustrated for a wheat crop growing in a semi-arid environment, for which detailed data on crop, site and weather conditions were available (Hochman, 1978).

3.4.2 Experimental details

The wheat crop was grown in an experiment to study the effects of water shortage during specific growth stages on crop performance and yield, as part of a research project on actual and potential production of semi-arid regions. The experiment was carried out in the central Negev desert of Israel (30° N, 34° E). The soil there is a uniform gray desert soil, of eolian origin, with a loamy sand texture. Field capacity, determined in the field, 48 hours after irrigation, is 0.225 $cm^3\ cm^{-3}$; wilting point, determined in the laboratory is 0.09 $cm^3\ cm^{-3}$; and total pore space 0.40 $cm^3\ cm^{-3}$.

Exercise 45

What is the maximum quantity of soil moisture, available for uptake by the crop (Equation 52)? If the soil is at wilting point at the end of the dry season and 25 mm of water is added, what is the wetting depth? How much water must be added to restore the potential rooting zone of 150 cm to field capacity?

Spring wheat, cv. Lachish, was sown on 30 October 1977; germination was completed in about 10 days. The crop was amply supplied with nutrients (nitrogen at 150 kg ha^{-1}, phosphorus at 90 kg ha^{-1} and potassium at 100 kg ha^{-1}). Soil-moisture measurements at emergence showed that the volumetric soil-moisture content in the deeper soil layers was 0.16 cm^3 cm^{-3}, due to residual moisture from a previously irrigated crop. The top soil (10 cm) was wetted to field capacity just prior to sowing to ensure proper germination. In the control treatment, sufficient moisture was applied for optimum supply, i.e. the root zone was irrigated to field capacity whenever available moisture as determined by gravimetric sampling fell below 30% of its maximum value. Rooting depth was determined at two-weekly intervals by examining soil cores. In the treatment illustrated in this example, water stress was allowed to develop between maximum tillering and anthesis by withholding irrigation.

Meteorological data, i.e. daily minimum and maximum temperatures, wet and dry bulb temperatures and wind run were recorded at a standard meteorological station, about 2 km from the experimental field. Rainfall was recorded at the site. Daily total global radiation was obtained from a meteorological station about 9 km from the experimental area. These data were used to

Table 40. Relevant input data for the wheat experiment

No of 10 day period	T_a (°C)	P (mm)	I_e (mm)	F_{gs} (kg $ha^{-1}d^{-1}$)	ET_0 (mm d^{-1})
1	15.2	0	0	251	1.85
2	12.9	0.1	0	257	1.76
3	10.4	7.4	0	225	1.47
4	8.4	19.6	0	263	1.14
5	8.8	1.6	0	257	1.23
6	7.9	0.2	0	258	1.41
7	12.4	0	0	260	2.22
8	11.5	0	0	297	2.27
9	12.3	0	0	338	2.87
10	11.7	2.2	81	344	2.72
11	15.9	0.1	96	402	3.65
12	13.6	3.9	0	374	4.25
13	13.2	0	0	437	4.40
14	17.0	4.2	60	464	5.26
15	19.6	0	0	457	5.56

calculate ten-day averages of air temperature, rainfall and potential gross CO_2 assimilation (Table 40). Total evaporative demand of the atmosphere is calculated on the basis of weather data, applying Penman's equation (Section 3.1). These data too are given in Table 40.

3.4.3 The actual calculation procedure

The calculation procedure is illustrated in Table 41, the bracketed letters in the text referring to the various lines of the table. Because this example refers to a semi-arid region with perma-dry conditions, i.e. without a groundwater table affecting moisture content in the rooting zone, capillary rise is absent. The values of the hydraulic conductivity are so low that transport from the root zone to deeper layers may safely be neglected, hence unsaturated flow above the groundwater table (cf. (CR – D) in Section 3.2) is neglected.

Line a: the first line of Table 41 specifies the initial conditions, i.e. those existing at the onset of the calculations. For wheat, the moment of emergence is chosen as the starting point. This moment is assumed to coincide with the transition from the situation where the seedlings grow from the reserves contained in the seed, to one where current assimilation provides the substrate for growth. At emergence the rooting depth of the seedlings equals 100 mm (Line a, Column 6). The total amount of moisture in the rooted depth (W_r, Column 17) is 22.5 mm of water. Hence, the volumetric moisture content in the root zone is (Column 18):

$$SM_r = 22.5/100 = 0.225\ \mathrm{cm^3\ cm^{-3}}$$

In Column 19, the matric head, S_r, in the rooting zone is calculated, applying Equation 27, solved for S:

$$S_r = \exp\left(\sqrt{(-\ln(SM_r/SM_0)/\gamma)}\right) \qquad (67)$$

exp stands here for e, the base of the natural logarithm, the value in brackets denoting the power; ln stands for natural logarithm.

For the soil in this experiment a value of 0.0189 for γ was adapted on the basis of measurements, hence:

$$S_r = \exp\left(\sqrt{(-\ln(0.225/0.40)/0.0189)}\right) = 250\ \mathrm{cm}$$

From this, the average hydraulic conductivity in the root zone, k_r, can be calculated with Equation 31, given in Section 3.2:

Table 41. Calculation scheme for water-limited production

1		2	3	4	5	6	7	8	9	10
Period		T_a	TSUM	DVS	PGASS	RD	ET_0	P	I_e	IM
a	0					100				
b	1	15.2	152	0.14	15	220	1.85	0.0	0.	0.0
c	2	12.9	281	0.25	30.8	340	1.76	0.1	0.	0.01
d	3	10.4	385	0.35	47.3	460	1.47	7.4	0.	0.74
e	4	8.4	469	0.43	78.9	580	1.14	19.6	0.	1.96
f	5	8.8	557	0.51	123.4	700	1.23	1.6	0.	0.16
g	6	7.9	636	0.58	170.3	820	1.41	0.2	0.	0.02
h	7	12.4	760	0.69	205.4	940	2.22	0.0	0.	0.0
i	8	11.5	875	0.80	255.4	1060	2.27	0.0	0.	0.0
j	9	12.3	998	0.91	300.8	1180	2.87	0.0	0.	0.0
k	10a	11.7	1103	1.0	282.1	1288	2.72	2.2	81.	9.24
l	10b	11.7	1115	1.02	251.0	1288	2.72	0.0	0.	0.0
m	11	15.9	1274	1.27	293.5	1288	3.65	0.1	96.	9.61
n	12	13.6	1410	1.48	235.6	1288	4.25	3.9	0.	0.39
o	13	13.2	1542	1.68	244.7	1288	4.40	0.0	0.	0.0
p	14	17.0	1712	1.94	222.7	1288	5.26	4.2	60.	6.42
q	15	19.6	1750	2.0	187.0	1288	5.56	0.0	0.	0.0

Table 41. (continued)

1		11	12	13	14	15	16	17	18	19	20
Period		E_m	E_a	T_m	T	dM_r	dW_r	W_r	SM_r	S_r	k_r
a	0							22.5	0.225	250	0.01
b	1	1.74	1.74	0.11	0.11	1.91	0.06	23.1	0.105	4640	0.0002
c	2	1.55	0.59	0.21	0.17	1.91	1.16	34.7	0.102	4929	0.00015
d	3	1.16	0.43	0.31	0.22	1.91	2.00	54.7	0.119	2997	0.0003
e	4	0.80	0.36	0.34	0.34	1.91	3.17	86.4	0.149	1379	0.0009
f	5	0.64	0.39	0.59	0.59	1.91	1.09	97.3	0.139	1769	0.0006
g	6	0.48	0.27	0.92	0.92	1.91	0.74	104.7	0.128	2375	0.0004
h	7	0.47	0.24	1.75	1.75	1.91	−0.08	103.9	0.110	3884	0.0002
i	8	0.32	0.13	1.94	1.69	1.91	0.09	104.8	0.098	5578	0.0001
j	9	0.32	0.11	2.55	0.64	1.91	1.16	116.4	0.098	5578	0.0001
k	10a	0.30	0.10	2.22	0.57	1.91	10.48	210.7	0.163	976	0.0015
l	10b	0.30	0.20	1.99	1.99	0.0	−2.19	208.5	0.162	1024	0.001
m	11	0.40	0.27	2.67	2.67	0.0	6.67	275.2	0.214	319	0.007
n	12	0.47	0.44	2.69	2.69	0.0	−2.74	247.8	0.192	532	0.003
o	13	0.48	0.40	2.47	2.47	0.0	−2.87	219.1	0.170	831	0.002
p	14	0.58	0.42	2.51	2.51	0.0	3.49	254.0	0.197	453	0.004
q	15	0.61	0.52	2.29	2.29	0.0	−2.81	248.4	0.193	497	0.004

Table 41. (continued)

1		21	22	23	24	25	26	27	28	29	30
Period		W_{nr}	GASS	MRES	ASAG	DMI	FR	IWRT	WRT	FL	IWLV
a	0	222.5							50		
b	1	203.4	15.0	1.5	13.5	9.5	0.44	4.2	92	0.56	5.3
c	2	184.3	24.9	2.9	22.0	15.4	0.33	5.1	143	0.67	10.3
d	3	165.2	33.6	5.1	28.5	20.0	0.26	5.2	195	0.80	12.0
e	4	146.1	78.9	7.9	71.0	49.7	0.205	10.2	297	0.53	26.3
f	5	127.0	123.4	15.3	108.1	75.7	0.16	12.1	418	0.46	34.8
g	6	108.9	170.3	26.7	143.6	100.5	0.12	12.1	539	0.39	39.2
h	7	88.8	205.4	41.8	163.6	114.5	0.09	10.3	642	0.30	34.4
i	8	69.7	222.5	59.0	163.5	114.5	0.055	6.3	705	0.18	20.6
j	9	50.6	75.5	75.2	0.3	0.2	0.03	0.2	707	0.08	0.0
k	10a	33.4	73.0	69.1	3.9	2.7	0.01	0.0	707	0.02	0.0
l	10b	33.4	251.0	43.6	207.4	165.9	0.0	0.0	707	0.0	0.0
m	11	33.4	293.5	45.0	248.5	198.8	0.0	0.0	707	0.0	0.0
n	12	33.4	235.6	62.7	172.9	138.3	0.0	0.0	707	0.0	0.0
o	13	33.4	244.7	74.8	169.9	135.9	0.0	0.0	707	0.0	0.0
p	14	33.4	222.7	87.0	135.7	108.6	0.0	0.0	707	0.0	0.0
q	15	33.4	187.0	96.8	90.2	72.2	0.0	0.0	707	0.0	0.0

Table 41. (continued)

1		31	32	33	34	35	36	37	38	39	40
Period		DWLV	WLV	FS	IWST	WST	FG	IWGR	WGR	LAI	TADW
a	0		50							0.1	50
b	1	0.0	103	0.0	0.0	0.0	0.0	0.0	0.0	0.21	103
c	2	0.6	200	0.0	0.0	0.0	0.0	0.0	0.0	0.40	206
d	3	1.7	303	0.14	2.8	28	0.0	0.0	0.0	0.61	354
e	4	0.0	566	0.265	13.2	160	0.0	0.0	0.0	1.1	721
f	5	0.0	914	0.38	28.8	448	0.0	0.0	0.0	1.8	1385
g	6	0.0	1306	0.49	49.2	940	0.0	0.0	0.0	2.6	2269
h	7	0.0	1650	0.61	69.8	1638	0.0	0.0	0.0	3.3	3311
i	8	6.4	1792	0.765	87.6	2514	0.0	0.0	0.0	3.6	4393
j	9	40.3	1389	0.61	0.0	2514	0.28	0.0	0.0	2.8	4395
k	10a	30.9	1111	0.19	0.5	2519	0.78	2.2	22	2.2	4420
l	10b	22.2	1089	0.0	0.0	2519	1.0	165.9	188	2.2	4586
m	11	21.8	871	0.0	0.0	2519	1.0	198.8	2176	1.7	6574
n	12	17.4	697	0.0	0.0	2519	1.0	138.3	3559	1.4	7957
o	13	13.9	558	0.0	0.0	2519	1.0	135.9	4918	1.1	9316
p	14	11.2	446	0.0	0.0	2519	1.0	108.6	6004	0.9	10402
q	15	8.9	428	0.0	0.0	2519	1.0	72.2	6148	0.8	10546

Table 41. (continued)

1		41	42	43
Period		TWD	TWDL	TDWD
a	0	100	100	
b	1	195	195	
c	2	349	343	6
d	3	549	526	23
e	4	1046	1023	23
f	5	1803	1780	23
g	6	2808	2785	23
h	7	3953	3930	23
i	8	5098	5011	87
j	9	5100	4610	490
k	10a	5127	4359	768
l	10b	5293	4503	790
m	11	7281	6273	1008
n	12	8664	7482	1182
o	13	10023	8702	1321
p	14	11309	9676	1433
q	15	11253	9802	1451

$$k_r = a \cdot (S_r)^{-1.4} \qquad (68)$$

The value of 'a' appeared to be equal to 22.6, hence:

$$k_r = 22.6 \times (250)^{-1.4} = 0.01 \text{ mm d}^{-1} \text{ (Column 20)}$$

The amount of moisture in the soil between the actual rooting depth and the potential rooting depth (in this case 1500 mm) is calculated as a separate state variable, W_{nr}, because this moisture becomes gradually available to the plants when the roots grow deeper. At emergence, the total amount in that zone equals 222.5 mm (Column 21), which is assumed to be evenly distributed.
At emergence, the dry weight of the root system, WRT, equals 50 kg ha^{-1} (Column 28) and that of the leaf blades, WLV, also 50 kg ha^{-1} (Column 32). From the latter value, the leaf area index, LAI, is calculated, assuming a constant specific leaf area of 20 m^2 kg^{-1}, hence:

$$LAI = 50 \times 20 \times 10^{-4} = 0.1 \text{ m}^2 \text{ m}^{-2} \text{ (Column 39)}$$

As in Section 2.3, four auxiliary state variables are calculated: the total amount of above–ground dry weight, TADW (Column 40), which is here equal to the weight of the leaf blades, the only above – ground organ present;

the total plant dry weight, TDW (Column 41), equal to the sum of leaf blades and roots (= 100 kg ha^{-1}). Finally the latter value is subdivided into live plant tissue, TDWL (Column 42) and dead plant tissue, TDWD (Column 43).

Line b: this line covers the first ten – day period of the growing period. Average air temperature during the period, T_a, equals 15.2 °C (Column 2). Integration over the ten – day period yields a temperature sum, TSUM, of 152 d °C (Column 3). The temperature sum is used to define the phenological development stage of the crop (Sections 2.2; 2.3). For spring wheat the temperature requirement between emergence and anthesis is 1100 d °C (van Keulen & Seligman, 1986). The development stage of the crop is thus calculated as the ratio between the accumulated temperature sum and 1100:

DVS = 152/1100 = 0.14 (Column 4)

Potential daily gross assimilation for the period, expressed in CH_2O is read from Table 40 as 251 kg ha^{-1} d^{-1}. The reduction factor for incomplete light interception, RA, is read from Table 42. For LAI = 0.1 the reduction factor equals 0.06. Potential gross assimilation in the absence of water stress is calculated as:

PGASS = 251 x 0.06 = 15 kg ha^{-1} d^{-1} (Column 5)

It is assumed on the basis of the experimental results that root extension proceeds at a rate of 12 mm d^{-1}, if the soil in which the roots grow is sufficiently wet. In this case the total potential rooting zone is wetted, thus that

Table 42. Reduction factor for gross Co_2 assimilation due to incomplete radiation interception ($f_h = 1 - e^{-0.6.LAI}$)

LAI	Reduction factor
0	0
0.25	0.14
0.5	0.26
1.0	0.45
1.5	0.59
2.0	0.70
2.5	0.78
3.0	0.84
3.5	0.88
4.0	0.91
5.0	0.95

condition is satisfied throughout. The rooted depth at the end of the ten – day period is 220 mm (Column 6).

Maximum evapotranspiration during the ten – day period, ET_0, is read from Table 40 and introduced in Column 7. For the present period it amounts to 1.85 mm d^{-1}. In the next columns the inputs of moisture into the system are defined. For this ten – day period, both precipitation (P, Column 8) and irrigation (I_e, Column 9) are absent, hence total infiltration (IM, Column 10) is also zero. Under the conditions prevailing in the area, loss of water by surface run – off has not been observed, hence in all cases total infiltration equals the sum of precipitation and irrigation. Maximum evaporation from the soil surface is derived from potential evapotranspiration, taking into account the reduction factor for the shading effect of the vegetation:

$$E_m = ET_0 \cdot (1 - RA) \qquad (69)$$

It is thus the complementary fraction of light interception by the vegetation. Hence:

$$E_m = 1.85 \times 0.94 = 1.74 \text{ mm d}^{-1} \text{ (Column 11)}$$

Actual evaporation is subsequently obtained from the average moisture content in the root zone. If the soil – moisture content is too low, the supply of water to the surface cannot meet the evaporative demand and actual evaporation falls short of the maximum. A detailed treatment of the process of soil evaporation is given by van Keulen (1975). In the present approach it is approximated by assuming a linear decline in evaporation rate between field capacity and air – dry soil, where evaporation ceases completely (Section 3.2). The moisture content of air – dry soil (SM_a), is estimated as one third of that at wilting point. For this soil it is 0.03 $cm^3\ cm^{-3}$. Hence:

$$E_a = E_m \cdot (SM_r - SM_a)/(SM_{fc} - SM_a) \qquad (70)$$

$$= 1.74 \times (0.225 - 0.03)/(0.225 - 0.03)$$
$$= 1.74 \times 1 = 1.74 \text{ mm d}^{-1}$$

Potential transpiration could, in principle, be obtained from potential gross assimilation and the transpiration coefficient, as outlined in Section 3.3. For most field situations it appears that potential evapotranspiration calculated according to Penman is a reasonable approximation of potential transpiration for a closed crop surface, apparently because stomatal control by CO_2 is the most common behaviour in field crops. Therefore 'Penman' (Column 7) is used as the basis for potential transpiration. To calculate maximum transpiration for crops with incomplete soil cover, the same reduction factor, RA, is used as for gross assimilation. Hence:

$$T_m = ET_0 \cdot RA = 1.85 \times 0.06 = 0.11 \text{ mm d}^{-1} \text{ (Column 13)}$$

Actual transpiration is only equal to the maximum value, if soil moisture is adequate to supply sufficient water to the plant roots. As explained in Section 3.2, a crop – specific critical soil – moisture content exists, above which roots can freely take up water from the soil. The critical soil – moisture content is determined by both the physical properties of the soil and the evaporative demand of the atmosphere. It is approximated by:

$$SM_{cr} = (1 - p) \cdot (SM_{fc} - SM_w) + SM_w \tag{71}$$

where

SM_{cr} is critical soil – moisture content ($cm^3\ cm^{-3}$)
p is soil water depletion fraction (Table 20)
SM_{fc} is soil – moisture content at field capacity ($cm^3\ cm^{-3}$)
SM_w is soil – moisture content at wilting point ($cm^3\ cm^{-3}$)

For potential transpiration equal to 1.85 mm d^{-1}, the value of p equals 0.86, hence:

$$SM_{cr} = 0.14 \times (0.225 - 0.09) + 0.09 = 0.109 \text{ cm}^3 \text{ cm}^{-3}$$

As the actual moisture content is well above the critical value, actual transpiration (T, Column 14) equals maximum transpiration.
Another 'input' into the plant – soil system is the amount of moisture that becomes available to the vegetation as a result of vertical extension of the root system. In the present approach, this is approximated by assuming that the total amount of moisture in that part of the potential rooting zone, where the roots have not yet penetrated, is evenly distributed. The amount, added to the plant – soil system as a result of root growth, is then calculated as:

$$dM_r = W_{nr} \cdot R_r / D_{nr} \tag{72}$$

where

dM_r is amount of moisture added to the soil plant system by root growth (mm d^{-1})
W_{nr} is total amount of moisture in the non – rooted part of the potential rooting zone (mm)
R_r is growth rate of the roots (mm d^{-1})
D_{nr} is thickness of the non – rooted part of the potential rooting zone (mm), hence: $D_{rm} - D_r$, with D_{rm} potential rooting depth and D_r the rooting depth (mm)

On the basis of Equation 72, the increase in available soil moisture by root growth is calculated as:

$dM_r = 222.5 \times 12/(1500 - 100) = 1.91 \text{ mm d}^{-1}$ (Column 15)

All processes influencing the water balance have been treated now, so that their effect on the rate of change in soil – moisture status of the root zone can be evaluated:

$$dW_r = IM + dM_r - E_a - T \qquad (73)$$

Hence:

$dW_r = (0 + 1.91 - 1.74 - 0.11) = 0.06 \text{ mm d}^{-1}$ (Column 16)

Total soil moisture in the root zone at the end of the ten – day period equals the value at the beginning (Column 17, line a) plus the rate of change multiplied by the length of the time interval:

$W_r = 22.5 + 0.06 \times 10 = 23.1$ mm (Column 17)

From the amount of moisture in the root zone, the average volumetric soil-moisture content (Column 18) is calculated, assuming it to be homogeneous. Hence:

$$SM_r = W_r/D_r \qquad (74)$$

where

SM_r	is volumetric soil – moisture content in the root zone ($cm^3\ cm^{-3}$)
W_r	is soil – moisture content in the root zone (mm)
D_r	is rooting depth (mm)

For the present ten – day period:

$SM_r = 23.1/220 = 0.105 \text{ cm}^3 \text{ cm}^{-3}$

The values for the matric head in the root zone, S_r, and the average hydraulic conductivity, k_r, are calculated according to Equations 67 and 68. These variables are included here for completeness sake, but they are not used, because unsaturated flow above the groundwater table can be neglected in this situation. In the remainder of this example Columns 19 and 20 will therefore not be treated.

The total amount of moisture in the non – rooted zone of the profile at the end of the period, is obtained by subtracting dM_r times Δt (Column 15) from the amount at the beginning of the decade (Line a, Column 21):

$$W_{nr} = 222.5 - 1.91 \times 10 = 203.4 \text{ mm (Column 21)}$$

In Column 22 actual gross assimilation, AGASS, is introduced. For this ten-day period it equals potential gross assimilation (Column 5) as there is no reduction due to water shortage, i.e. actual transpiration is equal to the maximum value. Maintenance respiration, i.e. the amount of energy invested in maintaining the existing cells and their structures, is calculated from the total live dry weight of the crop and the relative maintenance respiration rate, R_m. During the pre – anthesis phase of the crop the latter value is set at 0.015 kg CH_2O per kg dry matter per day, hence:

$$MRES = 0.015 \times 100 = 1.5 \text{ kg ha}^{-1} \text{ d}^{-1} \text{ (Column 23)}$$

The amount of assimilates available for increase in dry weight of the vegetation is the difference between actual gross assimilation and maintenance respiration:

$$ASAG = 15 - 1.5 = 13.5 \text{ kg ha}^{-1} \text{ d}^{-1} \text{ (Column 24)}$$

The rate of increase in dry weight of the vegetation is obtained from the amount of available assimilates, taking into account the losses associated with the conversion of primary photosynthates into structural plant material (growth respiration). For vegetative material of average composition, these losses amount to about 30% of the consumed carbohydrates (Penning de Vries, 1975). Thus:

$$DMI = 0.7 \times ASAG = 0.7 \times 13.5 = 9.5 \text{ kg ha}^{-1} \text{ d}^{-1} \text{ (Column 25)}$$

The assimilates are used to produce various plant organs at the same time (Sections 2.2 and 2.3). The partitioning of dry matter among the various organs is predominantly governed by the phenological stage of the crop. At the production level treated in this section, where at times production may be limited by moisture availability, the development of moisture stress in the plant may influence the partitioning pattern, according to the functional balance principle (Brouwer, 1963). The exact influence of water stress in the plant on the distribution of assimilates is difficult to quantify, however. Therefore, this effect has been neglected in the present approach.
The relation between phenological stage of the crop and the traction of assimilates diverted to the various organs is given in Table 43. The independent variable determining the fraction of assimilates partitioned to the root system

Table 43. Partitioning factors for dry matter to various plant organs as a function of development stage.

Development stage	f_r	f_l	f_s	f_g
0	0.50	0.50	0.0	0.0
0.1	0.42	0.58	0.0	0.0
0.2	0.33	0.67	0.0	0.0
0.3	0.26	0.60	0.14	0.0
0.4	0.20	0.52	0.28	0.0
0.5	0.15	0.43	0.42	0.0
0.6	0.10	0.34	0.56	0.0
0.7	0.07	0.23	0.70	0.0
0.8	0.04	0.12	0.84	0.0
0.9	0.02	0.06	0.42	0.50
1.0	0.0	0.0	0.0	1.0

(Column 26) is the development stage of the crop. For each ten – day period the relevant value is found as the average of the values at the end and at the beginning, as in Section 2.3. For the present period:

$$\text{DVS} = (0 + 0.14)/2 = 0.07$$

In Table 43 it is found that the fraction of dry matter partitioned to the root system at that value of DVS equals 0.44 (Column 26). The rate of increase in dry weight of the root system is thus:

$$\text{IWRT} = \text{FR} \cdot \text{DMI} = 0.44 \times 9.5 = 4.2 \text{ kg ha}^{-1} \text{ d}^{-1} \text{ (Column 27)}$$

The total dry weight of the root system at the end of the ten – day period is equal to its value at the beginning of the period (Line a, Column 28), augmented with the rate of increase times the length of the time interval, hence:

$$\text{WRT} = 50 + (4.2 \times 10) = 92 \text{ kg ha}^{-1} \text{ (Column 28)}$$

The fraction allocated to the leaf blades is also obtained from Table 43 at DVS is 0.07 and equals 0.56 (Column 29). The rate of increase in dry weight of the leaf blades follows from multiplication of this fraction with the rate of dry matter increase:

$$\text{IWLV} = \text{FL} \cdot \text{DMI} = 0.56 \times 9.5 = 5.3 \text{ kg ha}^{-1} \text{ d}^{-1} \text{ (Column 30)}$$

The weight of the leaf blades at the end of the period equals its value at the beginning plus the rate of increase multiplied by the length of the time interval:

$$WLV = 50 + 5.3 \times 10 = 103 \text{ kg ha}^{-1} \text{ (Column 32)}$$

In the early stages of crop development in spring wheat, only roots and leaf blades are produced (van Keulen & Seligman, 1986), thus the fraction allocated to the stems (FS, Column 33) is zero, and consequently the rate of increase in stem dry weight (IWST, Column 34) too. Total stem weight at the end of the period (WST, Column 35) is still zero. The crop is in its vegetative stage, hence no grain growth takes place (Columns 36, 37 and 38). The leaf area index at the end of the period is obtained from the dry weight of the leaf blades, taking into account the constant specific leaf area of 20 m^2 per kg dry matter and the surface area:

$$LAI = 103 \times 20 \times 10^{-4} = 0.21 \text{ m}^2 \text{ m}^{-2} \text{ (Column 39)}$$

The total canopy variables are finally obtained from the weights of the various plant organs at the end of the period. Total above – ground dry weight is the sum of the weights of the above – ground organs, i.e. leaf blades, stems and grain:

$$TADW = 103 + 0 + 0 = 103 \text{ kg ha}^{-1} \text{ (Column 40)}$$

Total dry weight of the vegetation is equal to above ground dry weight, augmented with the weight of the root system:

$$TDW = 103 + 92 = 195 \text{ kg ha}^{-1} \text{ (Column 41)}$$

At this moment no dead tissue is present (Column 43), hence total live dry weight (Column 42) equals total dry weight. These calculations complete the treatment of the first ten – day period.

Line c: the calculations in this period are in principle equal to those in the previous one. Therefore, they will be treated in less detail.

Average daily air temperature during this ten – day period equals 12.9 °C (Column 2), so the accumulated temperature sum at the end of the period is 281 d °C (Column 3) and the associated development stage is equal to:

$$DVS = 281 / 1100 = 0.25 \text{ (Column 4)}$$

Gross assimilation in the absence of water stress is obtained from the radiation determined value (Table 40) and the reduction factor for incomplete light interception, which is 0.12 at a LAI of 0.21 (Table 42). Hence:

$$PGASS = 257 \times 0.12 = 30.8 \text{ kg ha}^{-1} \text{ d}^{-1} \text{ (Column 5)}$$

Root extension proceeds unhampered, thus the rooting depth at the end of the period equals 340 mm (Column 6). Potential evapotranspiration, obtained from Table 40, equals 1.76 mm d^{-1} (Column 7). Rainfall during the period is 0.1 mm. To convert this value into a daily rate, it must be divided by Δt. Hence P = 0.01 mm d^{-1} (Column 8). No irrigation was applied (Column 9), thus the rate of infiltration is equal to 0.01 mm d^{-1} (Column 10). Maximum soil evaporation is calculated on the basis of potential evapotranspiration, taking into account the reducing effect of partial shading by the vegetation:

$$E_m = ET_0 \cdot (1 - RA) = 1.76 \times 0.88 = 1.55 \text{ mm d}^{-1} \text{ (Column 11)}$$

The reduction factor for soil evaporation due to soil – moisture content follows from:

$$(SM_r - SM_a)/(SM_{fc} - SM_a) = (0.105 - 0.03)/(0.225 - 0.03) = 0.38$$

Actual soil evaporation equals E_m x 0.38, i.e. 0.59 mm d^{-1} (Column 12). Maximum transpiration is the complementary fraction of potential evapotranspiration and equals:

$$T_m = ET_0 \cdot RA = 1.76 \times 0.12 = 0.21 \text{ mm d}^{-1} \text{ (Column 13)}$$

For the determination of actual transpiration the critical soil – moisture content has to be calculated by Equation 71, substituting 0.86 for p (Table 20):

$$SM_{cr} = (0.14 \times 0.135) + 0.09 = 0.109 \text{ cm}^3 \text{ cm}^{-3}$$

Actual volumetric soil – moisture content is 0.105 cm^3 cm^{-3} (Line b, Column 18), which is below the critical value. The reduction in transpiration is calculated from:

$$T/T_m = (Sm_r - SM_w)/(SM_{cr} - SM_w) \quad (75)$$

In this situation thus:

$$T = 0.21 \times (0.105 - 0.09)/(0.109 - 0.09) = 0.17 \text{ mm d}^{-1} \text{ (Column 14)}$$

To calculate the rate of water addition to the root zone as a result of root growth, Equation 72 is applied:

$$dM_r = 203.4 \times 12/(1500 - 220) = 1.91 \text{ mm d}^{-1} \text{ (Column 15)}$$

The rate of change in moisture content in the root zone follows from the balance:

$$dW_r = 0.01 + 1.91 - 0.59 - 0.17 = 1.16 \text{ mm d}^{-1} \text{ (Column 16)}$$

Total soil moisture in the root zone at the end of the ten – day period equals:

$$W_r = 23.1 + (1.16 \times 10) = 34.7 \text{ mm (Column 17)}$$

The average volumetric moisture content at the end of the ten – day period is then total moisture content divided by the rooting depth:

$$SM_r = 34.7 / 340 = 0.102 \text{ cm}^3 \text{ cm}^{-3} \text{ (Column 18)}$$

The moisture content in the non – rooted part of the profile equals:

$$W_{nr} = 203.4 - dM_r \cdot \Delta t = 203.4 - 19.1 = 184.3 \text{ mm (Column 21)}$$

Actual gross assimilation follows from the potential value, taking into account the reduction due to water shortage:

$$GASS = PGASS \cdot T/T_m = 30.8 \times 0.17/0.21 = 24.9 \text{ kg ha}^{-1} \text{ d}^{-1} \text{ (Column 22)}$$

The maintenance respiration rate is obtained from total live dry weight and the relative maintenance respiration rate:

$$MRES = 0.015 \times 195 = 2.9 \text{ kg ha}^{-1} \text{ d}^{-1} \text{ (Column 23)}$$

The amount of assimilates available for increase in dry matter is the balance between actual gross assimilation and maintenance respiration:

$$ASAG = 24.9 - 2.9 = 22.0 \text{ kg ha}^{-1} \text{ d}^{-1} \text{ (Column 24)}$$

The rate of dry – matter increase of the vegetation, taking into account growth respiration, equals:

$$DMI = 22.0 \times 0.7 = 15.4 \text{ kg ha}^{-1} \text{ d}^{-1} \text{ (Column 25)}$$

The fraction of dry – matter increase partitioned to the root system is obtained from Table 43 at a value of the development stage midway between the beginning and the end of the ten – day period, hence $(0.25 - 0.14)/2 + 0.14 = 0.195$. FR equals 0.33 at that value (Column 26) and the rate of increase in dry weight of the root system follows from:

$$IWRT = FR \cdot DMI = 0.33 \times 15.4 = 5.1 \text{ kg ha}^{-1} \text{ d}^{-1} \text{ (Column 27)}$$

and

$$WRT = 92 + 5.1 \times 10 = 143 \text{ kg ha}^{-1} \text{ (Column 28)}$$

The fraction allocated to the leaf blades also follows from Table 43 and equals 0.67 (Column 29), hence:

$$IWLV = 0.67 \times 15.4 = 10.3 \text{ kg ha}^{-1} \text{ d}^{-1} \text{ (Column 30)}$$

In the present approach it is assumed that leaf blades deteriorate when the vegetation suffers from water shortage, in an attempt to reduce transpirational losses. The rate of decline is proportional to the relative transpiration rate, T/T_m. At severe moisture stress, leaf blades are assumed to deteriorate at a maximum rate of 0.03 kg dry matter per kg leaf blade dry – matter present per day.

The rate of decline is thus calculated as:

$$DWLV = RDR \cdot WLV \tag{76}$$

with:

$$RDR = 0.03 \cdot (1 - T/T_m) \tag{77}$$

For the present ten – day period:

$$RDR = (1 - 0.17 / 0.21) \times 0.03 = 0.0057 \text{ d}^{-1}$$

and

$$DWLV = 0.0057 \times 103 = 0.6 \text{ kg ha}^{-1} \text{ d}^{-1} \text{ (Column 31)}$$

Leaf – blade dry weight at the end of the period equals:

$$WLV = 103 + (10.3 - 0.6) \times 10 = 200 \text{ kg ha}^{-1} \text{ (Column 32)}$$

At development stage 0.195, the fraction allocated to stem and grain is still zero, hence Columns 33 till 38 all contain zeros. The leaf area index at the end of the ten – day period follows from the dry weight of the leaf blades:

$$LAI = 200 \times 20 \times 10^{-4} = 0.40 \text{ m}^2 \text{ m}^{-2} \text{ (Column 39)}$$

Total above – ground dry matter equals the weight of the live leaf blades plus the weight of the dead leaves, hence:

$TADW = 200 + 6 = 206$ kg ha^{-1} (Column 40)

Total plant dry weight is obtained by adding root weight to the previous value:

$TDW = 206 + 143 = 349$ kg ha^{-1} (Column 41)

Total live dry weight equals 343 kg ha^{-1} (Column 42) and the amount of dead tissue is 6 kg ha^{-1} (Column 43). These calculations complete the treatment of the second ten – day period.

In the periods 3 to 9 no basically different processes take place, therefore these periods are not treated in detail here. From the 8th period onwards, a serious water shortage develops, because irrigation is purposely withheld in this treatment. The onset of moisture stress was somewhat delayed because of fairly heavy rainfall in the 4th period. The result of the water shortage is that growth practically ceases after the 8th period and that green leaf area declines drastically (Table 41).

Line k: this is the 10th ten – day period of the growing period (Column 1). Average air temperature during the period is 11.7 °C (Column 2). Temperature sum at the end of the period equals:

$TSUM = 998 + 117 = 1115$ d °C

and the associated development stage:

$DVS = 1115/1100 = 1.02$

That is a point beyond anthesis and, because the temperature relations for development are different before and after anthesis, this period is subdivided into two. The first part covers the period till anthesis, its duration being determined by the remainder of the required temperature sum:

$(1100 - 998)/11.7 = 9$ d

Hence, the temperature sum at the end of period 10a equals 1103 d °C and the development stage at the end of the period is 1.0, i.e. anthesis. Gross assimilation in the absence of moisture stress is calculated in the usual way and yields 254 kg ha^{-1} d^{-1} (Column 5). Root growth still proceeds and the rooting depth at the end of the period equals 1288 mm.

It should be noted here, that according to the partitioning factors shown in Table 43, the contribution to the roots is negligible during this period. However, if root extension proceeds, assimilates are necessary. It is likely therefore, that the moisture shortage, developing in this period, affected the distribution pattern to allow root growth. In the present schematized set – up, however, this effect is not taken into account, and the data of Table 43 are used indiscriminately. Potential evapotranspiration rate during the period equals 2.72 mm d^{-1}, showing that gradually time is moving towards a warmer and drier spring.

It was the intention to end the moisture stress at anthesis, therefore a heavy irrigation of 81 mm was applied during this period to restore the water content in the root zone to a value well above the critical level. As there was also a small rain shower, total infiltration amounted to 9.24 mm d^{-1} (Column 10).

For the calculation of the maximum rate of soil evaporation it is assumed that the dried leaves remain on the vegetation and act effectively in radiation interception. Thus, in calculating the shading effect, the maximum value of the leaf area index (LAI = 3.6) is applied. For this period:

$$E_m = ET_0 \cdot (1 - RA) = 2.72 \times 0.11 = 0.30 \text{ mm d}^{-1} \text{ (Column 11)}$$

Actual soil evaporation falls short of the maximum, because of the low moisture – content in the root zone, according to Equation 70:

$$E_a = E_m \cdot (0.098 - 0.03) / (0.225 - 0.03) = 0.10 \text{ mm d}^{-1} \text{ (Column 12)}$$

The maximum rate of transpiration is calculated in the usual way from the potential evapotranspiration rate and amounts to 2.22 mm d^{-1} (Column 13). The actual rate of transpiration is derived from the maximum value taking into account the moisture content in the root zone. The critical volumetric soil – moisture content follows from Equation 71 at 0.121 cm^3 cm^{-3}. The reduction in transpiration rate is obtained from Equation 75:

$$T = 2.22 \times (0.098 - 0.09)/(0.121 - 0.09) = 0.57 \text{ mm d}^{-1} \text{ (Column 14)}$$

The rate of increase of moisture in the root zone as a result of root growth amounts to 1.72 mm d^{-1} (Column 15). Total soil moisture in the root zone at the end of the period follows from realization of the calculated rates over the time interval of nine days and amounts to 210.7 mm. From this, the volumetric soil-moisture content in the root zone is calculated as 0.163 cm^3 cm^{-3} (Column 17). The actual assimilation rate amounts to 73 kg ha^{-1} d^{-1} after correcting for the transpiration deficit (Column 22). The rate of maintenance respiration is 69.1 kg ha^{-1} d^{-1}, calculated on the basis of total live dry weight of the vegetation (Column 23). The difference of 3.9 kg ha^{-1} d^{-1} (Column 24) allows for a rate of increase in dry matter of 2.7 kg ha^{-1} d^{-1} (Column 25). This amount is distributed among roots, leaf blades, stems and grain in accordance with the partitioning factors given in Table 43.

In this period, as in the preceding ten – day period, some of the assimilates are diverted to the grain, although anthesis has not yet been reached. This partitioning pattern has been chosen to account for the fact that during the last part of the pre – anthesis phase, a substantial part of the assimilates is stored in the stem as reserves, which after grain set are translocated to the growing grain (van Keulen & Seligman, 1986). In the present schematized calculation procedure these assimilates are designated grain weight directly. The remainder of this period follows the same rules as the ones that have been discussed in detail.

Line l: to follow through the ten – day subdivision, the one remaining day from the 10th period is treated in a separate line. After anthesis the development scale runs from 1 at anthesis to 2 at maturity, or in fact at the end of the grain filling period. The temperature requirement for that period is, for this and most other spring wheat varieties, 650 d °C (van Keulen & Seligman, 1986).

The development stage in the post – anthesis period is thus calculated as:

DVS = (TSUM – 1100)/650 + 1 (Column 4)

Extension growth of the root system is assumed to cease at anthesis, because no more assimilates are invested in growth of the root system (Column 26). Treatment of the Columns 6 till 22 is not essentially different from that detailed in the preceding lines, therefore the calculations are not repeated in detail at this point. A few remarks may serve to draw attention to specific points:

– the reduction factor for soil evaporation due to shading by the vegetation is kept constant at 0.11 (Column 11), corresponding to the maximum leaf area index of 3.6 m^2 m^{-2}.
– after cessation of root extension, the 'input' of moisture into the root zone as a result of root growth remains zero (Column 15). Transport of moisture between the root zone and the non – rooted part of the profile, caused by developing potential gradients is neglected. Thus the soil moisture stored in the non – rooted zone remains constant (Column 21).
– the relative maintenance respiration rate is changed at anthesis from 0.015 kg kg^{-1} d^{-1} to 0.01. The main reason for this change is that in general the nitrogen concentration in the vegetative material declines rapidly after anthesis, due to translocation of nutrients to the developing grain. Moreover, the dry matter accumulating in the grain requires less energy for maintenance. Hence, from anthesis onwards: MRES = 0.01 x TDWL
– the conversion efficiency of primary photosynthates into structural plant dry matter, set at 0.7 for vegetative material of average composition, is changed to 0.8 for grain dry matter. Again, the major reason for this change is the fact that the nitrogen concentration of the grain is lower than that of the vegetative material during the pre – anthesis phase.
– leaves have a limited life – time and even under optimum growing condi-

tions part of the earlier formed leaves deteriorate. This deterioration increases rapidly after anthesis, mainly because of the translocation of nitrogen and other essential growth substances to the developing grain. In the present model this process is taken into account by assuming a relative death rate of the leaf blades of 0.02 kg kg^{-1} d^{-1}.

Exercise 46
Calculate the weight of leaf blades 10 days after anthesis, starting from an initial value of 2000 kg ha^{-1}. Employ first the value of 0.02 kg kg^{-1} d^{-1} and a time step of one day. Apply subsequently the time step of the model.
What do you notice? What is the reason?

From Table 41 it follows, that under the prevailing environmental conditions, the post – anthesis phase lasts 43 days, the calculated grain yield amounts to 6.2 t ha^{-1} and total above – ground dry matter to 10.5 t ha^{-1}.

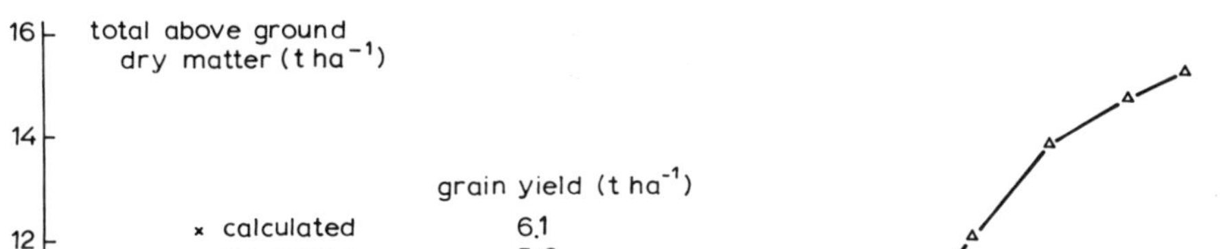

Figure 38. Comparison of measured and calculated dry-matter accumulation of wheat growing under temporary water shortage. The calculated dry-matter accumulation of the control is given for comparison.

Exercise 47
Calculate the transpiration coefficient for this example. Use both total dry matter and above – ground dry matter.

3.4.4 Comparison with the experiment

In Figure 38 the calculated growth curve is compared to the measured one, showing satisfactory agreement in terms of total dry – matter production. The calculated grain yield is however higher than the measured one, for which there is no obvious reason. It is clear from the graph, that the difference is equal to the difference in dry – matter accumulation after anthesis, provided that interpolation between the measured points on either side of the anthesis date is permitted.

4 CROP PRODUCTION AS DETERMINED BY NUTRIENT AVAILABILITY

4.1 Crop yield and nutrient requirements

H. van Keulen

4.1.1 Introduction

In the previous chapters it has been shown that plants need energy (supplied by the sun), carbon dioxide and water to produce organic material. However, plant tissue contains not only oxygen, hydrogen and carbon, but also other elements like nitrogen, phosphorus and sulphur in proteins, and potassium accompanying organic anions. Many other elements are found in small quantities as constituents of enzymes.

These elements must be taken up by the root system from the soil. In many cases the soil in its natural situation does not supply sufficient plant nutrients to satisfy the demand of the crop. The yield level obtained is then determined by the amount of the limiting element that can be absorbed by the vegetation.

In the foregoing chapters a quantitative treatment has been given of the influence of energy and water on the agricultural production process. In this chapter the influence of plant nutrients will be treated, with this difference, however, that the dynamic aspect, i.e. the change with time, is not considered.

4.1.2 Nutrient supply and crop response

Since the discovery, during the middle of last century, that inorganic ions are needed by growing crops, an enormous number of fertilizer experiments have been carried out at different locations, with various elements and with many crops. In general these trials yield information of the type presented in Figure 39, i.e. the measured yield is presented as a function of the amount of an element supplied. This graph, referring to experiments carried out at IRRI, in the Philippines (Tanaka et al., 1964), shows clearly one of the difficulties encountered in interpreting these experiments: the results are extremely variable.

It should be realized, therefore, that to obtain a yield response to fertilizer application two conditions must be fulfilled:

- the fertilizer that is applied to the soil (or in the case of bunded rice to the water) must be taken up by the crop.
- after uptake by the crop it must be utilized to produce the required plant material, i.e. grains in the case of rice and maize, roots in the case of cassava.

Both processes may be hampered by external or internal conditions. A presen-

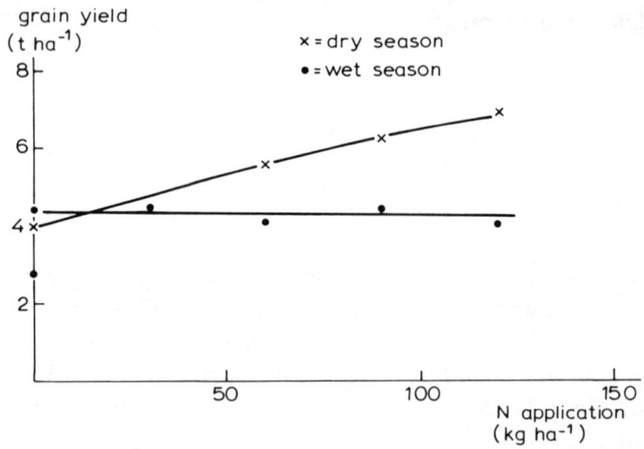

Figure 39. Relation between nitrogen application and grain yield for bunded rice grown at IRRI, Los Banos, the Philippines in the dry season and the wet season (Tanaka et al., 1964).

tation such as that in Figure 39 gives no clue to the relative importance of both in the final response. For useful interpretation of fertilizer experiments, yield determinations must be accompanied by chemical analysis of the harvested material, so that the uptake of the element by the vegetation can be calculated and its distribution in the plant.

When in fertilizer experiments both yield and chemical composition have been determined, graphical presentation of the results as suggested by de Wit (1953) facilitates interpretation. The method is illustrated in Figure 40, using the same data as in Figure 39.

Figure 40 consists of three graphs: in the first quadrant, Quadrant a, the relation is given between the economic yield (grain) and the total uptake of the element in the above ground parts of the vegetation, that is both in grain and in straw. This value is calculated by multiplying the amounts of grain and straw harvested by their respective nitrogen concentrations.

Exercise 48

Calculate the total uptake of nitrogen for the following experiment:

N application ($kg\ ha^{-1}$)	Yield ($kg\ ha^{-1}$)		Nitrogen concentration (%)	
	grain	straw	grain	straw
0	4400	4500	1.12	0.44
50	5450	5200	1.29	0.40
100	6200	6250	1.24	0.52
150	6350	6400	1.43	0.55

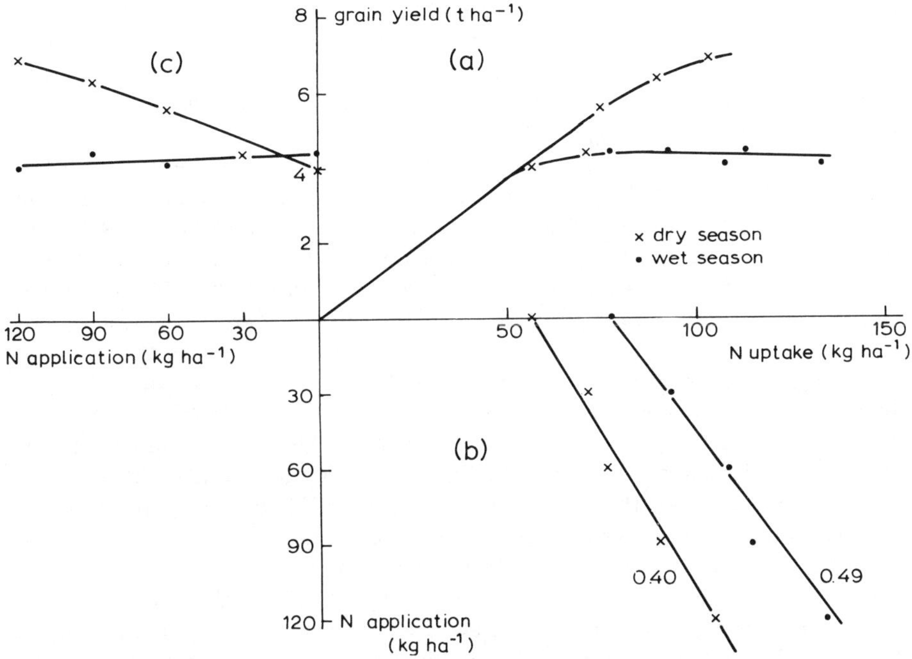

Figure 40. The relation between nitrogen uptake and grain yield; the relation between nitrogen application and nitrogen uptake, and the relation between nitrogen application and grain yield for bunded rice grown at IRRI, Los Banos, the Philippines, in the dry season and the wet season (Tanaka et al., 1964). Numbers in lower half of the graphs denote recovery fraction of applied fertilizer.

Quadrant b shares the uptake axis with Quadrant a, while on the vertical axis the amount of fertilizer applied is given in downward direction. Quadrant c shows the relation between fertilizer application and yield, i.e. that given in Figure 39. The three graphs are not independent, as one can always be constructed from the other two through elimination of one variable.

The lower end of the curve in Quadrant a passes through the origin, i.e. at zero uptake no yield is to be expected. Because uptake refers to total uptake in both grain and straw, theoretically a situation could exist where only vegetative material is produced, which contains some nitrogen, so that zero yield is associated with a small nitrogen uptake. For all practical purposes, however, that phenomenon may be neglected and the easily obtained origin may be considered part of the curve. At low levels of nitrogen uptake, a proportional relation exists between total uptake and grain yield. This proportionality reflects the existence of a minimum nitrogen concentration, both in grain and in straw. In the grain, no further accumulation of dry matter can take place beyond the point where nitrogen has been diluted to its minimum concentration. In the straw, which in the reproductive phase looses nitrogen to the developing grains, a residual non – remobilizable level of nitrogen remains. The fact that each unit of nitrogen taken up, yields a constant amount of grain also indicates that the ratio between grain yield and total dry matter yield (the harvest index) is not strongly influenced in that range.

A different situation may arise under arid and semi – arid conditions. Moisture limitation during the post – anthesis phase may hamper crop assimilation due to stomatal closure and accelerated leaf senescence, and thus interfere with the growth of the storage organs. The consequence of such behaviour is a very unfavourable harvest index. An additional effect is that not all the nitrogen in the vegetative plant parts can be remobilized and translocated, but material dies with a high residual nitrogen content. Moreover, the nitrogen content of the storage organs is high, because nitrogen is incorporated preferentially in the storage organs and is subsequently not diluted (van Keulen, 1977; van Keulen & van Heemst, 1982). The art of nitrogen fertilizer application in such situations is to apply the fertilizer judiciously so that the moisture is just depleted when the seed ripens. In semi – arid regions, too liberal an application at the early growth stages must be avoided, because that leads to abundant vegetative growth with the associated high water use and the risk of moisture shortage during the reproductive stage. These intricate interrelations are treated and simulated in detail elsewhere (van Keulen & Seligman, 1986). The combination of a poor harvest index and a high protein content in the harvested plant material, decreases the efficiency of nitrogen utilization considerable, even under limited nitrogen supply. In such situations it is, therefore, much more difficult to predict the effect of nitrogen uptake on economic yield quantitatively.

A similar situation may arise when a high proportion of the total nitrogen uptake takes place after anthesis. Such a process hardly influences the harvest index, but it leads to high protein contents in the storage organs and incomplete remobilization of the proteins from the vegetative material. However, such a situation rarely occurs under conditions of limited nitrogen supply, except if temporary water shortage during the pre – anthesis phase is followed by abundant water supply during the post – anthesis phase. In the situations described here, the yield – uptake curve must thus be applied cautiously.

Exercise 49

Calculate the harvest index for the treatments of the experiment presented in Exercise 48.

What is the mathematical relation between the harvest index and the grain/straw ratio?

With increasing uptake of nitrogen the yield – uptake curve deviates from the straight line, reflecting an increase in nitrogen concentration in the harvested products. The efficiency of nitrogen utilization, expressed as grain yield per unit nitrogen taken up, thus decreases. However, the quality of the harvested material, in terms of its protein content increases.

Finally, the yield curve reaches a plateau level, where further uptake of nitrogen is not reflected in higher yields. At that point, nitrogen availability is no longer the determinant factor for growth and production. The level of the plateau in any actual situation is determined by the growth factor that is constraining. If there is a sufficient supply of the other major inorganic elements, such as phosphorus and potassium, and the other chemical properties of the soil are also favourable (pH, salinity), the yield plateau is equal to that calculated for Production Situation 2, i.e. for situations where either radiation or water may be at times limiting. If by means of proper drainage and irrigation facilities also the moisture status of the soil can be maintained around its optimum value, the yield plateau is that of Production Situation 1, i.e. potential production determined by radiation and temperature only. The influence of other limiting factors on the level of the plateau will be treated in more detail in Subsection 4.1.2. The yield – uptake curve will extend to a point where the vegetation has taken up so much nitrogen that the maximum concentration in the tissue is maintained throughout the plants life cycle.

Various analyses of yield – uptake curves have shown that the relationship is independent of the type of fertilizer applied, provided the fertilizer is only effective through its main acting element (van Keulen & van Heemst, 1982; van Keulen, 1982; van Keulen, 1977; de Wit, 1953).

The relation in Quadrant b appears to be linear over the complete range of applications presented, which is characteristic for the majority of fertilizer experiments for nitrogen. Of course, when very high amounts of fertilizer are applied, a situation may be created where the vegetation is continuously 'saturated' with nitrogen, i.e. it is always at its maximum concentration, so that uptake does not increase any more at the highest application levels. For most practical situations, however, the relation in Quadrant b may be characterized by two parameters: the intercept with the uptake axis, representing the uptake from the unfertilized soil, and the slope with respect to the vertical, representing the proportion of the applied fertilizer taken up by the above – ground plant material. This is referred to as the recovery fraction.

The nitrogen uptake at zero fertilizer application shows wide fluctuations, partly because of soil characteristics, partly because of variations in environmental conditions like temperature and rainfall. Furthermore, management influences are of importance such as crop rotation and water management.

The fertilizer recovery fraction may vary between 0.1 and 0.8: the efficiency of uptake is influenced by such factors as fertilizer type, method and timing of application, environmental conditions, etc.

Exercise 50
In a certain situation, the uptake of nitrogen at zero N application is 20 kg ha^{-1} and at a fertilizer rate of 100 kg ha^{-1} the nitrogen uptake is 60 kg ha^{-1}. What is the recovery fraction of the applied nitrogen?

The recovery fraction is smaller than one, because many processes are competing for the applied nitrogen: uptake by plants, immobilization by bacteria, volatilization, leaching and denitrification. The linearity of the application rate – uptake relation suggests strongly that all these processes proceed at a rate which is proportional to the concentration of mineral nitrogen (NO_3^-, NH_4^+) in the soil solution, so that they can be described as first – order reactions.

It is obvious from this presentation that the curves in Quadrant c may vary widely under different conditions, because both uptake without fertilizer application and recovery of applied fertilizer show wide variability. In order to suggest methods for improvement in a given situation it is, however, necessary to know all three relations depicted in Figure 40.

Exercise 51
Construct the three quadrant diagram for the experiment given in Exercise 48.

4.1.3 Yield – uptake relations

Initial slope

Rice

In Figure 41, a number of representative examples are given for yield – uptake curves of nitrogen on rice, representing a range of cultivars, environmental conditions, fertilizer treatments and management practices. The plateau level in the various examples presented is not necessarily the yield level of Production Situation 2, because at the higher rates of N application other elements could have been in short supply. However, information about this is, in general, lacking. For our purpose that is immaterial because only the initial slope of the yield – uptake curve is considered here, and points on this slope materialize only if elements other than nitrogen are not limiting. Examination of the various curves shows that the initial slope, expressed in kg grain (at a moisture content of 0.15 kg kg^{-1}) per kg nitrogen taken up varies in the examples between 51 (Figure 41a) and 80 (Figure 41b). These variations reflect differences in grain/straw ratios in the various experiments, resulting from

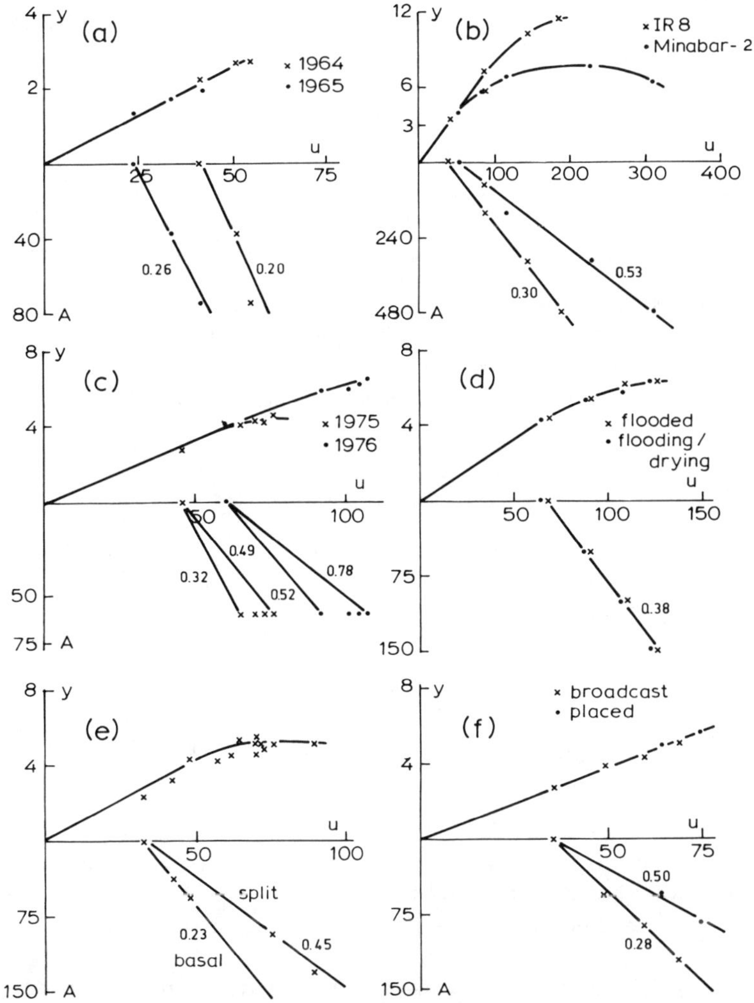

Figure 41. The relation between total nitrogen uptake (u, in kg ha^{-1}) and grain yield, y (t ha^{-1}), at a moisture content of 0.15 kg kg^{-1} and the relation between nitrogen application, A (kg ha^{-1}), and nitrogen uptake of bunded rice. Numbers in lower half of the graphs denote recovery fractions of applied fertilizer. a. India (Majumdar, 1973). b. Peru (Sanchez et al., 1973). c. USA (Reddy & Patrick, 1978). d. Philippines (Khind & Ponnamperuma, 1981). e. Thailand (Koyama et al., 1973). f. Indonesia (Ismunadji & Sismiyati, 1976).

differences in growing conditions or cultivar characteristics. The experiment of Figure 41a was carried out in India with a local tall indica cultivar, Raghusail, producing a relatively abundant vegetative apparatus, hence the average grain/straw ratio of 0.53. On the other hand, the improved short straw cultivar IR8 growing under the conditions of the coastal plain in Peru, ended up with a grain/straw ratio of 1.3 (Figure 41b).

Exercise 52
Calculate the harvest index for the examples of Figures 41a and 41b.

In both situations, however, and in the other examples in Figure 41, the minimum nitrogen concentration in the grains was around 0.01 kg kg^{-1}, whereas in the straw the residual N concentration was around 0.004 kg kg^{-1}. On the basis of these parameters the initial efficiency, E_{in}, can be calculated as:

$$E_{in} = 1/(0.01 + s/g \times 0.004) \quad (78)$$

where

s is weight of straw (kg ha^{-1})
g is weight of grain (kg ha^{-1})

Exercise 53
Calculate the 'theoretical' efficiencies for the two examples of Figures 41a and 41b.

In an extensive analysis of yield – uptake curves for nitrogen on rice (van Keulen, 1977), it was shown that the initial efficiencies are always within the range set by the two extremes in Figure 41. Also in pot trials similar values are obtained.

In Figure 42, a similar analysis is presented for the effect of phosphorus on rice. Because less sufficiently detailed field data are available where phosphorus is the limiting factor for rice production, some results of pot trials have also been included. The minimum phosphorus concentrations in grain and straw found in various experiments are 0.0011 and 0.0005 kg kg^{-1}, respectively. Applying the same reasoning as in the case of nitrogen and, assuming a value of one for the grain/straw ratio, leads to an initial efficiency of 625 kg grain per kg phosphorus taken up. The slopes in Figure 42 vary between 410 and 625 kg kg^{-1}, the variability being again mainly the result of variations in the grain/straw ratio.

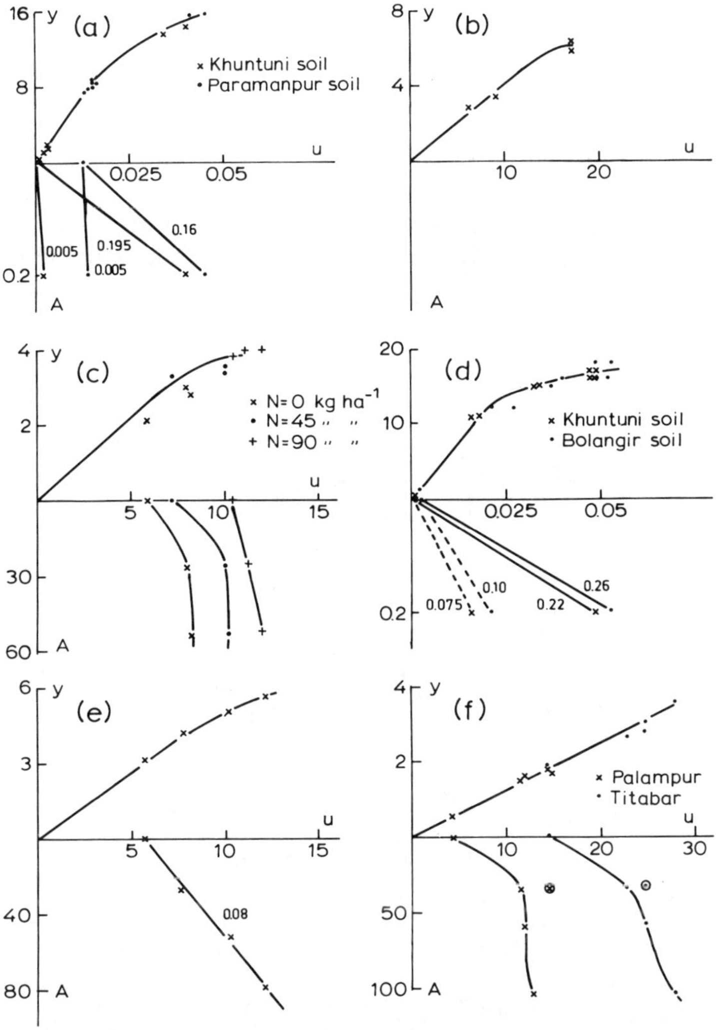

Figure 42. The relation between total phosphorus uptake, u (kg ha^{-1}), and grain yield, y (t ha^{-1}), at a moisture content of 0.15 kg kg^{-1} and the relation between phosphorus application, A (kg ha^{-1}), and phosphorus uptake for bunded rice. Numbers in lower half of the graphs denote recovery fractions of applied fertilizer (in a and d, u, y and A in g pot^{-1}). a. pot experiment, India (Dash et al., 1982). b. Mali (Traoré, 1974). c. Nigeria (Bredero, 1966). d. pot experiment, India (Sadanandan et al., 1980). e. India (Agarwal, 1980). f. India (Motsara & Datta, 1971).

Exercise 54

Calculate the grain/straw ratio of the experiment presented in Figure 42b, where the initial efficiency is 410 kg grain per kg P taken up.

In Figure 43 yield – uptake curves for potassium on rice are presented for situations where a clear effect of increased potassium uptake on yield was observed. The number of sufficiently detailed experimental reports found in

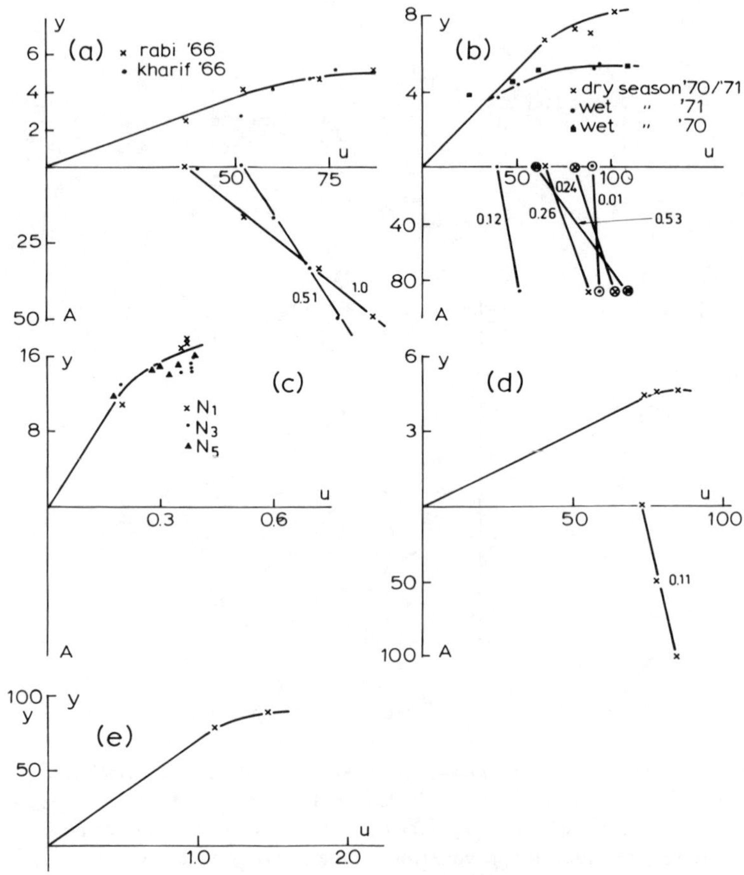

Figure 43. The relation between total potassium uptake, u (kg ha^{-1}), and grain yield, y (t ha^{-1}), at a moisture content of 0.15 kg kg^{-1} and the relation between potassium application, A (kg ha^{-1}), and potassium uptake for bunded rice. Numbers in lower half of the graphs denote recovery fractions of applied fertilizer (in c and e, u, y and A in g pot^{-1}). a. India (Mahapatra & Panda, 1972). b. Senegal (Beye, 1974). c. pot experiment, India (Esakkimuthu et al., 1975). d. pot experiment, India (Agarwal, 1980). e. India (Patnaik & Gaikawad, 1969).

the literature is still smaller than for phosphorus. The variation in initial slope for this element is larger than for either nitrogen or phosphorus. One reason is that at maturity the larger part of the potassium is found in the straw. Therefore, variations in the grain/straw ratio have a more pronounced effect on the slope of these curves, whereas also losses of potassium due to inevitable leaf loss at the end of the growing period will have greater effect. Another reason for the variability may be the fact that the element has a double function in the plant. For one part it is an essential element for certain physiological functions, for another part it serves as a positive charge, accompanying organic and inorganic anions during transport through the plant. In the latter function it may be almost completely replaced by other positive ions, if present in sufficient amounts.

However, the minimum concentrations in the vegetative material at maturity seem to be around 0.008 kg kg^{-1} and in the grains between 0.0025 and 0.005 kg kg^{-1}. The combination of these two values with a grain/straw ratio of one, yields an initial slope of 55 – 80 kg grain per kg potassium absorbed. The examples presented in Figure 43 all fall within this range.

Maize

Yield – uptake curves for nitrogen on maize are presented in Figure 44. Despite the fact that maize is a C_4 species, which are reported to be more economical with respect to nitrogen (Brown, 1978), the initial slopes found here vary between 55 and 70 kg kg^{-1} and appear, thus, not to be different from those for rice. It is true that under low nitrogen availability the residual N concentration in the vegetative structures may drop to values as low as 0.002, which is lower than found for rice, but normal variations in grain/straw ratio mask this effect. Average grain/straw ratios for the experiments of Figure 44 vary between 0.85 and 1.33.

In Figure 45 yield – uptake curves for phosphorus on maize are given. The value of the initial efficiency varies between 420 and 590 kg kg^{-1}, which is again very similar to the values found for rice, suggesting that the minimum P concentrations in the tissue are identical for this graminae as well.

For potassium on maize the number of suitable experiments found in the literature was even more limited than for rice. Whatever was available is presented in Figure 46, which again shows a wide variability in initial efficiencies. In this case it may well be that in some of the experiments presented the potassium – supplying capacity of the soil in the absence of fertilizer application was so high that the minimum levels in the tissue were not really attained.

Cassava

Yield – uptake curves for the three macro – elements on cassava are presented in Figures 47, 48 and 49, respectively. For nitrogen, the initial efficiency in the examples presented here varies between 34 kg tuber dry matter per kg N taken up (Figure 47a) to 175 kg kg^{-1} (Figure 47b). These variations seem a

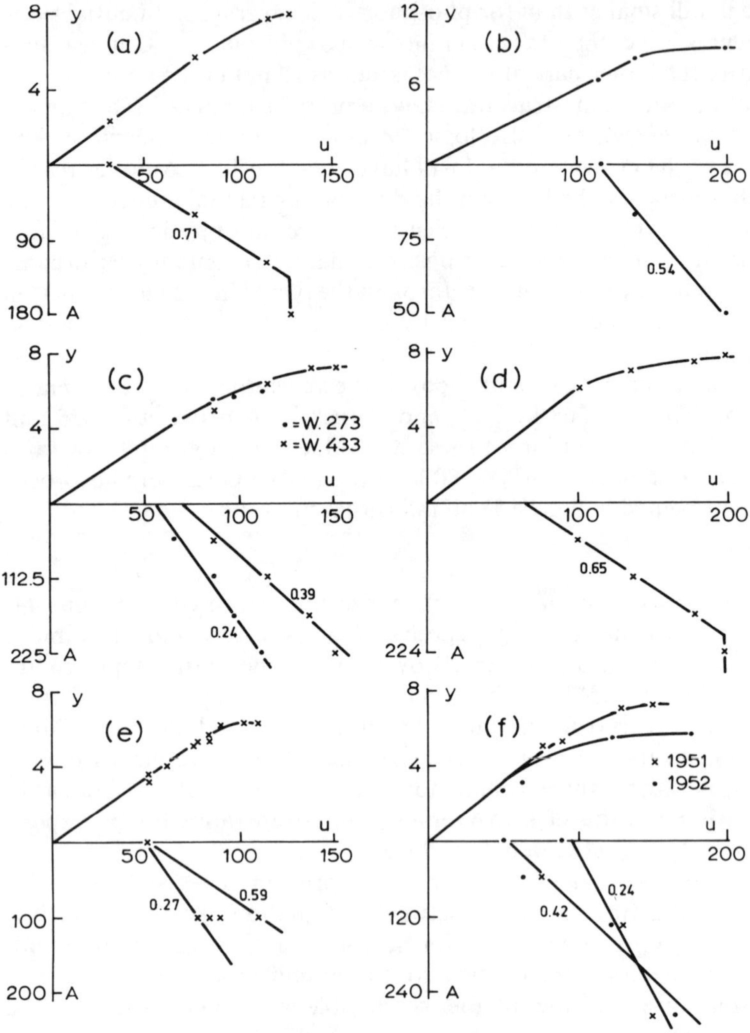

Figure 44. The relation between total nitrogen uptake, u (kg ha^{-1}), and grain yield, y (t ha^{-1}), at a moisture content of 0.15 kg kg^{-1} and the relation between nitrogen application, A (kg ha^{-1}), and nitrogen uptake for maize. Numbers in lower half of the graph denote recovery fractions of applied fertilizer. a. Nigeria (Balasubramanian & Singh, 1982). b. USA (Olson, 1980). c. USA (Jung et al., 1972). d. USA (Rabuffetti & Kamprath, 1977). e. Brasil (Grove et al., 1980). f. USA (Flynn et al., 1957).

little too high if the reasoning followed for rice and maize also applies to this crop. However, the ratio of root weight to top weight varies between 0.8 and 1.75, which is a much greater variation than in the case of grains. Part of this

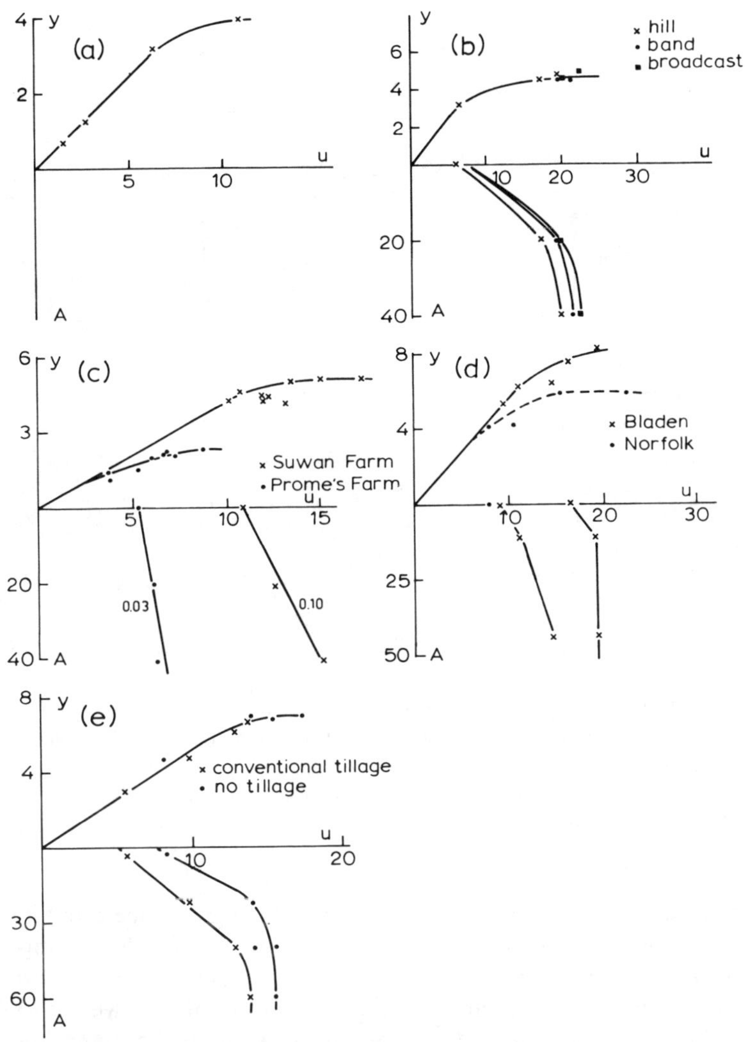

Figure 45. The relation between total phosphorus uptake, u (kg ha^{-1}), and grain yield, y (t ha^{-1}) at a moisture content of 0.15 kg kg^{-1}, and the relation between phosphorus application, A (kg ha^{-1}), and phosphorus uptake for maize. Numbers in lower half of the graphs denote recovery fractions of applied fertilizer. a. Mali (Traoré, 1974). b. Nigeria (Kang & Yunusa, 1977). c. Thailand (Suwanarit, 1975). d. USA (Krantz et al., 1949). e. USA (Moschler & Martens, 1975).

variation may be due to the fact that leaves fallen from the plant in the later growth stages have not been taken into account in the final dry weight (Nijholt, 1936). That has also its consequences for the total amount of N utilized,

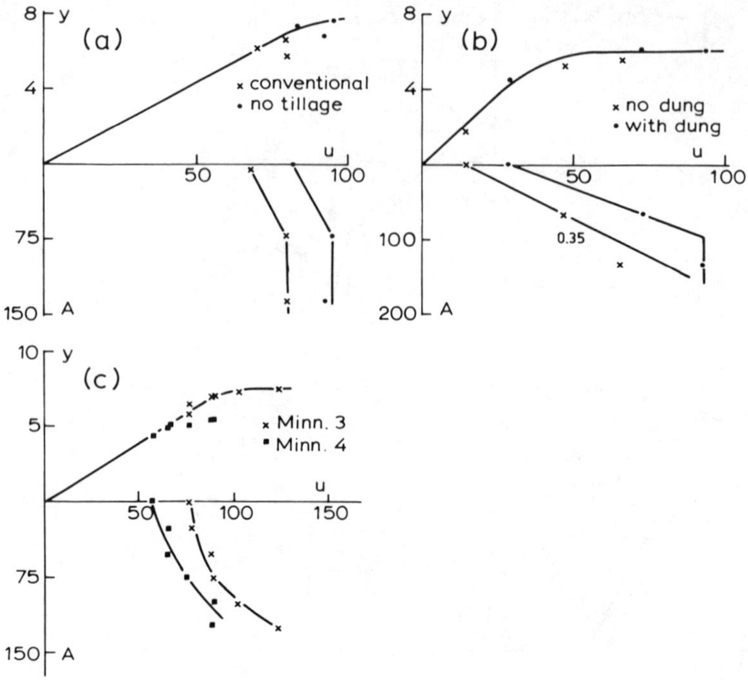

Figure 46. The relation between total potassium uptake, u (kg ha^{-1}) and grain yield, y (t ha^{-1}) at a moisture content of 0.15 kg kg^{-1} and the relation between potassium application, A (kg ha^{-1}) and potassium uptake for maize. Numbers in lower half of the graphs denote recovery fraction of applied fertilizer. a. USA (Moschler & Martens, 1975). b. France (Loué, 1963). c. USA (Hanway et al., 1962).

because the nitrogen irreversibly incorporated in those leaves, is neglected in the analysis. Moreover, the growth habit of cassava is much more indeterminate than that of the grains, i.e. harvesting time of the roots varies between 6 and 24 months after planting. With increasing age, the ratio of root weight to top weight also increases (Section 2.2). From various sources the minimum nitrogen concentration in the roots at the age of 12 months is estimated at around 0.003. For the above – ground plant parts there is quite some variability in the reported nitrogen concentrations, the variability being also related to the status of the leaves that were collected. Values range from 0.0065 to about 0.01 kg N per kg dry matter. In the first case, probably more older leaves were incorporated in the analysis. Combination of the extreme values of the relevant parameters given so far will yield the range of initial efficiencies that may be expected.

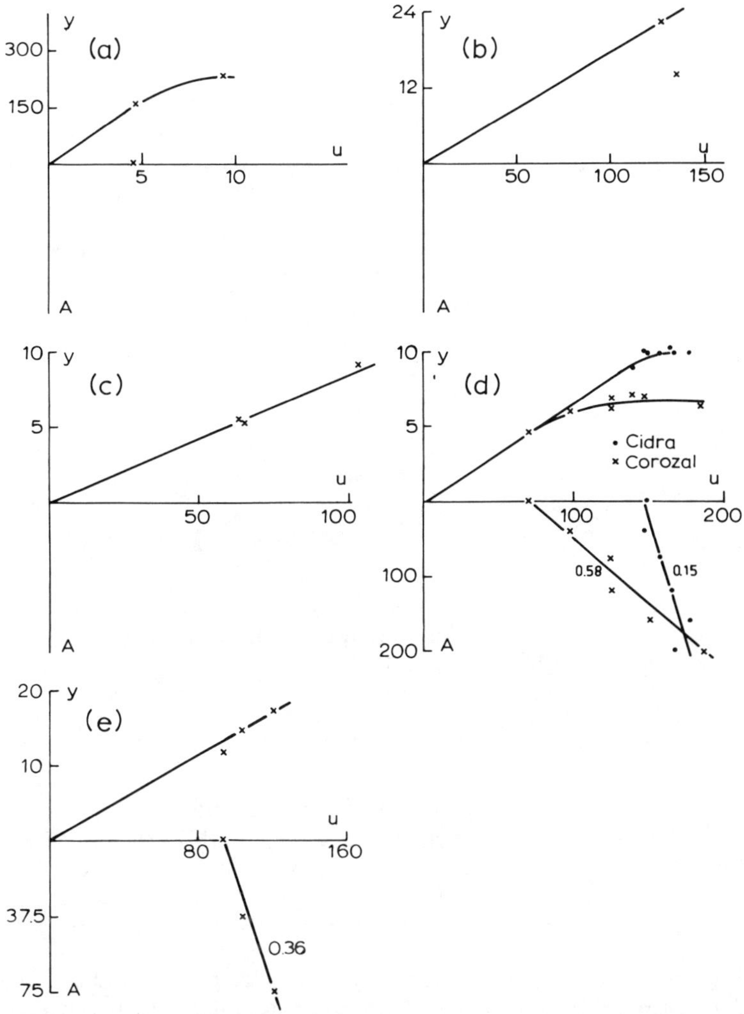

Figure 47. The relation between total nitrogen uptake, u (kg ha^{-1}) and root dry matter yield, y (t ha^{-1}), and the relation between nitrogen application, A (kg ha^{-1}), and nitrogen uptake for cassava. Numbers in lower half of the graphs denote recovery fraction of applied fertilizer (in a, u, y and A in g pot^{-1}). a. Argentine (Orioli et al., 1967). b. Indonesia (Nijholt, 1936). c. Malaysia (Kanapathy, 1974). d. Puerto Rico (Fox et al., 1975). e. Thailand (Somsak, 1974).

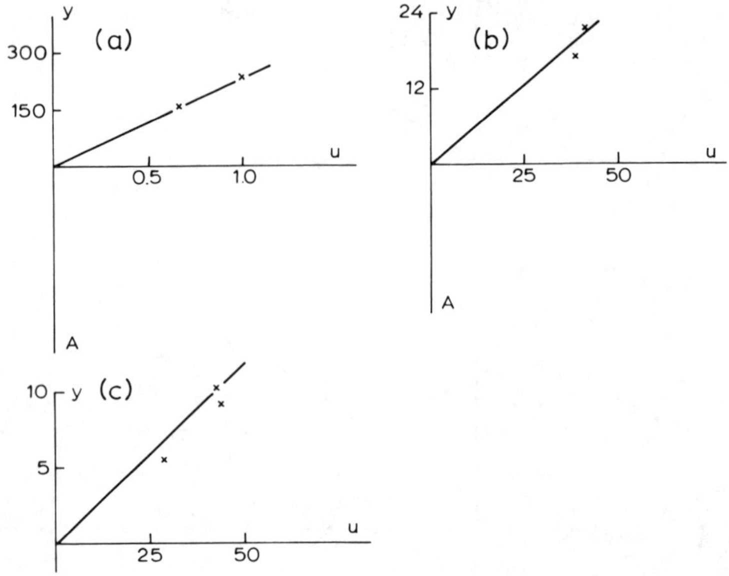

Figure 48. The relation between total phosphorus uptake, u (kg ha^{-1}) and root dry matter yield, y (t ha^{-1}), for cassava (in a, u, y and A in g pot^{-1}). a. Argentine (Orioli et al., 1967). b. Indonesia (Nijholt, 1936). c. Malaysia (Kanapathy, 1974).

Exercise 55

Calculate the minimum and maximum value of the initial efficiency for nitrogen on cassava on the basis of the numerical values given in this section. What equation is applied?

The reasoning followed for nitrogen applies also to phosphorus and potassium. In Figure 48 the initial efficiency of phosphorus utilization varies between 230 and 510 kg tuber dry matter per kg P taken up. For potassium (Figure 49) the values range between 40 and 100 kg tuber dry matter per kg K absorbed.

On the whole, it must be said that the lack of quantitative data and the indeterminate growth habit of cassava are a disadvantage for the treatment of nutrient response of this crop. However, the indicative values given here enable a first approximation of the nutrient requirements of the crop for different production situations.

The ceiling yield level

As mentioned in Subsection 4.1.1, the ceiling yield level of the yield – up-

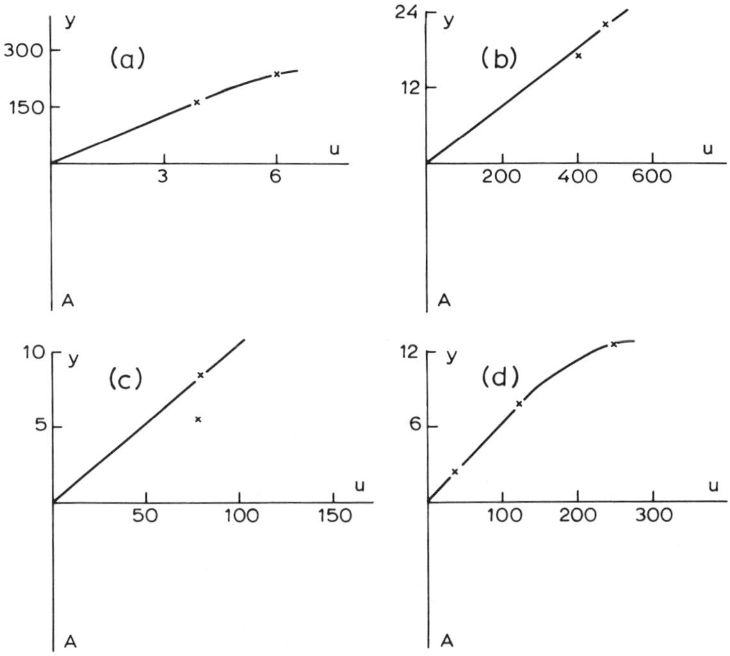

Figure 49. The relation between total potassium uptake, u (kg ha^{-1}), and root dry matter yield, y (t ha^{-1}) for cassava (in a, u, y and A in g pot^{-1}). a. Argentine (Orioli et al., 1967). b. Indonesia (Nijholt, 1936). c. Malaysia (Kanapathy, 1974). d. Madagascar (Dufournet & Goarin, 1957).

take curve is determined by the growth factor that is in short supply. By definition this plateau is beyond the point where the factor considered, i.e. the uptake of a particular nutrient element influences growth and production.

In Figure 50 some experimental data are summarized that illustrate the influence of various growth factors on the level of the plateau. Figure 50a shows the influence of the radiation level in a potential production situation (Production Situation 1) for bunded rice in the Central Plain of Thailand. The 'dry season' crop maturated during a period with an average daily irradiance of about 22 MJ m^{-2}, whereas average daily irradiance during maturation of the 'wet season' crop amounted to about 17.5 MJ m^{-2}.

Figure 50b illustrates the effect of moisture supply on the plateau level for maize grown in Missouri, USA. In 1951, a total of 625 mm of rain fell during the growing period of maize, whereas in 1953 it was only 306 mm during the same period. The experiment in question was a fertilizer experiment and not an irrigation experiment. Therefore it can be concluded that the plateau level is that of Production Situation 2 in 1953, but in 1951 it could be either that of Production Situation 1 or Production Situation 2.

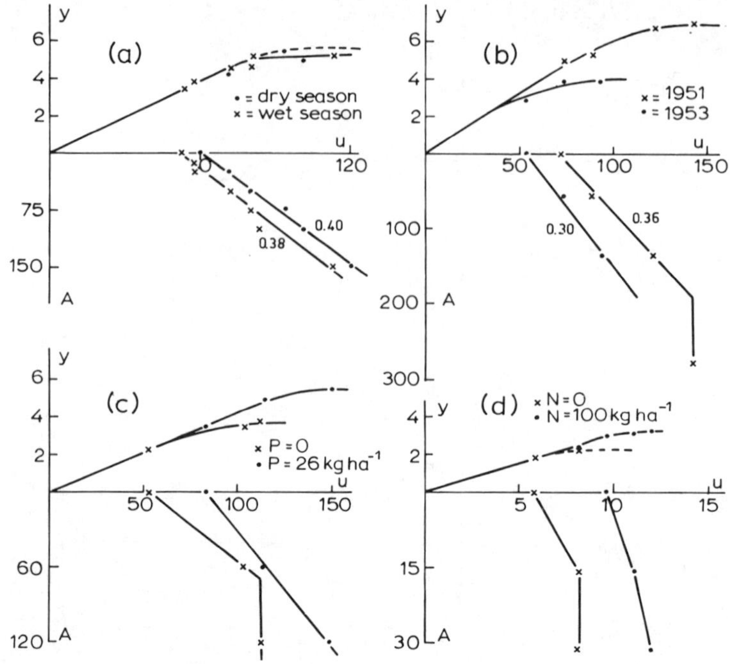

Figure 50. The relation between nutrient uptake, u (kg ha^{-1}), and yield, y (t ha^{-1}) and the relation between nutrient application, A (kg ha^{-1}) and nutrient uptake. Numbers in lower half of the graphs denote recovery fraction of applied fertilizer. a. Rice, Thailand. y is grain yield at a moisture content of 0.15 kg kg^{-1}, u and A are expressed in kg N ha^{-1} (Walcott et al., 1977). b. Maize, USA. y is grain yield at a moisture content of 0.15 kg kg, u and A are expressed in kg N ha^{-1} (Flynn et al., 1957). c. Grain sorghum, USA. y is grain yield at a moisture content of 0.15 kg kg, u and A are expressed in kg N ha^{-1} (Roy & Wright, 1973; 1974). d. Rice, Nigeria. y is grain yield at a moisture content of 0.15 kg kg, u and A are expressed in kg P ha^{-1} (Bredero, 1965).

Figure 50c refers to grain sorghum in the USA and illustrates the effect of phosphorus application on the level of the plateau. If no phosphorus is applied, the plateau level is a grain yield of about 3500 kg ha^{-1}, whereas at a phosphorus application rate of 22 kg ha^{-1} the plateau level increases to about 5500 kg ha^{-1}. On the basis of this information it is not possible to conclude whether the latter level indeed represents Production Situation 3, but it does not seem unlikely.

The reverse effect is illustrated in Figure 50d referring to bunded rice grown in Nigeria. Application of phosphorus in the absence of nitrogen application leads to an increase in yield, but the effect is rather small, because soon nitrogen becomes the limiting factor. At a nitrogen application rate of about

100 kg ha^{-1} the yield level increases with increasing P uptake to about 3200 kg ha^{-1}. This seems still low for that situation, but on the basis of the available information it is not possible to determine what the yield – limiting factor could have been.

4.1.4 Application rate — uptake relations

Single elements

If the yield potential of a given crop is to be estimated from the amount of a particular nutrient absorbed by the vegetation, it is necessary to estimate uptake as a function of application rate. These relations will be discussed for each nutrient separately, distinguishing the two relevant parameters: uptake at zero fertilizer application (zero level) and the slope with respect to the vertical (the recovery fraction).

Nitrogen

The intersect of the application rate – uptake curve with the uptake axis represents the inherent fertility of the soil for the element, for a given crop: the amount available without fertilizer application in the current cropping season. Its value in a particular case is partly a soil characteristic, determined by the mineral composition of the soil and its organic matter content and quality. This is illustrated in Figure 47d, where the soil at the Cidra site supplied twice as much nitrogen in the unfertilized situation as the soil at the Corozal site. The chemical analyses of the two soils (Fox et al., 1975) show that the total N content of the top soil at Cidra is about 1.3 times that at Corozal, at about the same organic matter content. It must be assumed, therefore, that the quality of the organic material, particularly its C/N ratio, is more favourable at the former site. The amounts of nitrogen taken up at zero fertilizer application are in both cases high for oxisols. This might be attributed to the fact that the experimental sites had fertilizer applications previous to the cassava experiment, which may have improved the nitrogen fertility.

Environmental factors, especially temperature and precipitation, also have a distinct effect on the zero N level. Temperature influences microbial activity (c.f. van Veen, 1977) and thus for instance the rate of mineralization, which is in general higher at higher temperatures. On the other hand higher temperatures may lead to greater losses of the mineralized N by increased denitrification (again a microbial process) or by favouring volatilization of ammoniacal compounds (Bouwmeester & Vlek, 1981).

Rainfall may have various effects on the nitrogen balance: it supplies nitrogen to the soil from atmospheric sources and through its effect on the moisture balance in the soil, it influences the rate and duration of mineralization (van Veen, 1977), as well as the magnitude of losses through denitrification and leaching. Such differences in environmental conditions are presumably at the basis of (sometimes large) differences in zero N level between years, as in Figures 41a and 41c. In these cases it is not possible to pinpoint the exact process responsible for the differences, because the reports do not provide sufficient details.

Some other examples, mostly referring to experiments in the temperate zone, are given in Figure 51. The first one refers to the first spring cut after an early nitrogen application to permanent pasture in the Netherlands. The spring in 1961 was warm with almost normal precipitation, whereas in 1962 it was a cold spring with rainfall about 25% above normal. In the latter year the low temperatures, hampering microbial activity and the high rainfall, favouring denitrification and leaching, resulted in a zero N level of less than half that of the previous season.

In Figure 51b, the effect of precipitation *per se* is illustrated for a winter wheat crop in Germany. In one of the treatments of this experiment winter rainfall was intercepted by protective shelters, which increased the uptake of nitrogen in the non – fertilized situation by about 15 kg ha^{-1}, presumably because leaching of the mineralized nitrogen beyond the rooting zone was prevented (and maybe denitrification was suppressed).

Land reclamation also improves the availability of native nitrogen as illustrated in Figure 51c, which refers to an experiment with winter wheat in one of the reclaimed polders in the Netherlands. In the well – drained plot, where the watertable was below 30 cm throughout the growing period, appreciably more native N is available to the crop than in the plot where under natural drainage the watertable was less than 10 cm below the surface from October till April. In the latter situation, the high soil – water content and the associated low oxygen concentration hamper organic matter decomposition and favour denitrification. In many cases, therefore, land reclamation has a triple effect on crop yield, as it improves concurrently the plateau level, the uptake at the zero N level and the efficiency of nitrogen – fertilizer utilization (Section 7.1).

Management practices can also have an influence on the uptake of nitrogen from natural sources. Figure 51d shows that the introduction of a leguminous crop in a rotation increased the uptake at zero – fertilizer application, in this case by about 25 kg ha^{-1}. The magnitude of the gain depends on the growing conditions for the legume, the effectiveness of nodulation and nitrogen fixation and the conditions for decomposition in the subsequent crop growing season.

Figure 51e illustrates the effect of weed control on nitrogen availability at zero – fertilizer application. When weeds are allowed to compete with the crop, part of the available nitrogen is absorbed by these unwanted plants and is thus lost for the crop. In this particular case, the difference in nitrogen uptake by the crop amounts to about 45 kg ha^{-1}.

The effect of a rather common phenomenon in growth of rice under rained conditions is illustrated in Figure 51f: that of temporary drying of the field when rain is insufficient to maintain a water layer. In the situation depicted

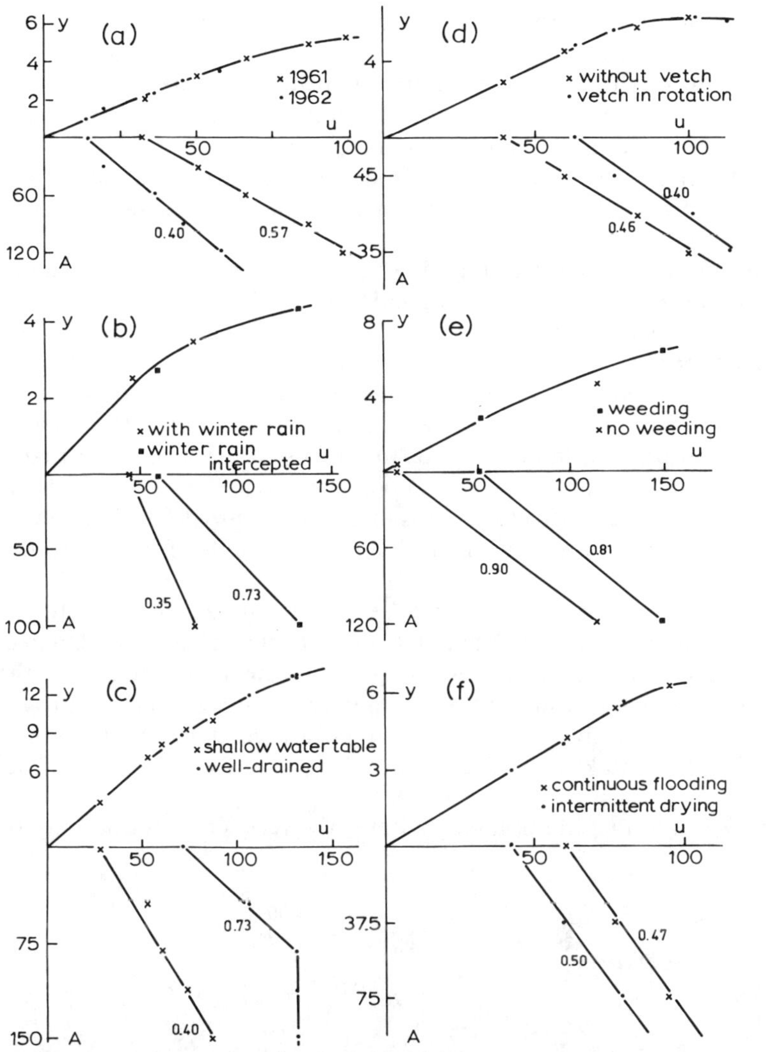

Figure 51. The relation between total nitrogen uptake, u (kg kg^{-1}), and yield, y (t ha^{-1}) and that between nitrogen application (A in kg ha^{-1}) and nitrogen uptake. Numbers in lower half of the graphs denote recovery fraction of applied fertilizer. a. Permanent pasture, the Netherlands. y is total dry matter (Oostendorp, 1964). b. Winter wheat, Western Germany. y, is grain yield at a moisture content of 0.15 kg kg^{-1}. c. Winter wheat, the Netherlands. y is total dry matter (Sieben, 1974). d. Rice, USA. y is grain yield at a moisture content of 0.15 kg kg^{-1} (Williams et al., 1972). e. Maize, Nigeria. y is grain yield at a moisture content of 0.15 kg kg^{-1} (Kang et al., 1977). f. Rice, USA. y is grain yield at a moisture content of 0.15 kg kg^{-1} (Patrick et al., 1967).

here, intermittent drying substantially reduces the availability of native nitrogen, as compared to situations of continuous flooding. The most likely explanation is that during the periods of soil drying nitrification takes place, leading to an abundance of nitrates in the soil. During subsequent flooding, reduced conditions build up in the root zone and the resulting denitrification leads to the loss of nitrogen from the system. Poor water control is in many cases the cause of substantial losses of nitrogen from the soil – plant system of bunded rice.

In conclusion, it may be stated that despite the existence of a great deal of knowledge about the factors influencing nitrogen availability at zero – fertilizer application, reliable quantitative predictions for concrete situations are very difficult. For many situations, an alternative is the analysis of yield data at low input levels, from which nitrogen uptake may be derived (Section 4.2).

The slope of the application – uptake curve represents the efficiency of uptake of the fertilizer, or in other words the fraction of that fertilizer that is recovered in the (mostly above – ground) plant material. This variable determines how much of the expensive input is effectively utilized and is therefore of prime importance for decisions on economically feasible fertilizer application rates.

In general there is a positive correlation between the uptake at zero – fertilizer application and the recovery fraction in a given situation (cf. Brockman et al., 1971), because the processes rendering the nutrient unavailable to the plants act in identical ways on nitrogen from natural sources and on nitrogen applied as fertilizer. In the examples presented in Figures 41, 44 and 47, in most cases a higher uptake of native N coincides with a higher recovery fraction.

In addition to the factors discussed before, the recovery fraction may be

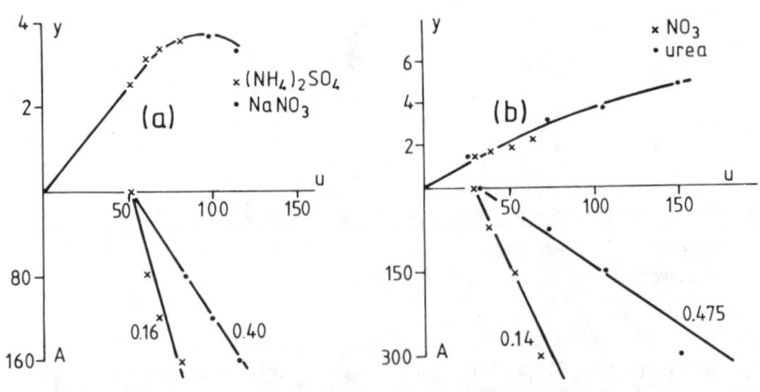

Figure 52. The relation between total nitrogen uptake, u (kg ha^{-1}) and yield, y(t ha^{-1}), and that between nitrogen application, A (kg ha^{-1}), and nitrogen uptake. Numbers in lower half of the graphs denote recovery fraction of applied fertilizer. a. Winter wheat, the Netherlands. Netherlands. y is grain yield at a moisture content of 0.15 kg kg^{-1} (Lehr, 1959). b. Natural rangeland, Mali. y is total dry matter (Penning de Vries et al., 1980).

influenced by the type of fertilizer applied, as illustrated in Figure 52a for a winter wheat crop growing in one of the reclaimed IJsselmeer polders. The nitrate fertilizer is absorbed 2.5 times as efficiently as the ammoniacal fertilizer. This is due to losses through ammonia volatilization on the lime–rich gritty clay soil (pH 8.2) in these polders.

In contrast to that, the recovery of nitrate fertilizer is less than one third of that of urea for the situation depicted in Figure 52b. This refers to an experiment on a heavy clay soil in the Sahelian region, which received run–off water from surrounding areas and was flooded for some time during the growing season. The low recovery of nitrate must be due to denitrification and/or leaching. The much higher recovery of urea indicates that nitrification must have been slow under these conditions, so that nitrogen remained in the system in ammoniacal form.

Fertilizer recovery is also influenced by timing and method of application. The effect of method of application is illustrated in Figure 41f, which refers to an experiment with bunded rice on Java, Indonesia. Placement of urea fertilizer directly into the reduced soil layer, where oxygen is absent, prevents the transformation of ammonium ions, formed after hydrolysis of urea into nitrates (nitrification), and the subsequent loss through denitrification. When urea is broadcast onto the layer of standing water, nitrification takes place in the aerobic environment, after which the nitrates formed enter the reduced soil layer, either by mass flow or by diffusion. In that environment, rapid denitrification follows, which results in a much lower availability of nitrogen for the vegetation.

Split application of the total amount of nitrogen fertilizer generally results in higher recovery fractions (Figure 41e), because it leads to a shorter average residence time of the element in the soil and hence to lower losses. The same effect plays a role in comparisons between early application and later applications. In the latter situation, the rate of crop growth is higher, which increases the demand for nitrogen and hence the rate of nutrient uptake.

Differences in recovery fraction between different cultivars (Figures 41b and 44c) may also be related to the same phenomenon. Minabar–2 is a traditional tall cultivar that produces much more vegetative material than the improved short straw cultivar IR8 (grain/straw ratio 0.9 vs. 1.5). The recovery fractions in practice vary typically between 0.1 and 0.8, with a value of 0.5 as an 'acceptable norm'. If the recovery fraction is much lower than that, too much of the expensive nitrogen fertilizer is lost and achieving much higher values may require too much investment in sophisticated management practices.

Phosphorus

In many parts of the world, crop production under natural conditions is limited by phosphorus availability (e.g. Penning de Vries et al., 1980). Therefore phosphorus fertilizer application is necessary to achieve potential yields.

In contrast to nitrogen, a major part of the total phosphorus store in a soil may be present as inorganic compounds of low solubility, notably aluminium, iron and calcium compounds. There are, however, large variations in the relative proportions of the inorganic and organic forms of phosphorus in different soils.

The inherent fertility level of the soil for phosphorus (zero P level) may vary as a result of variations in mineralogical composition as well as in quantity and quality of their organic matter content. Such differences are illustrated in Figure 45c, where Prome's farm yields almost twice as much P as Suwan's farm. That may be related to the difference in pH, because the former soil is slightly alkaline, the latter acidic. Under low pH the concentration of Al ions tends to be higher, leading to formation of insoluble phosphates when fertilizer is applied.

As for nitrogen, different environmental conditions may result in differences in the zero P level. This is partly related to the conditions affecting decomposition of organic material, which supplies part of the phosphorus, and partly to the inorganic P cycle in the soil, which is affected by temperature and soil moisture conditions (Beek, 1979).

There is an interesting example of the interaction between nitrogen and phosphorus in Figure 42c, where increased N availability improves uptake of native phosphorus even though phosphorus is the limiting element for production. This aspect will be treated in more detail in Subsection 4.1.4.

The relation between application rate and uptake is much more complex for phosphate fertilizers than for nitrogen, as is clear from Figures 42 and 45. In some situations the relation seems to be a straight line (Figure 45c), but the recovery fraction is so low in this case that not much importance can be attached to these results.

The complexity of the relation may be expected because, contrary to the situation with N, the kinetics of the reactions of phosphorus between the soil solution and the solid phase of the soil are not of first – order (Beek, 1979). Phosphate ions are effectively removed from the soil solution by absorption, precipitation and immobilization, so that the concentration in the soil solution is more or less constant and not proportional to the amount of fertilizer applied. The recovery of applied phosphorus fertilizer is therefore in general low and often decreases with increasing rates of application, because the longer residence time in the soil leads to larger proportions being immobilized by precipitation or incorporation in the organic fraction (Figures 42c, 45b). In such situations, where recovery is low, a more efficient utilization of the fertilizer may be achieved by concentrating the fertilizer in a limited soil volume, for instance by placement (de Wit, 1953).

A 'special' case is encountered in Figure 42f, where the recovery fraction first decreases with increasing application rate and subsequently increases at higher application rates. A possible explanation for this phenomenon is that with addition of phosphates to the soil precipitation takes place, which increases with increasing concentration of phosphates. At the highest application rate, all free cations that may form insoluble phosphates (Fe^{3+}, Al^{3+}, Ca^{2+}) are exhausted. The phosphate concentration in the soil solution will then rise, resulting in increased uptake of the fertilizer.

The results presented here clearly illustrate that the relation between phosphorus application rate and uptake by a crop may be highly variable and that it is difficult to predict the effect of phosphorus application without detailed knowledge of the local conditions.

Potassium

The potassium content in plant tissue is generally high as may be deduced from the uptake – yield curves. In areas with less intensive agriculture, however, yields are predominantly limited by availability of nitrogen and phosphorus, hence potassium shortage is not widespread. Moreover, at harvest, most of the potassium is in the crop residues that are partly or completely left in the field or returned to the field later on. That is probably the main reason that so few relevant experimental results could be collected.

The potassium – supplying capacity of soils is mainly determined by their mineralogical composition. The main sources of the element are minerals like micas and feldspars, which upon weathering release large amounts of potassium. Therefore clay soils, which contain more of these minerals, are generally superior in the supply of native potassium to sandy soils.

The reactions involved in the supply of potassium from the soil store to the plant are equilibrium reactions whose rate is governed by the relative concentrations of the various forms of potassium: ions in the soil solution, ions adsorbed in exchangeable form on negatively – charged clay or organic matter particles and potassium chemically bound in the clay minerals (van Diest, 1978). Thus, conditions changing these relative concentrations influence the zero K level. Weather plays an important role: higher temperatures and more favourable moisture conditions favour weathering of minerals and increase potassium availability. Differences between years may therefore be considerable (Figures 43a and b).

It has also been observed that crops differ considerably in their ability to extract potassium from a given soil (van Keulen & van Heemst, 1982), which could be related to the different potassium requirements of the crops, but also to the length of the growing season and hence the time available for uptake.

Two main processes influence the efficiency of uptake of potassium fertilizer. In sandy soils potassium ions are very mobile and may be easily lost by leaching. Clay soils, on the other hand – especially those having a high proportion of a special type of clay minerals (illite, vermiculite) – may fix potassium. In that case, potassium is absorbed at specific sites in the clay lattice, rendering it unavailable for plant uptake. The result is that, especially at lower application levels, the recovery of fertilizer may be very low (cf. Figure 43b) but increases gradually at higher levels of application, when the absorption sites are more or less saturated (van Keulen & van Heemst, 1982).

Where the processes of leaching and fixation are not of great importance, recoveries of potassium fertilizers are generally high, and, as in the case of nitrogen, independent of the amount applied (Figures 43a and 46b).

Levelling off of the application – uptake rate at high levels of application (Figure 46b) must be attributed to active exclusion of potassium ions by the plant at high concentrations in the solution. When the concentration in the tissue has reached its maximum value, further uptake is apparently inhibited. This will, however, only happen at very high levels, because 'luxury' consumption does apparently occur quite often (Figures 43b and 46b).

4.1.5 Interaction between elements

In Subsection 4.1.2 it was shown that the response of a crop to a specific element is also influenced by the supply of other elements. For example, increased uptake of nitrogen may not lead to higher yields if the supply of phosphorus is insufficient (Figure 50c). However, the uptake of a specific element may also be affected by the supply of other elements, as is illustrated for example in Figures 50c and d. It shows, that the application of phosphorus increases the uptake of nitrogen in the absence of nitrogen application, and vice versa. Similar reactions have also been observed in Sahelian natural pasture (Penning de Vries & van Keulen, 1982) and in other experiments with bunded rice (de Wit, 1957).

For an explanation of this phenomenon, it is necessary to consider the dynamics of the two elements in the plant tissue. It has been argued that at maturity species – characteristic minimum and maximum element concentrations exist in the various plant organs, provided that other elements are not in short supply. This is not only the case at maturity, but at each point in the plants life cycle such characteristic minimum and maximum concentrations may be established. In general there is a ratio of about four to one between the maximum and the minimum concentrations of one element at a particular phenological stage. Both values decline during ageing of the plant. This gradual decrease in characteristic concentration is related to the physiological functioning of the plant. In the early stages of plant growth, protein – rich material – mainly in the leaves – is being produced, whereas later on more supporting tissue, such as cell walls is formed, consisting for the larger part of structural carbohydrates with less nitrogenous compounds. The specific function of nitrogen and that of phosphorus in the plant are strongly related: both elements are found in the nucleic acids; nitrogen is an essential component of the enzymes, whereas phosphorus plays a role in the molecules that are responsible for energy transfer in enzymatic processes. It may be expected, therefore, that a correlation exists between the functional concentration of nitrogen in the plant and that of phosphorus. That is borne out by experimental results obtained in the field (Penning de Vries & van Keulen, 1982) and under controlled conditions (Dijkshoorn, personal communication). As the maximum and minimum concentrations of each of the elements N and P vary by a factor of four, the ratio of their concentrations (the P/N ratio) could in theory vary 16 fold. However, only about a four – fold variation is found experimentally. The minimum and maximum value of the P/N ratio for various grasses and dicotyledons varied between 0.04 for situations where a relative shortage of phosphorus exists to about 0.15 in situations where a relative nitrogen shortage exists. When these limiting values are approached, absorption of the element with the relative surplus is inhibited, even though it may be abundantly available. This relative shortage should not be confused with an absolute shortage of the element, i.e. the P/N ratio may be at its minimum value, while the nitrogen concentration in the tissue is also at its absolute minimum. Thus application of the element with the relative shortage (phosphorus in Figure 50c) increases the uptake of the other element (nitrogen), which in turn leads to a proportional increase in production.

Under conditions where both elements are available in sufficient amounts, the P/N ratio in the tissue is often close to 0.1, which may be considered the optimum value. A first approximation of the phosphorus requirements of plants may thus be set equal to one – tenth of their nitrogen requirements.

Exercise 56

In an experiment, the following observations were made at a certain point in time.

	Element concentration ($kg\ kg^{-1}$)	
	N	P
Field 1	0.02	0.0018
Field 2	0.02	0.0008
Field 3	0.02	0.003

In which field do you expect that P fertilizer application will increase the uptake of N? In which field will application of N increase uptake of P? Explain your answer.

4.2 Nutrient demand and fertilizer requirements

P.M. Driessen

In Section 4.1 the effects of the main plant nutrients nitrogen, phosphorus and potassium on crop yield and the relations between nutrient supply and nutrient uptake were discussed using a three quadrant presentation. If these principles are applied to actual cropping situations, four practical questions must be answered:

- how can a nutrient shortage be recognized in a certain situation?
- how much of that nutrient does the unfertilized soil supply?
- what is the recovery of a certain fertilizer – nutrient if applied in a specific way?
- how much fertilizer must be applied to realize Production Situation 2, where crop yield is solely determined by weather conditions and water availability?

As is evident from the discussion of the principles of nutrient demand and supply, such questions pertaining to a specific cropping situation can only be answered with the aid of information collected in field experiments and from chemical analyses. The purpose of this section is to show how these questions can be answered on the basis of minimal information. For practical reasons, the discussion will be limited to situations in which the availability of nitrogen and/or phosphorus determines crop performance; situations where yields are limited by potassium shortage are less common.

4.2.1 Recognition of nutrient limitation

The easiest way to recognize nutrient stress in a crop is through chemical analysis of such biomass components as straw and seed, or – if the deficiency is serious enough to manifest itself in the physical appearance of the crop – through identification of specific deficiency symptoms such as discolouration or necrosis of plant organs. The information needed for correct diagnosis can only be collected with the backing of an adequately equipped plant analytical laboratory. This is the major reason that such data are scarce. Soil chemical data are more commonly available as they are routinely collected in soil surveys. Unfortunately, there is not a generally valid quantitative correlation between soil analytical data and crop performance. Soil analysis data can at best give an indication of likely element deficiency or of the occurrence of unfavourable soil conditions which would conceivably obstruct the normal functioning of plants. This subject will be further elaborated in Section 5.3.

A convenient way to recognize nutrient limitation is by comparing actual plant production in the field with the calculated production in Production Situation 2. Obviously, such calculations must be done for a crop species and variety with similar properties as the one grown in the field and for weather conditions and a water regime the same as that in the field. Likewise, the actual field production must have been obtained under conditions that apply to Production Situation 2, viz. in a weed – free environment, with adequate control of pests and diseases and with optimum harvesting methods. In addition, chemical and physical soil conditions must be such that they have no adverse effect on plant performance. The outcome of such a comparison can be either of two possibilities:

- one is that the actual production is close to the calculated production. In that case growth is not limited by a defiency of nitrogen or mineral elements.
- the other is that the actual production is clearly lower than the one calculated for Production Situation 2. For identification of the element that is in short supply it is then inevitable to perform a chemical analysis of crop components.

If, for example, the nitrogen concentrations of the analysed plant parts are distinctly higher than the minimum values as given in Section 4.1, it may be concluded that nitrogen availability is not the limiting factor and the concentrations of other nutrients must be checked. Phosphorus shortage is then a likely candidate. If, on the other hand, the nitrogen concentrations of the analysed plant parts approach their minimum values, it may be concluded that nitrogen shortage limits crop production. It is then worthwhile to consider application of a nitrogen fertilizer. The improved plant growth resulting from this N application will also increase the demand for other elements such as phosphorus and potassium. If this additional demand cannot be met by the soil, i.e. if the effect of N application remains below expectations, mineral elements must be applied in addition to nitrogen.

In practical plant nutrition research, nutrient demands are not identified sequentially, but simultaneously. This is done in fertilizer experiments, which involve cropping a number of identical fields, arranged in a randomized design and each planted to the same variety and fertilized in a similar fashion but with different combinations of N, P and K fertilizers. The advantage of such experiments is that possible effects of nitrogen application, mineral element(s) application, or combinations thereof, become apparent after only one cropping season with one weather and one water regime which facilitates their interpretation. Obviously, the number of plots included in a particular experiment depends on the number of types/combinations and doses of fertilizers tested; it is often considerable. Moreover, the experiments must be done with a number of replications to rule out misinterpretation of the situation due to local anomalies or human failure. Each experiment includes at least one unfertilized plot, the 'control plot'. The function of this control plot deserves particular attention, as it is instrumental in answering the second practical question posed at the beginning of this section: How much of a nutrient does the unfertilized soil supply?

Exercise 57

In Section 2.3 (discussing Production Situation 1, where water, nitrogen and mineral elements are optimally supplied) an example was given in which the production of the high yielding rice variety IR8 was calculated for a bunded experimental site near Paramaribo, Suriname. Suppose that on this same station an unfertilized field planted to IR8 (on the same date) produced a total above – ground dry weight of 2000 kg ha^{-1} with a grain – straw ratio of 1.0. The total quantity of nitrogen contained in the above – ground production was 14 kg ha^{-1}.

- Identify the total above – ground biomass production and the grain yield calculated for Production Situation 1 in Section 2.3; calculate the production of straw in this situation.
- What would the above – ground biomass production be at Production Situation 2? (Remember: bunded rice fields)
- Do you think that fertilizers must be used for high production on that location?
- If so, is nitrogen the first element that has to be supplied? (Recall that minimum nitrogen concentrations in rice grain and straw are around 0.01 and 0.004 kg kg^{-1}, respectively)

4.2.2 Nutrient uptake from unfertilized soils

As unfertilized soils can normally still support a crop, they cannot be entirely devoid of nutrients. Influx of elements with wind, rain or irrigation water is a possibility, but more commonly these nutrients originate from sources in the soil itself. Nitrogen is predominantly supplied through microbial breakdown of soil organic matter, which – depending on its botanical origin and genetic history – normally has a nitrogen concentration between 0.01 and 0.05 kg kg^{-1}. Mineral nutrients are also supplied in this process but, with the possible exception of phosphorus, only in rather negligible quantities. The bulk of the mineral elements originates from weathering rock fragments.

The rate at which indigenous nutrients are supplied depends partly on soil characteristics such as organic matter content and composition, soil mineral composition and soil pH, but also on a score of exogenous factors that are often highly variable (Section 4.1). An example is soil temperature, subject to daily and seasonal fluctuations, and also influenced by weather conditions, vegetation cover and the action of man. To complicate things even more, not all the nutrients released are available for uptake by plant roots, as they may leach out of the root zone, precipitate as low – solubility compounds, etc.

This complexity makes it virtually impossible to predict the uptake of nutrients from an unfertilized soil on the basis of theoretical considerations only. For the time being, the quantity of the growth limiting element taken up by a certain crop from the unfertilized soil can best be established by simply dividing the control yield by the slope of the yield – uptake curve for that element.The growth limiting element can be identified, as explained, by determining element concentrations in the plant tissue. The slope of the yield – uptake curve is established on the basis of the ratio of economic product and crop residues and their minimum element concentrations.

If fertilizer trials are continued at an experiment station over a number of years, it is not uncommon that the 'base uptake', i.e. the quantity of nutrients taken up from unfertilized plots, increases in successive years. To understand this, it must be realized that the location of the control plot changes each year, because the individual fields of the experiment are laid out in a randomized design. Nutrients applied, but not completely taken up in one year may to some extent remain in the soil and increase the level of soil – supplied nutrients in subsequent experiments. The same happens in practical farming where fertilizers are used: natural soil fertility improves in the course of the years and base uptake reaches a new (higher) level. The magnitude of this improvement depends on both cultivation practices and environmental conditions. This may be illustrated by two extreme examples: in the case of nitrogen application to bunded rice, carry – over effects are often low because most of the nitrogen that is not directly taken up is lost through denitrification and/or leaching. A completely different situation exists in natural pastures in semi – arid regions, where almost no nitrogen is lost from the soil except by plant uptake.

Phosphorus is less mobile in the soil – plant – atmosphere system than nitrogen. Therefore, losses of phosphorus from the system are normally lower than losses of nitrogen and, consequently, the carry – over of phosphorus is higher. In practical farming, phosphorus is sometimes applied in high doses, which promise a satisfactory P supply over a number of years. However this effect should not be overestimated, as a large part of the fertilizer not taken up in the first year is transferred to forms that are far less available to plants.

Exercise 58

Table 44, from 'A Summary Report on Yield Response of some Thai Rice Varieties to Varying N and P_2O_5 Combinations' (Thai Rice Department, 1956), presents average yields of some Thai lowland rice varieties obtained in

Table 44. Average yields (kg ha^{-1}) of Thai lowland rice varieties at different levels of N and P combinations at Bangkhen Experiment Station.

Rates of nitrogen applied (kg ha^{-1})	Rates of phosphorus applied (kg ha^{-1}) P-0	P-37.5	P-75.0	P-150.0
A. Crop year 1952-1953				
N-0	1214.4			
N-37.5		2107	2232	2493
N-75.0		2211	2394	2551
N-150.0		2389	2484	2727
B. Crop year 1953-1954				
N-0	1529.4			
N-37.5		2706	3048	3164
N-75.0		3074	3220	3303
N-150.0		3159	3665	3649
C. Crop year 1954-1955				
N-0	1755.6			
N-37.5		3547	3686	3649
N-75.0		4188	4166	4291
N-150.0		4682	4647	4691

three consecutive seasons at Bangkhen Experiment Station.

- Identify the control plot yields obtained in the three consecutive crop years.
- Can you explain why the control yields obtained in the three consecutive years are different?
- Experience with traditional Thai rice varieties has shown that their rough grain* – straw ratio is normally close to 0.5 ('long straw varieties'). The minimum nitrogen concentrations of rough grain and straw are 0.0078 and 0.0036, respectively.

Calculate with these numbers the slope of the rough grain yield N uptake curve and use this slope to calculate the nitrogen uptake from the control plots in the three consecutive crop years.

---

* Rough grain pertains to unhusked rice with 12 percent water; one kilogram of rough rice is equivalent to 0.8 kg husked rice (Section 5.4).

4.2.3 *The recovery of nutrients applied in fertilizers*

In Section 4.1 it was shown that the increase in nutrient uptake following fertilizer application is generally proportional to the quantity of that nutrient added, at least in the case of nitrogen and potassium. It was also shown that there are exceptions to this rule, e.g. in the case of immobilization or fixation of the added element by the soil. In this section not all possible exceptions will be treated, but element recovery in a normal situation where application – uptake relations are linear in the relevant range will be discussed. It was also explained in Section 4.1 that the slope of the application – uptake relation, i.e. the quantity of a certain element taken up from a fertilized soil minus the quantity taken up from the same but unfertilized soil divided by the quantity of that element contained in the applied fertilizer, is called the recovery fraction, or in a mathematical notation:

$$R_x = (u_{f,x} - u_{o,x})/A_x \tag{79}$$

where

R_x is the recovery fraction of element x (kg kg^{-1})
$u_{f,x}$ is the uptake of nutrient x from fertilized field (kg ha^{-1})
$u_{o,x}$ is the uptake of nutrient x from control field (kg ha^{-1})
A_x is the application of nutrient x to fertilized field (kg ha^{-1})

Theoretically, R_x ranges in value from close to 0 to close to 1.0; it expresses the efficiency with which a certain fertilizer is used.

The actual recovery of an element depends on the competitive position of the plant relative to processes in the plant – soil – atmosphere system that render the element unavailable to the plant. Taking nitrogen as an example, volatilization of ammoniacal nitrogen, leaching of nitrogen in nitrate form, denitrification to gaseous N forms and immobilization of nitrogen by soil microbes are processes lowering the recovery of applied fertilizer N. It follows, therefore, that a low recovery can be improved by either increasing the uptake activity of the plant roots or by decreasing the impact of the competing processes, or both. In the case of nitrogen, about one – tenth of the nitrogen taken up is needed in the root system, so that the maximum N recovery in the above – ground parts of a crop would be of the order of 0.9 kg kg^{-1}. In practice, it is often difficult to reach a higher recovery than 0.5 kg kg^{-1}. If the actual N recovery is well below this value it is worthwhile to try to improve the situation. It is this aspect, viz. indication of the need for adaptation of variety and/or cultural practice, that makes it so important to quantify the recovery of applied fertilizer nutrients in the analysis of a cropping situation.

The recovery of a certain element can be calculated from fertilizer experiments, if the biomass components are analysed for that element. It was explained why such complete experiments are rather scarce. Their number and regional distribution may be sufficient to indicate whether in a certain region a situation exists or has been created where shortage of a certain element is prominent, but complete experiments are nearly always too few in number to allow sufficiently reliable estimation of the recovery of that element from applied fertilizers. In practical farming, nutrient recovery depends not only on plant properties and environmental conditions but also on management factors, such as the type of fertilizer used and the timing and mode of fertilizer application. If, for instance, a high dose of a nitrogen fertilizer is broadcast at the beginning of the cropping season, when the nitrogen demand of the crop is still low, high losses of nitrogen from the root zone can be expected and, consequently, low N recovery values. Adapted application methods and good timing of fertilizer application may lead to increased N recovery in such situations, as will be explained later.

Where complete experimental information is scarce, the limited amount available must be used as efficiently as possible. If it is known which element is most likely limiting the growth of a certain crop in a particular situation, a first estimate of nutrient recovery can be made without further chemical analysis of plant components. This is possible because it may be assumed then that the limiting element in the crop is diluted to its minimum concentration in the various plant parts. The simplest experiment that still yields the required information, consists of a control plot and a fertilized plot whose production components, viz. economic yield and crop residues, are separately weighed. It must be realized that the reliability of only one recovery value, calculated in this way, is low and that such results are only indicative. The results of properly conducted experimental series, normally published as average yields per treatment with specified standard deviations, are a sound basis for recovery calculations.

Exercise 59

In Table 44 average yields of rough rice are given for three consecutive years of N – P fertilizer experiments at Bangkhen Experiment Station, Thailand.

- From the average yield values for 1954 – 1955, it can be observed that at a given level of nitrogen application different P applications do not result in significant yield differences. Which conclusion can be drawn with regard to the P supply to the crop in this situation?
- What has caused this situation? (Note: look also at the effects of P application on yields in 1952 – 1953 and 1953 – 1954). Is it logical that the same situation has apparently not been reached for the soil's nitrogen status?
- Calculate the nitrogen recovery fractions realized in 1954 – 1955 for the fertilizer combinations N 37.5 – P 37.5, N 75.0 – P 37.5 and N 150.0 – P

37.5. Assume the same grain – straw ratio and minimum N concentrations as in Exercise 58.

Earlier in this section it was argued that unsatisfactory recovery of fertilizer nutrients can be remedied by adaptation of crop characteristics, environmental conditions or management, alone or in combination. Agricultural research has developed crop varieties that can recover nutrients and realize an acceptable yield level under conditions that would be prohibitive for unimproved varieties, for example acidity – tolerant varieties and varieties which can successfully be grown in a brackish environment. Better nutrient recovery is not, or only marginally, involved in the success of the modern high yielding rice varieties (HYV's) developed at the International Rice Research Institute in the Philippines. The key to the success of these varieties is their high grain – straw ratio, which is of the order of 1.0; HYV's are so – called 'short straw varieties'. The high grain – straw ratio is associated with a high initial efficiency of the yield – uptake relation, i.e. a high increment in grain yield per kg nitrogen taken up. Their short posture allows uptake of considerable quantities of nitrogen with less risk of lodging, which quickly follows too luxuriant vegetative growth of traditional long – straw rices. In combination, these two effects result in very high grain yields per hectare, provided that the crop is grown under favourable environmental and management conditions that allow a high nutrient recovery. Under unfavourable conditions such as nutrient shortage or heavy weed infestation, traditional rice varieties, which have evolved in the course of centuries of natural selection by rice farming communities, may well perform better.

Increasing nutrient recovery by adaptation of the environment has been practiced as long as agriculture exists. Bunded rice fields are an example: nutrient recovery is promoted by the artificially created conditions in the puddled top layer of a wet rice field. Often, manipulation of the environment involves management measures. In the management sphere, too, wetland rice cultivation offers many practical examples of how nutrient recovery can be manipulated and improved. Illustrative of the great practical significance of good management is the mode of nitrogen fertilizer application to bunded rice. In many rice areas, urea is used as a nitrogen source and broadcast on the flooded field at the time of transplanting. The average recovery fraction of the fertilizer N is then often only 0.2, or even lower. This low recovery may be understood as follows: the favourable temperature conditions and oxygen-rich environment of the shallow water layer on top of the rice field promote rapid transformation of urea N to ammonium ions (NH_4^+) and subsequently to nitrate ions (NO_3^-) . The nitrate ions are highly mobile and move downward into the soil with percolating water or by diffusion. The oxygen in the waterlogged soil is rapidly depleted by microbes decomposing soil organic matter, and subsequently some bacterial species use the nitrate ions as an oxygen source. Nitrate N is then converted to gaseous forms (N_2 or N_2O) and escapes to the atmosphere. A good way to combat this nitrogen loss and improve the nitrogen recovery is placement of the urea directly into the low-oxygen layer. There urea N is only converted to NH_4^+ ions that are, in contrast with anions like NO_3^- to some extent protected against leaching because they are retained by the soil through adsorption at the surfaces of negatively charged clay and organic matter particles (the process of ion adsorption will be discussed in some detail in Section 5.3). Placement of urea in the puddled surface layer may well bring the recovery of nitrogen to a value of 0.5 or higher. In general, losses of fertilizer N can be reduced and recovery improved, if only small doses of fertilizer are given at a time, so that most of the nutrient applied can be absorbed by the roots in a relatively short time. Therefore, repeated application of smaller doses spread over the growing season ('split application') is often superior to application of all fertilizer in one dressing.

Exercise 60

The Technical Division of the Thai Rice Department in Bangkok has conducted fertilizer experiments with bunded rice to investigate the recovery of nitrogen from different types of fertilizer, applied at different moments and in different ways (Lusanandana et al., 1966). Nitrogen application was 60 kg ha^{-1} on all fields. In addition, each field received phosphorus at a rate of 26 kg ha^{-1} and potassium at a rate of 50 kg ha^{-1} at the beginning of the growing season to eliminate the possibility of P or K shortages.
Table 45 presents essential information on the different treatments, as well as grain and straw yields and the results of chemical plant analyses.

- Examine carefully the different treatments tested in this experiment. Note that four different nitrogen fertilizers were used, applied either at transplanting or at the time of ear primordium initiation, and at three modes of application.
- Calculate for each treatment the P/N ratio at the time of primordium initiation. Answer the following questions:
  Do the P/N ratios indicate that nitrogen availability is the growth – limiting factor in this experiment? Are there indications that N fertilization at transplanting (16 July) has a measurable beneficial effect on production at the time of primordium initiation (21 September)?
- Calculate the grain – straw ratios at harvest time. Are the rice varieties used in the experiment modern HYV's or traditional long – straw varieties?
  What does this mean for the slope of the yield-N uptake curve? Check your answer by calculating this slope with the results obtained for one or more of the treatments.

Table 45. Nitrogen and phosphorus concentration at primordium initiation, and in grain and straw at harvest; and dry weight of grain and straw of Thai rice varieties at different fertilizer treatment.

Treatment	Harvest at primordium initiation (21 September 1966)			Final harvest (22 November 1966)			
	dry weight	N	P	grain weight	N	straw weight	N
	(g)	(kg kg^{-1})	(kg kg^{-1})	(kg ha^{-1})	(kg kg^{-1})	(kg ha^{-1})	(kg kg^{-1})
A Ammonium sulphate 16 July 1966; in rows at 5 cm depth	8.51	0.0186	0.0012	3335	0.0086	12410	0.0032
B Urea, 16 July 1966; in rows at 5 cm depth	8.32	0.0178	0.0013	3325	0.0089	11825	0.0032
C Ammonium nitrate 16 July 1966; in rows at the surface	7.26	0.0168	0.0012	2685	0.0088	9585	0.0036
D Sodium nitrate 16 July 1966; in rows at the surface	5.31	0.0156	0.0014	2300	0.0091	5785	0.0036
E. Ammonium sulphate 21 September 1966; broadcast on surface	7.17	0.0281	0.0014	3015	0.0099	8840	0.0040
F Urea, 21 September 1966; broadcast on surface	5.72	0.0255	0.0013	2900	0.0094	7555	0.0039
G Ammonium nitrate, 21 September 1966; broadcast on surface	5.74	0.0271	0.0015	2775	0.0088	7185	0.0038

Table 45. (continued)

Treatment	Harvest at primordium initiation (21 September 1966)			Final harvest (22 November 1966)			
	dry weight	N	P	grain weight	N	straw weight	N
	(g)	(kg kg^{-1})	(kg kg^{-1})	(kg ha^{-1})	(kg kg^{-1})	(kg ha^{-1})	(kg kg^{-1})
H Sodium nitrate, 21 September 1966; broadcast on surface	5.35	0.0251	0.0014	2740	0.0091	6790	0.0035
K Control (blanket) dressing only				1930	0.0091[a]	4085	0.0036[a]

[a] Estimated value.
(Source: Lusanandana et al., 1966).

– Calculate the total nitrogen uptake for each of the treatments and for the control (on the basis of above – ground parts). Then, calculate the nitrogen recovery realized in each of the eight treatments and answer the following questions:

a. Are the recoveries realized under Treatments A and B satisfactory? If so, can you explain why (for both treatments)?
b. Compare the recoveries realized in Treatments C and D.
Is there reason to expect high losses of fertilizer nitrogen in these treatments? Can you explain the difference in N recovery between Treatment C and Treatment D?
c. Compare the N recoveries realized in Treatments D and H, where sodium nitrate was applied at different stages of the growth cycle. Can you explain the difference in nitrogen recovery if you consider for each treatment the competitive position of the plant relative to other processes removing nitrogen from the root zone?
d. It has been stated in this section that N recovery from urea, broadcast at the time of transplanting is often lower than 0.2. Consider the N recovery realized under Treatment F and explain why you have found a different recovery value.

4.2.4 The nutrient requirement from fertilizer

The quantity of fertilizer that must be applied to realize Production Situation 2, can now be established. It has been discussed in which way the limiting element can be identified. In Section 4.1 the minimum concentrations of the three elements in the relevant plant parts are given. The total uptake of a certain element required for maximum production can therefore be calculated by multiplying the dry weights of economic product and crop residue calculated for Production Situation 2, with the respective minimum nutrient concentrations. Part of the total required uptake is covered by the natural soil fertility or base uptake. This base uptake is established through analysis of the production obtained on an unfertilized field. The difference between the required uptake at Production Situation 2 and base uptake must be supplied as fertilizer nutrient, the recovery of which is established for a given situation by making use of published fertilizer experiments.

This recapitulation demonstrates that the quantity of fertilizer nutrient needed to create Production Situation 2, i.e. the 'fertilizer nutrient requirement', D_x, can be quantified by subtracting from the calculated total uptake under Production Situation 2, $u_{m,x}$, the uptake under unfertilized (control) conditions, $u_{o,x}$, and subsequent division by the recovery fraction, R_x. In a mathematical notation:

$$D_x = (u_{m,x} - u_{o,x})/R_x \qquad (80)$$

Earlier in this section it was shown that the same nutrient can be supplied by different fertilizer materials. In the case of nitrogen, urea, ammonium fertilizers and nitrate fertilizers have been mentioned, but others, both chemical and 'natural' (e.g. farmyard manure) are available. Table 46 lists some common commercial fertilizers and their concentration of pure nutrient. From this table it may be deduced that a calculated nitrogen fertilizer requirement of 100 kg ha^{-1} can be met with an application of 100/0.21 kg ammonium sulphate, but also with 100/0.45 kg urea, provided that nitrogen recovery is the same from both sources. This difference partly explains the popularity of urea as a nitrogen source, especially when transport costs are high.

Exercise 61

In Exercise 57 an experiment with a HYV near Paramaribo, Suriname was treated. Assume that in that experiment nitrogen availability is the growth determining factor at any level of production. Assume furthermore that urea is used as a fertilizer and that the recovery fraction from this N source is 0.55. Dry – grain yield of an unfertilized field was 1000 kg ha^{-1}.

- Calculate the total N uptake required to realize the production calculated for Production Situation 2. Grain weight and total above – ground dry weight at maturity are given in Subsection 2.3.2.
- Calculate the slope of the yield – N uptake relation and the nitrogen uptake if fertilizers are not used.
- Calculate the nitrogen requirement of the crop and decide how many 25 kg bags of urea you must reserve for one hectare. (Use Table 46).

So far the concept of nutrient requirement was illustrated using nitrogen as the limiting element and Production Situation 2 as the pursued production target. It is well possible, however, that a farmer is not interested in producing the maximum yield of Production Situation 2, e.g. if there is no market for all the produce. He may then aim at a lower yield and adapt his fertilizer use accordingly.

It is also possible that crop growth is not limited by nitrogen supply but by the supply of some other nutrient. Experience has shown that in many such cases phosphorus supply is the limiting factor. Then, the phosphorus requirement can be calculated with Equation 80, i.e. by quantifying the total phosphorus uptake needed to realize the production target, subtracting from that amount the uptake of P from an unfertilized field and dividing by the P recovery fraction. The calculation procedure is identical for any nutrient but

Table 46. Common commercial N and P-fertilizers and their nutrient concentration.

Nitrogen fertilizers	*N concentration (kg kg⁻¹)*
Ammonium sulphate	0.21
Urea	0.45
Ammonium nitrate	0.33
Sodium nitrate	0.16
Calcium nitrate	0.155
Calcium ammonium nitrate	0.205
Phosphorus fertilizers[a)]	***P concentration (kg kg⁻¹)***
Superphosphate	0.08
Triple superphosphate (T.S.P.)	0.20
Rock posphate[b)]	0.16

Data from De Geus, 1973; ILRI, 1972; Jacob and V. Uexküll, 1958.

a) Phosphorus contents are often expressed in P_2O_5; 1 kg P_2O_5 corresponds with 0.44 kg P.
b) Phosphorus contents of phosphate rocks vary greatly; the P-content given refers to Christmas Island Rock Phosphate (C.I.R.P.)

the behaviour of the different nutrients in the soil – plant – atmosphere system is not. Unlike nitrogen, phosphorus is in general not lost from the system in significant amounts. Its chemistry in the soil is, on the whole, more complicated than that of nitrogen and that is often reflected in non – linear application-uptake curves. Levelling off at higher application rates, as signalled in Section 4.1 and illustrated in Figure 45, is not uncommon. It should be realized that in situations with non – linear application – uptake relations, nutrient recovery is not independent of the fertilizer application rate anymore! Most problems with phosphorus recovery are related to the low solubility of many P compounds. Low P solubility may be due to a secondary reaction in the soil solution, e.g. precipitation of phosphorus from a soluble fertilizer as insoluble calcium, iron or aluminium phosphates, but there are also P fertilizers that are themselves hardly soluble such as rock phosphates. P recovery from rock phosphate is therefore commonly less than 0.05 (Penning de Vries & van Keulen, 1982), but it does have the advantage that the effect may persist for quite some years even after only one application. Obviously such low recoveries are inadequate to make rock phosphate a suitable fertilizer for realizing Production Situation 2. By treating rock phosphate with sulphuric acid, the fertilizer industry produces so – called superphosphate which has a higher P concentration than rock phosphate and is more readily soluble. Consequently, P recovery from superphosphate is higher than from rock phosphate. A recovery of 0.3 may be obtained on some soils, but on soils that strongly immobilize phosphorus, recovery from superphosphate is still very low. Phosphorus immobilization in soils is complex and difficult to predict. As a first approximation, Table 47 lists a number of common soil groups arranged according to the phosphorus recovery that may be achieved if they are cropped with an annual crop, optimally supplied with water and nitrogen.

In production situations where water may at times be in suboptimum supply, recovery may be lower due to the reduced activity of plants under water stress. The uncertainty with respect to phosphorus availability in soils makes it generally advisable to be generous in the estimation of P requirements. Especially as too much phosphorus does not do any harm and what is not taken up on short notice is added to the phosphorus stock of the soil and may be used by a subsequent crop. This does not apply to soils that strongly immobilize phos-

Table 47. Indicative P recovery fractions of superphosphate applied to a grain crop. The soils are arranged according to decreasing P recovery.

Class	Recovery fraction (kg kg^{-1})	Soil type
I	0.30-0.15	— quartzitic sandy soils — organic soils/peats — young, neutral, coarse and medium-textured alluvial soils
II	0.15-0.08	— weakly acid to neutral alluvial clay soils of intermediate age — weakly acid to neutral soils with a thick black organic surface layer — weakly or medium acid well structured clay soils — churning heavy clay soils — neutral to weakly alkaline, calcareous soils
III	0.08-0.02	— permanently waterlogged mucky soils, low in bases — old acid red and/or yellow soils, rich in iron and aluminium — very acid leached 'podzol' soils, rich in iron — strongly acidified soils in pyrite-containing marine sediments, rich in aluminium — young brown or black volcanic soils, rich in allophanes

phorus. With them the availability of fertilizer – P decreases dramatically in the course of time.

Exercise 62

In Exercise 61 the nitrogen requirement of a HYV rice near Paramaribo, Suriname, was calculated under the assumption that nitrogen availability determined crop growth. In this exercise it is assumed that the availability of phosphorus is the limiting factor at any level of production.

- Calculate the total P uptake under Production Situation 2. Grain weight and above – ground dry weight at maturity are given in Subsection 2.3.2. Minimum P concentrations in grain and straw dry matter are 0.0011 kg kg^{-1} and 0.0005 kg kg^{-1}, respectively.
- Calculate the slope of the yield – P uptake relation and the phosphorus uptake if fertilizers are not used. Assume a dry – grain yield on the control plot of 1000 kg ha^{-1}.
- Calculate the P requirement of the crop and the quantity of rock phosphate that must be applied for realization of Production Situation 2. Assume that P recovery from rock phosphate is 0.02 kg kg^{-1} (Consult Table 46).
- Calculate how much triple superphosphate (TSP) would have to be applied to realize Production Situation 2. Assume a P recovery from TSP of 0.15 kg kg^{-1} (Consult Table 46).
- Assume that there is a rice glut and that the production target is lowered to 0.8 times the production calculated for Production Situation 2. How much TSP would have to be applied then? Explain why this quantity is less than 0.8 times the amount needed to realize the production calculated for Production Situation 2.

So far it was assumed that only one nutrient is in short supply: the nitrogen requirement was discussed for a situation with phosphorus sufficiency or the phosphorus requirement for a situation where nitrogen supply was optimal. It is, of course, possible to create such conditions, e.g. by applying a generous P dressing at the beginning of cropping in a situation where nitrogen supply is known to be limiting. There would be little risk involved in a high P application, but the same cannot be said of a high blanket dressing of nitrogen if phosphorus is known to be in short supply. Excessive nitrogen losses could be the undesirable result. It is therefore necessary to establish the need for nitrogen fertilizers, if the P concentration of the plant is known to be limiting, and the P fertilizer requirement in the reverse case. It was shown in Section 4.1 that relative nutrient shortage is witnessed by the ratio of element concentrations in the plant tissue. In the case of phosphorus and nitrogen, the P/N ratio varies between a maximum of about 0.15 and a minimum of about 0.04. This knowledge can be put to use: if, for instance, the availability of phosphorus is limiting, the overall P concentration of an unfertilized grain crop with a grain straw ratio of 1.0 has a minimum value of:

$$\frac{1 \times 0.0011 + 1 \times 0.0005}{2} = 0.008 \text{ kg kg}^{-1}$$

Its overall N concentration cannot be higher than 0.0008/0.04 = 0.02 kg kg^{-1}. It is also unlikely that the nitrogen concentration is much lower than 0.02, because N uptake is not dictated by nitrogen availability but is entirely a function of P availability. Addition of P would probably result in further nitrogen uptake without any addition of nitrogen fertilizer (see also Figure 50 in Section 4.1).

Assume that in a particular situation the dry – grain yield and dry – straw production of an unfertilized crop are 1000 kg ha^{-1} and that for Production Situation 2 the calculated grain yield equals 5000 kg ha^{-1} and the straw yield also 5000 kg ha^{-1}. Let TSP be used as a P fertilizer and assume a recovery of 0.1 kg kg^{-1}. From this information the fertilizer phosphorus requirement, D_P, can be established according to Equation 80:

$$D_P = ((5000x0.0011 + 5000x0.0005) - (1000x0.0011 + 1000x0.0005))/0.1$$

Hence, D_P amounts to 64 kg ha^{-1}, which can be satisfied with TSP at a rate of 320 kg ha^{-1}.

To quantify an accompanying N fertilizer application to ensure optimum nitrogen supply at the high production level of Production Situation 2, the same reasoning is followed. However, it would be unrealistic to estimate N uptake from an unfertilized field on the basis of minimum N concentrations, because the real (overall) N concentration of the control material has already been estimated. Let ammonium sulphate be the selected N fertilizer with an N recovery (established under conditions of mineral element sufficiency) of 0.5 kg kg^{-1}. Then the maximum possible fertilizer nitrogen requirement amounts to:

$$D_N = \frac{5000 \text{ x } 0.01 + 5000 \text{ x } 0.004) - (2000 \text{ x } 0.02)}{0.5} = 60 \text{ kg ha}^{-1}$$

This requirement can be satisfied with an ammonium sulphate application of 286 kg ha^{-1}. Compared with a calculation based on assumed minimum N concentrations in an unfertilized situation, this represents a reduction in calculated fertilizer input by no less than 114 kg ha^{-1}.

It should be realized that the so calculated N requirement guarantees N sufficiency at a production as calculated for Production Situation 2, but that there is still a likely overestimation of the N requirement. The total N uptake from an unfertilized field (now estimated at 2000 x 0.02 = 40 kg N ha^{-1}) was limited by P deficiency and could conceivably have been higher in a situation with P sufficiency. In other words: natural soil fertility may contribute more than 40 kg N ha^{-1} if the TSP has been applied as calculated.

The practical consequences of this – mild – overestimation are normally not prohibitive. It appears that, in general, the stock of readily available nitrogen is rapidly exhausted and that withholding N fertilization for only one or two years creates already a situation where P is in sufficient supply and N is limiting. In normal soils that do not strongly immobilize phosphorus, a comparatively small maintenance application of P fertilizer will suffice to maintain P sufficiency once a favourable phosphorus level has been established. To what extent a certain soil immobilizes P, how high the maintenance dose of a certain P fertilizer should be, and for how long that maintenance dose would probably ensure P sufficiency, can only be judged (in a rather qualitative way) if analyses of soil chemical conditions and soil mineralogical composition are available. Some attention will be paid to this in Section 5.3. In practice, maintenance applications are largely a matter of experience, gained from years of fertilizer experiments and transferred to the farming community by agricultural extension services.

Above, a situation was considered in which P availability limits plant growth in an unfertilized situation (sufficient N available to realize the control yield but no guaranteed N sufficiency at Production Situation 2). There is no need to argue that the same reasoning applies – with the respective differences considered – to a situation in which the control yield is dictated by nitrogen shortage and P sufficiency at Production Situation 2 is uncertain.

Exercise 63

In Exercise 60, an experiment with traditional rice varieties at Bangkhen Experiment Station, Thailand, was considered. In that experiment different N fertilizers were applied at various moments and with different application techniques. To make sure that the effects of nitrogen application were not blurred by P shortage, superphosphate was given to all fields at the time of transplanting. The P application of this blanket dressing was 26 kg ha^{-1}. The value of that P application can now be judged (refer to Exercise 60 and Tables 45 and 47).

- Calculate the total N uptake from the control plot and divide this value by the total dry weight of grain and straw to establish the overall N concentration of the dry matter produced under control conditions. Assume a moisture content of 0.12 kg kg^{-1} in the field produce.
- Was the control yield limited by nitrogen shortage? If so, approximate the maximum possible P concentration of the dry control material at harvest

by multiplying the overall N concentration with the appropriate P/N ratio.
- The total P uptake by the control plot can now be estimated.
- The crop under treatment A (Table 45) produced the highest amount of plant matter, viz. 3335 kg grain plus 12410 kg straw. Assume an average moisture content of 0.12 kg kg^{-1} and calculate the total P uptake, which is minimally required to realize the dry matter production in Treatment A (minimum P concentrations of dry grain and straw are 0.0011 and 0.0005 kg kg^{-1}, respectively).
- Bangkhen Experiment Station is situated on soils belonging to the Rangsit Soil Series. This soil has, under inundated conditions, properties which place it in the upper half of P recovery class II in Table 47. Calculate the P requirement for the production as obtained under Treatment A. Assume a P recovery from superphosphate of 0.15 kg kg^{-1}.
- Was the blanket phosphorus dressing of 26 kg ha^{-1} a realistic one?

5 THE COLLECTION AND TREATMENT OF BASIC DATA

5.1 Introduction

P.M. Driessen

In the preceding chapters of this book, three production situations have been discussed that differ with respect to the number of land qualities that influence plant performance. In each situation, the influence of the land quality or land qualities concerned (i.e. intercepted solar radiation, water availability, nutrient status) on plant performance is described in a number of functional relations. All functional relations that play a role in a given production situation are included in the crop production model for that situation. It is important to realize that a model cannot predict crop production. The production estimate that is computed is entirely determined by the user of the model who defines the crop characteristics and the production environment in a set of basic data; a model merely makes the consequences of the user's data selection visible.

It follows that the quality and reliability of the generated production estimates cannot surpass the quality of the available basic data and therefore basic data must be sufficiently accurate and complete. Basic data used in the present model can be grouped into four categories:

- basic weather data
- basic plant data
- basic soil/land data
- basic management data.

These same headings are used in Tables 48, 49, and 50, in which the minimum basic data requirements for production calculations under Production Situations 1, 2 and 3 are summarized.

Data within one category are not necessarily all of the same nature. Three different types of data can be distinguished. The first type concerns data that are characteristic for a certain location or land use. Such data do not change in the course of a crop cycle. They are 'CONSTANTS'. Examples of these are such soil physical characteristics as total pore space and saturated hydraulic conductivity, and plant characteristics such as the threshold temperature for development or the maximum rate of CO_2 assimilation of single leaves. The second type pertains to data that do vary with time, but their variation is independent of the crop production process. The variables themselves do, however, influence system behaviour. They are therefore called 'FORCING VARIABLES'. Rainfall, temperature and irradiance are examples of forcing variables. The third type are again variables, but variables of this type change in value in an endogenous way, i.e. as a result of the calculations. Their values must be initialized by the user at the beginning of the computations, after

Table 48. Basic data requirements for the simulation of crop performance in Production Situation 1. Variety, location and crop calendar must be known.

Symbol	Description	Source
Weather data		
H_g	total radiation on clear day ($J\ m^{-2} d^{-1}$)	Table 2
H_a	measured total radiation ($J\ m^{-2} d^{-1}$)	Meteorological Serv
T_a	measured air temperature (°C)	Meteorological Serv
Plant data		
k_e	extinction coefficient for visible light	normally $k_e = 0.8$
SLA	specific leaf area ($m^2\ kg^{-1}$)	Table 51
F_{cl}	gross assimilation rate on clear day ($kg\ CO_2 ha^{-1} d^{-1}$)	Table 1
F_{ov}	gross assimilation rate on overcast day ($kg\ CO_2 ha^{-1} d^{-1}$)	Table 1
F_g	maximum rate of gross CO_2 assimilation of a single leaf ($kg\ CO_2 ha^{-1} h^{-1}$)	C_3 crops: $F_g = 40$ C_4 crops: $F_g = 70$
E_g	conversion efficiency kg dry matter ($kg^{-1} CH_2O$)	Table 4
R_m	relative maintenance respiration rate ($kg\ CH_2O\ kg^{-1}$ dry matter d^{-1})	Table 4
TU_{pre}	temperature sum before anthesis (d °C)	This volume and a mic literature
TU_{post}	temperature sum after anthesis (d °C)	This volume and a mic literature
T_O	threshold temperature (°C)	Table 7 and agronc terature
WLVI	leaf dry weight at beginning of the first interval ($kg\ ha^{-1}$)	This volume and agı
WI	total dry weight at beginning of the first interval ($kg\ ha^{-1}$)	nomic literature

which they remain invariate for the duration of one time interval. Thus they reflect the state of the system at any moment, which explains why they are termed 'STATE VARIABLES'. Examples of state variables are the soil — moisture content and the weight of the leaf mass. A rate of change calculated for one time interval is used to update the variable value at the end of the calculations. The updated value is then used as an input for the calculation of system behaviour during the next interval.

Table 49. Additional basic data requirements for the simulation of crop performance in Pı Situation 2.

Symbol	Description	Source
Weather data		
P	gauged rainfall rate (cm d^{-1})	Meteorological service
E_0	potential evaporation rate (cm d^{-1})	Meteorological service Section 2.1
ET_0	potential evapotranspiration rate (cm d^{-1})	Meteorological service Section 2.1
Plant data		
p	soil water depletion factor	Table 20
RDI	rooting depth at beginning of first time interval (cm)	This volume and agron literature
Soil/land data		
∅	slope angle of the land (°)	Contour maps
$z_t(i)$	groundwater depth at beginning of first time interval (cm)	This volume and maps/ surements
$SS_t(i)$	surface storage at beginning of first time interval (cm)	This volume and field c servations
SM_0	total pore space (cm^3 cm^{-3})	Table 16 + soil reports
γ	texture-specific geometry factor (cm^{-2})	Table 16 + soil reports
S_0	standard sorptivity (cm $d^{-1/2}$)	Table 18 + soil reports
A	transmission zone permeability (cm d^{-1})	Table 18 + soil reports
k_0	saturated hydraulic conductivity (cm d^{-1})	Table 17 + soil reports
a	texture-specific constant ($cm^{2.4}$ d^{-1})	Table 17 + soil reports
α	texture-specific constant (cm^{-1})	Table 17 + soil reports
ψ_{max}	texture-specific suction limit (cm)	Table 17 + soil reports
$\psi(i)$	matric suction at beginning of first time interval (cm)	This volume and field observations
Management data		
I	rate of water release at headworks	User-defined/ Irrigation authority
E_d	field application efficiency factor	Table 19
E_f	field canal efficiency factor	Table 19
E_i	conveyance efficiency factor	Table 19
d	surface roughness (cm)	User defined/this volume
σ	clod/furrow angle (°)	normally: $\sigma = 30$
DD	drain depth (cm)	Irrigation authority
L_d	drain spacing (cm)	Irrigation authority
r_d	drain radius (cm)	Irrigation authority

Table 50. Minimum basic data set, needed for estimation of the nutrient requirement (Production Situation 3).

Symbol	Description	Source
Plant data		
N_Y	minimum nitrogen concentration in marketable product (kg kg^{-1})	This volume and agronomic literature
$N_{(P-Y)}$	minimum nitrogen concentration in crop residue (kg kg^{-1})	This volume and agronomic literature
N_A	actual nitrogen concentration in plant tissue (kg kg^{-1})	Chemical analysis
Management data		
Y_c	control yield of fertilizer experiment (kg ha^{-1})	Agronomic literature
Y_{Ax}	yield obtained on experimental plot fertilized with A kg ha^{-1} of nutrient x	Agronomic literature
A_x	application of nutrient x in fertilizer (kg ha^{-1}).	Agronomic literature

Exercise 64

Carefully examine the input data listed in Tables 48 and 49 and specify for each entry whether it is a constant, a forcing variable or a state variable.

Explain why the input data listed in Table 50 (for Production Situation 3) cannot be classified in constants, forcing variables and state variables. Read once more the introduction to Section 4.1 before attempting to answer this question.

It has been argued before that the quality of the model results cannot be better than the quality of the input data. If complete, accurate and sufficiently detailed basic data are lacking, even the most sophisticated simulation exercise becomes futile. Perfect basic data sets are rarely available. In most practical situations, data sets are incomplete or partly of unknown quality.The gaps must then be filled with approximate data or 'default values'. In some instances, standard values can be used that are, for all practical purposes, not too far from the truth. Examples have been presented in the preceding chapters (e.g. F_g for a C_3 crop equals 40 kg ha^{-1} h^{-1}; all texture – related soil parameters). Where indicative standard data cannot be used, approximate data must be estimated by the user of the model. This can, in some instances, be done with an interpolation or extrapolation routine (e.g. weather data for areas between meteorological stations). In other cases, the help of a specialist with local experience, who can make reliable estimates by interpreting a qualitative or semi – quantitative description of the production environment, is required. Soil survey reports are examples of such descriptions from which semi – quantitative information can be inferred.

The models presented so far are essentially mathematical descriptions of processes that take place in plant production and the effects of specific environmental conditions on these processes. Clearly plant production is a complex affair and an in – depth analysis of all factors and processes involved is beyond the scope of this book. Attention has therefore been focussed on only the most important aspects of plant production, which have been described in a number of necessarily simple mathematical relations. (Gross simplification is not the exclusive hallmark of the approach presented here; some aspects of plant production are still poorly understood and must be described in an oversimplified fashion in any model). These necessary simplifications are another reason why the fit between observed and simulated plant performance is normally less than perfect. However, such structural imperfections are likely to become less disturbing in the future as better methodologies are developed.

There is also a possibility that the generated production figures deviate from observed figures for reasons that cannot be attributed to inaccuracies in the data base or the model itself. Production could be affected by factors that are not considered in the simulation procedure. The possible occurrence of extreme temperatures, storms, specific soil disorders or endemic diseases are but a few examples. For this reason it is prudent to apply the suggested production calculations only to the regions where the crop concerned is actually grown. Even then, the calculation results should not be regarded as an accurate prediction of crop yields but rather as a useful quantitative approximation of the productive capacity of land under conditions as defined.

In the following sections, the availability, determination and relevance of a number of important weather, soil and plant data will be discussed. In addition, some attention will be given to major environmental disorders that might affect plant performance or preclude production altogether.

5.2 Meteorological data

H. van Keulen and H.D.J. van Heemst

For estimating the production capacity of a region, the prevailing weather conditions are of major importance, a fact that has been made patently clear in the preceding chapters. In this section the relevant variables are treated in a schematic way. For a more detailed treatment of the processes involved, the reader is referred to textbooks on the subject (cf. Grace, 1983; Monteith, 1973).

Weather data are collected at weather stations, run mostly by the meteorological service of a country. Very often these services are not in the first place, or not at all, concerned with the use of their data in the field of agro – meteorology. Good contacts with the local service should promote cooperation, especially if the agro – meteorologist or agronomist is able to clearly explain his needs.

Weather data reported by the meteo stations should never be taken at face value, but should be judged using common sense and, if possible, the instrumentation and the method of collecting the data should be checked by the user.

5.2.1 Radiation

Solar energy is the primary source of energy for all terrestrial life. It is trapped by the green pigments of the plant and converted into chemical energy in the process of assimilation (Section 2.1). For the calculation of (potential) assimilation and (potential) evapotranspiration a reasonably accurate estimate of solar radiation is indispensable. Total global radiation received at the earth's surface may be measured by solarimeters, which provide accurate data on daily totals of global radiation. Unfortunately, such data are rather scarce and often use must be made of records on sunshine duration. For conversion of these data, Angströms equation may be used (Section 3.1). As is explained there, the coefficients applied are location specific, so if, even for a limited period of time, data on both measured radiation and sunshine duration are available, it may be worthwile to derive a specific equation.

If sunshine duration is also not available, it may be possible to obtain a first approximation of sunshine duration from the degree of cloudiness as estimated by an observer several times a day. It should be realized that especially for strongly fluctuating conditions such estimates may not be very accurate, but for the purposes of the present approach they may be useful. The average cloudiness weighted according to time can then be set equal to the fraction overcast as defined in Section 2.1, and Tables 1 and 2 can be used directly.

5.2.2 *Air temperature*

In general daily maximum and minimum air temperature are recorded at screen height. For the present approach, the arithmetic average of the two may serve as a reasonable estimate of the mean temperature sensed by the vegetation. This is due to the fact that most of the temperature – dependent relations in the model are essentially linear in the relevant range. If, however, maximum or minimum temperature are outside the linear range, a more accurate value of the dependent variable may be obtained by averageing hourly values. If hourly temperature values are not available they can be estimated from minimum and maximum temperature by a procedure developed by de Wit et al. (1978), which calculates a daily temperature regime from measured minimum and maximum temperature under the assumption of a sinusoidal daily wave.

5.2.3 *Air humidity*

Many different types of instruments are used to determine the humidity of the atmosphere, but many of these are either not very good or easily misused. The actual concentration of moisture in the air is given by either absolute humidity, which is the mass of water vapour per unit volume of air, W_w, or by the specific humidity, the mass of water per unit mass of the moist air.

In agro – meteorology, air humidity is either expressed by the dew point, T_d, the vapour pressure of the atmosphere, e_a, the relative humidity, h_r, or the wet bulb temperature, T_w. Without going into the details of measurement, relations between the various variables are presented here.

The dew point of a volume of air with temperature T_a and vapour pressure e_a is the temperature at which e_a would be the saturated vapour pressure. The relation between e_a and T_d may be approximated by the expression:

$$e_a = 6.11 \text{ x } e^{(17.4 \text{x} T_d/(T_d + 239))} \qquad (81)$$

with T_d expressed in °C and e_a in mbar.

The vapour pressure e_a (mbar) is often measured directly. It can also be deduced from the absolute humidity by applying the ideal gas law, which gives the expression:

$$e_a = T_k/0.217 \text{ x } W_w \qquad (82)$$

where

T_k is absolute temperature (K), equal to T_a + 273
W_w is absolute humidity in kg m^{-3}

The relative humidity of the air is the ratio between the actual vapour pressure of the air and the saturated vapour pressure at air temperature, e_s. The latter, expressed in mbar, may be approximated by:

$$e_s = 6.11 \times e^{(17.4 \times T_a/(T_a+239))} \tag{83}$$

The wet bulb temperature, T_w, is the temperature of a wet surface exposed to atmospheric conditions, but shielded from radiation. As a result of evaporation of the water, the temperature of the wet bulb will be lower than air temperature (Section 3.1) and the temperature difference is a measure of air humidity:

$$e_a = e_w - \gamma(T_a - T_w) \tag{84}$$

where

e_w is saturated vapour pressure at wet bulb temperature (mbar), equal to $e_w = 6.11 \times e^{(17.4 \times T_w/(T_w + 239))}$
γ is psychrometer constant (mbar °C^{-1}) (= 0.66)

5.2.4 *Wind speed*

Wind speed is recorded with integrating anemometers and very often these are read once daily, so that total daily wind run is obtained. As a rule of thumb it may be assumed that the wind speed during the day is twice that at night (de Wit et al., 1978).

If wind speed is not measured at the standard heigth of 200 cm, as required in the Penman equation (Section 3.1), the wind speed at 200 cm may be calculated, assuming a logarithmic wind profile. If wind speed was measured at a height H, the equation reads:

$$U_{200} = U_H \cdot \ln(200/z_o)/\ln(H/z_o) \tag{85}$$

with z_o the roughness length of the surface. As most meteorological observations are carried out over a short green grass cover, z_o may be approximated as 2 cm. The equation then reduces to:

$$U_{200} = 4.61\ U_H/\ln(H/2.) \tag{86}$$

5.2.5 *Precipitation*

A rain gauge, to record the amount of rainfall is probably the most widely spread meteorological instrument in use and consequently rainfall data are generally available in much higher density than any of the other variables. The accuracy of the measurements depends very much on the type of rain gauge used, its position and the prevailing wind. Variations of not less than 15% are common. Generally, daily values of precipitation are available, which for the present approach is sufficiently detailed. It should be realized, however, that proper estimation of, in particular, surface run – off requires greater detail, as rainfall intensity is the determining factor. To obtain such data a self – recording rain gauge is necessary, of which not many are in operation. If run – off plays an important role it may therefore be necessary to measure that component of the water balance directly and to introduce it in the model as a forcing variable.

5.2.6 Potential (evapo)transpiration

As has been discussed in preceding chapters, a reasonably accurate estimate of potential (evapo)transpiration is obtained by application of the Penman equation, or the Penman – Monteith equation. Application of that equation requires, however, availability of such data as total global radiation and wind speed, which are not always available. In some situations data may be available from a so – called evaporation pan, a container with water, from which the daily water loss is determined. Various types of such evaporation pans exist, and are in operation around the world. One of the earliest types was the so – called BPI sunken pan, which was lowered into the soil, with its rim approximately at surface level; it produced reasonable results. For unknown reasons that type was largely abolished in favour of the Class A pan, which is set up at some height above the soil surface. This type, in particular, reacts differently from an extended water layer because exchange of heat may also take place through the pan walls, and because there is influence from the rim, as the water surface is generally some – variable – distance below the rim. An additional disadvantage is that the presence of such a body of water, especially in not well – guarded situations, invites attention from men and animals, so that not all the changes in water level are due to atmospheric processes. However, despite these disadvantages, if only pan – evaporation data are available they may be applied as a first approximation of the evaporative demand by multiplying the reading by a factor of 0.7.

5.3 Soil data

P.M. Driessen

5.3.1 Soil physical data

Soil physical data can be obtained from two main sources: field observations and laboratory analyses. The two types of data are complementary.
Field data are collected in soil surveys and presented in soil survey reports. Such reports contain descriptions of representative observation sites and soil profiles. Soil profile descriptions specify for each individual layer, or 'horizon', in a soil, the depth of the horizon, its colour, texture, structure, porosity, mottling and foreign inclusions. Two profile descriptions from Sierra Leone are presented in Tables 51 and 52, as examples.

A profile description is thus a field record of observed (and measured) soil characteristics. The information that such a profile description provides is at best semi – quantitative but nonetheless valuable as it is indicative of important physical and chemical soil conditions. Examples are soil texture estimates ('field texture') which are indicative of a score of physical soil parameters – see Tables 16, 17 and 18 – and soil colour estimates, which contain implicit information on the contents and distribution of soil organic matter, oxides, etc. The interpretation of such evidence is not always easy and is best entrusted to a skilled soil scientist with ample local experience.

Profile descriptions are normally supplemented by analyses of soil samples from one or more of the horizons distinguished in the profile description. Such analysis data are always quantitative and reproducible but, here again, data interpretation is not always easy. One difficulty is that the sample material is disturbed and sometimes modified prior to analysis. An example of this is the drying, sieving and decalcification of soil material prior to the determination of its particle size distribution in the laboratory. (This is one reason why field texture estimates do sometimes differ from texture class estimates based on laboratory analysis).

In all cases, soil data are determined on the basis of a limited quantity of soil material. As soils are rarely homogeneous, even accurately determined quantitative data give only an indication of the properties of an entire soil body. An additional difficulty is presented by the translation of soil data into land characteristics, because a tract of land includes normally more than one soil type. This regional variability among soils introduces a need for aggregation of individual soils in soil (map) units. The map unit criteria, and therewith the range and significance of the provided soil information, are commen-

Table 51. Profile description of Profile P9, Masuba, Sierra Leone. Described by R. Miedema and A. A. Thomas on March 20, 1968

Location	Topographic map of Sierra Leone, scale 1:50000, sheet 43, coordinates $HE27_8$-87_2
Physiography	Lower part of streambed near valley edge.
Relief	Slope 0 to 3 percent
Vegetation	Farm with cassava, Kandi trees and weeds, and many wild oil palms.
Drainage	Moderately well drained.
Parent material	Gravel-free, transported alluvial/colluvial material.
Hor. code: A_p Hor. depth: 0-19 cm Lab. No.: S29810	Very dark grayish brown (10YR 3/2); sandy clay loam; weak fine to medium angular blocky; very hard; common macro- and many mesopores; few distinct fine charcoal mottles; many coarse, medium, and fine roots; many large and medium ant holes; clear, smooth boundary to horizon below.
Hor. code: B_1 Hor. depth: 19-57 cm Lab. No.: S29811	Pale brown (10YR 6/3); sandy clay loam; weak fine to medium angular blocky; very hard; many macro-and mesopores; few distinct fine charcoal mottles; common fine distinct reddish-yellow to strong brown (7.5YR 5.5/8) to red (2.5YR 5/8) iron mottles; common coarse, many medium and fine roots; less than 10% uncoated, nodular, coarse, porous, red, hardened plinthite glaebules[a], with few quartz grains; many large and medium ant holes; gradual, smooth boundary to horizon below.
Hor. code: B_2 Hor. depth: 57-170 cm Lab. No.: S29812	Very pale brown (10YR 6.5/3); sandy clay loam; weak fine angular blocky; firm; many macro- and mesopores; many distinct fine and medium yellowish-red (5YR 5/8) and reddish-yellow (7.5YR 6/8) iron mottles; common distinct fine and medium charcoal mottles; few coarse, common medium, and many fine roots, less than 10% gravel, and one quartz stone; common worm holes with dark coatings.

Soil physical and soil chemical data.

Classification: 'Plinthic' Udoxic Dystropept

Horizon code	A_p	B_1	B_2
Depth of horizon (cm)	0-19	19-57	57-170

Table 51. (continued)

Fraction of entire sample	>2 mm ($g\,g^{-1}$)	0.0009	0.018	0.020	
Particle-size distribution	<2 mm (%)				
Total sand	2.0-.05 mm	64.0	59.6	60.4	
Total silt	.05-.002 mm	12.6	13.6	13.9	
Total clay	$<.002$ mm	23.4	26.8	25.7	
Bulk density ($g\,cm^{-3}$)		1.1	1.3	1.3	
Moisture: SM_{333} ($cm^3\,cm^{-3}$)		0.16	0.15	0.16	
SM_{16000} ($cm^3\,cm^{-3}$)		0.10	0.10	0.10	
Organic carbon ($g\,g^{-1}$)		0.0125	0.006	0.005	
Total P (ppm)		n.d.[b]		n.d.	n.d.
Total CaO (%)		0.133		0.078	0.078
Total Fe_2O_3 (%)		2.67		3.13	3.18
Total K_2O (%)		1.361		1.277	1.200
Available K ($kg\,ha^{-1}$)		67		73	121
P-Bray No. 1 ($kg\,ha^{-1}$)		9.0		3.4	3.4
C.E.C. ($me(100\,g)^{-1}$)		5.71		3.79	3.14
exchangeable Ca	$me(100\,g)^{-1}$	0.47		0.16	0.21
exchangeable Mg	$me(100\,g)^{-1}$	0.37		0.16	0.16
exchangeable K	$me(100\,g)^{-1}$	0.03		0.02	0.01
exchangeable Na	$me(100\,g)^{-1}$	0.08		0.08	0.08
exchangeable A1	$me(100\,g)^{-1}$	0.99		1.09	1.05
Base saturation (%)		16.6		11.1	14.6
pH-H_2O		4.7		4.6	4.8

Source: Odell et al., 1974.

a laterite gravel

b n.d. is not determined

Table 52. Profile description of Profile Kpuabu 1, Manowa, Sierra Leone
Description after Sivarajasingham.

Location	Kpuabu Cocoa Experiment Station; about 450 feet (137 m) from the Kenema-Joru road on the road to the Station Office, and about 150 feet (46 m) on the right-hand side from the Station Office road.
Physiography	Accordant, flat-topped hill of the dissected lateritic upland.
Relief	Upper, convex 5 percent slope.
Vegetation	Cocoa plantation under many tall trees of original secondary vegetation; good grass cover.

Table 52. (continued)

Drainage	Moderately well drained.
Parent material	A thin layer of gravel-free material over a thick, very gravelly layer of locally transported material.
Hor. code: A_1 Hor. depth: 0-25 cm Lab. No.: S28558	Very dark grayish brown (10YR 3/2); sandy clay loam; moderate medium and fine subangular blocky; porous; friable, slightly sticky, slightly plastic; common fine, few medium, and very few coarse roots; clear, smooth boundary to horizon below.
Hor. code: A_3 Hor. depth: 25-53 cm Lab. No.: S28557	Dark brown (10YR 3/3); very gravelly sandy clay; 70% yellow-coated, nodular, coarse and medium, dense, red and yellow, hardened plinthite glaebules[a]; weak fine subangular blocky aggregates with no strong interface; friable, sticky, slightly plastic; few fine and very few medium roots; gradual, smooth boundary to horizon below.
Hor. code: B_{21} Hor. depth: 53-89 cm Lab. No.: —	Dark yellowish brown to yellowish brown (10YR 4/4-5/6); very gravelly sandy clay; 60% yellow-coated and uncoated, round, fine, dense, red and black, hardened plinthite glaebules[a]; weak fine subangular blocky aggregates with no strong interface; friable, sticky, slightly plastic; very few fine and medium roots; gradual, smooth boundary to horizon below.
Hor. code: B_{22t} Hor. depth: 89-178 cm Lab. No.: S28556	Strong brown (7.5YR 5/8); very gravelly clay; 75% yellow-coated, nodular, coarse, dense, red and yellow, hardened plinthite glaebules[a]; weak fine subangular blocky aggregates with no strong interface; porous; friable, sticky, slightly plastic; very few fine and medium roots.
Remarks	This soil is very gravelly and droughty and would be expected to be unsuitable for cocoa. The cocoa planted in 1959 appears as stunted trees of very poor health, with many vacant patches because of low survival rate of the planted seedlings. Management is very good, but shade appears to be excessive.

Soil physical and soil chemical data.

Classification: Orthoxic Palehumult (or Typic Umbriorthox)

Table 52. (continued)

Horizon code		A_1	A_3	B_{22t}
Depth of horizon (cm)		0-25	25-53	89-178
Fraction of entire sample	>2 mm (g g^{-1})	0	0.70	0.75
Particle-size distribution	<2 mm (%)			
Total sand	2.0-.05 mm	57.4	49.2	38.9
Total silt	.05-.002 mm	11.8	12.4	13.3
Total clay	<.002 mm	30.8	38.4	47.8
Bulk density (g cm^{-3})		1.2[c]	1.4	1.5
Moisture: SM_{333} (cm^3 cm^{-3})		0.19	0.17	0.22
SM_{16000} (cm^3 cm^{-3})		0.13	0.13	0.18
Organic carbon (g g^{-1})		0.027	0.019	0.007
Total P (ppm)		390	370	n.d.[b]
Total CaO (%)		0.076	0.074	0.069
Total Fe_2O_3 (%)		8.10	8.99	11.19
Total K_2O (%)		0.169	0.180	0.215
Available K (kg ha^{-1})		10.0	3.4	1.1
P-Bray No. 1 (kg ha^{-1})		5.6	2.2	0
P-Bray No. 2 (kg ha^{-1})		9.0	4.5	2.2
C.E.C. (me(100 g)$^{-1}$)		11.79	9.29	6.00
exchangeable Ca (me(100 g)$^{-1}$)		0.01	0.07	0
exchangeable Mg (me(100 g)$^{-1}$)		0.10	0.08	0.08
exchangeable K (me(100 g)$^{-1}$)		0.10	0.07	0.06
exchangeable Na (me(100 g)$^{-1}$)		0.04	0.05	0.04
exchangeable Al (me(100 g)$^{-1}$)		3.06	2.50	0.67
Base saturation (%)		2.1	2.9	3.0
pH-H_2O		4.3	4.5	4.8

Source: Odell et al., 1974.
[a] laterite gravel
[b] n.d. is not determined
[c] Estimated

surate with the map scale, which is dictated by the number of observations per unit surface area. In other words: there are detailed soil maps and less detailed ('semi – detailed' and 'reconnaissance') soil maps and the information that can be extracted from a soil survey report depends on the observation density and aggregation level of the soil map units.

The following sequence of taxonomic aggregation levels is widely used in soil science:

Soil	Orders
Soil	Suborders
Great	Soil Groups
Soil	Subgroups
Soil	Families
Soil	Series
(Soil	Types)

In this hierarchy, soil orders are the least specific with regard to soil properties, soil types are the most specific. At the lowest, most detailed level, soil types are distinguished within soil series on the basis of texture, a single characteristic (USDA, 1975). Most detailed soil surveys define only soil series, without subdivision in types, because a series is commonly already homogeneous with regard to texture. Detailed soil maps and reports describing individual soil series are valuable sources of soil physical and chemical information. Less detailed soil inventories provide less specific information and are consequently less useful for quantitative crop production analysis.

Even if reliable quantitative data are available, they cannot always be used in an indiscriminate way. Much of the soil physical information required for the modelling of a crop's production environment is related to the geometry of the soil matrix and correlated with the matrix texture. Quantitative correlations have been established for standard soils from the Netherlands with a normal mineral composition for Dutch conditions. A different situation may exist in the tropics and this may affect the relation between soil texture and soil physical characteristics. It is also possible that the matrix geometry itself differs from the typical situations studied in the Netherlands. This is the case in compacted soils or in soils with a very loose matrix, in soils with very high organic – matter content or in very gravelly or stony soils. The most important factors that disturb the tabulated relation between matrix texture and soil physical parameters (Tables 17, 18 and 19) will be discussed in the following. When such deviations occur, measured relations and variables must be used instead of tabulated standard values.

The relation between matrix geometry and physical soil properties

– The influence of the packing density

Normally, soil is a three – phase system. It is composed of a solid component (minerals and/or soil organic matter), a liquid component (soil moisture) and a gaseous component (soil air):

$$V_t = V_s + V_l + V_g \qquad (87)$$

where

V_t is the sample volume (cm^3)
V_s is the volume of solid soil materials (cm^3)
V_l is the volume of soil moisture (cm^3)
V_g is the volume of soil air (cm^3)

The fraction of the soil volume not occupied by the solid phase is the total pore space, SM_0 ($cm^3\ cm^{-3}$), which may be expressed as:

$$SM_0 = (V_t - V_s)/V_t \tag{88}$$

The actual value of SM_0 is largely determined by the 'bulk density' (BD, expressed in $g\ cm^{-3}$) of the soil material. The bulk density is basically the weight per unit volume of dry soil material:

$$BD = W_t/V_t \tag{89}$$

where

W_t is the dry sample weight (g)

In 'normal' soils, bulk density values vary between 0.9 and 1.5 $g\ cm^{-3}$. The solid component, V_s, has a weight that is almost identical to W_t, because the weight of a unit of soil air is negligible when compared with the weight of a similar unit of solid soil material. The weight of one volume unit of the solid component is called the specific density (SD, expressed in $g\ cm^{-3}$) of the soil material; it may be expressed as:

$$SD = W_t/V_s \tag{90}$$

Combination of Equations (88), (89) and (90) yields:

$$SM_0 = 1 - BD/SD \tag{91}$$

The overall specific density of composite (i.e. mineral and organic) soil material depends to some extent on the mineral composition of the solid phase, but is much more influenced by the soil organic – matter content. Soil minerals normally have a specific density of approximately 2.60 $g\ cm^{-3}$; the specific density of organic matter ranges between 1.30 and 1.50 $g\ cm^{-3}$, with a typical value of 1.43 $g\ cm^{-3}$. Consequently, the compounded specific density of an organic matter containing soil material can be approximated by:

$$SD = 1/((1 - O_m)/2.6 + O_m/1.43)$$

or

$$SD = 3.72/(1.43 + 1.17 \times O_m) = 1/(0.38 + 0.31 \times O_m) \quad (92)$$

where

O_m is the weight fraction of organic matter in soil ($g\ g^{-1}$)

In some soil survey reports, the organic matter content of the soil is expressed as the carbon content of the matrix material (C_m, expressed in $g\ g^{-1}$). As most soil organic matter consists for approximately 0.55 $g\ g^{-1}$ of carbon, the overall specific density can also be established with:

$$SD = 3.72/(1.43 + 2.13 \times C_m) = 1/(0.38 + 0.57 \times C_m) \quad (93)$$

Compression of soil material does not affect SD, but increases BD. Consequently, it lowers SM_0 (Equation 91). Total pore space values, which deviate from values typical for soils within a given texture class, are associated with a-typical hydraulic soil properties ($k_\psi - \psi$ and $SM_\psi - \psi$ relations) and this directly affects plant performance in situations where water availability to the plant is suboptimal.

Exercise 65
Reconstruct the SM_0 values of the three soil horizons distinguished in profile P9 (Table 51).
Compare the calculated SM_0 values with the indicative SM_0 value for sandy clayloam as tabulated in Table 16 (Section 3.2).

– High soil organic matter contents

Equation 92 demonstrates that the presence of soil organic matter decreases the overall specific density of soil material. However, it lowers the bulk density even more, because organic matter cements individual soil particles together to form loosely structured aggregates. The overall result of organic-matter accumulation in a soil is therefore generally an increase in total pore space. The quantity of organic matter in a soil is determined by the supply rate of plant litter and by the rate at which this litter is decomposed by soil organisms. Equilibrium between supply and decomposition of organic material may be approached in soils under a natural vegetation cover and in agricultural soils long cultivated under steady management. High organic-matter contents build up where the decomposition of organic material is reduced by factors that hamper or preclude the activity of soil organisms, such as low soil temperature, low soil pH, prolonged periods of waterlogging or the formation of toxins or compounds with a high resistance to microbial attack. In general, tabulated soil physical parameters cannot be applied if organic carbon contents exceed 0.04 kg kg^{-1} in more than one – third of the rootable surface soil.

Exercise 66

Explain why the upper 19 cm layer of soil profile P9 has a lower BD and higher SM_0 than the two subsurface horizons of this profile.

The bulk density of high moor peats varies commonly between 0.05 and 0.15 g cm^{-3}; the mineral content of such peats is low, say between 0.01 and 0.05 g g^{-1}. Calculate V_s of such peats.

– Dominance of swelling and shrinking clay minerals

Certain types of clay minerals, known as 'smectite' clays, shrink in volume upon drying. This causes the formation of deep soil cracks during droughts. The clays swell again in a subsequent wet period, which results in the destruction of voids and pores. Smectite – rich 'Vertisols', also known as 'Regur' or 'Black Cotton Soils', are notorious in this respect. Drying of their topsoil causes the formation of a thin granular surface layer of hard clay aggregates on top of dense (BD around 1.55 g cm^{-3}) polygonal structures, separated by deep, wide cracks. Swelling of the clays in wet periods causes smearing of the surface soil during tillage and shearing phenomena at some depth. Clearly, this affects $SM_\psi - \psi$ and $k_\psi - \psi$ relations. The mathematical description of the water regime of Vertisols is therefore virtually impossible.

– High contents of sesquioxides in the clay fraction

Many genetically old, red and yellow soils in the humid tropics have a clay fraction ($<$ 0.002 mm) dominated by aluminium oxides and iron oxides ('sesquioxides'). Such soils are classified as 'Oxisols' or bear a name in which the prefix 'ox' occurs. Profile P9, classified as a 'Plinthic Udoxic Dystropept', is an example of such a soil rich in sesquioxides (Table 51). The oxides cement individual clay particles together to stable sand – sized aggregates that have fine pores inside. In the low suction range, these fine pores remain filled with water and do not influence the $SM_\psi - \psi$ relation, which resembles that of a sand. In the high suction range, coarse soil pores are all empty. A normal sand with solid particles has then lost virtually all of its water because very few fine pores occur between the coarse sand grains. A cemented Oxisol, however, has fine pores inside the cemented aggregates. This explains why its $SM_\psi - \psi$ relation resembles that of a loam soil in the high suction range. Soils high in sesquioxides have typically compound $SM_\psi - \psi$ relations; in water – balance calculations for Oxisols one has to approximate the required $SM_\psi - \psi$ combinations by linear interpolation between measured values.

Exercise 67

Calculate SM_{333} and SM_{16000} for a sandy clay loam (use Equation 27 and Table 16; Section 3.2). Compare the calculated values with values measured for profile P9 (Table 51).

Calculate SM_{333} for profile P9, using SM_0 as calculated in Exercise 65 and γ = 0.035 cm^{-2} (a value appropriate for medium coarse sand). Compare the result with the measured SM_{333} value (Table 51).

Calculate SM_{16000} for profile P9, using SM_0 as calculated in Exercise 65 and γ = 0.017 cm^{-2} (a value appropriate for loam soils). Compare the result with the measured SM_{16000} value (Table 51).

– Slaking soils

A low structure stability of the uppermost soil layer may lead in some soils to dispersion of the individual soil particles in times of excessive wetness and to subsequent formation of a hard surface crust upon drying. This phenomenon is known as 'slaking' of the soil; it is associated with clogging of pores and it lowers the infiltration capacity and surface storage capacity of water. Slaking is particularly prominent in fine sandy soils that are low in organic matter and calcium. Slaking is aggravated by unnecessary tillage operations, which enhance the destruction of aggregates and intensify the breakdown of soil organic matter.

– Gravelly and stony soils

Soil texture determinations – both in the laboratory and in the field – disregard the presence of any component with a diameter in excess of 2 mm. In some soils, such inclusions make up a considerable part of the matrix. The presence of gravel reduces the volume of soil material that can be explored by the roots and consequently the quantity of water that the soil can supply to a crop. Normally, gravel contents are estimated in the field and expressed as a fraction of the volume of each soil horizon. An example is the soil description of profile Kpuabu 1, presented in Table 52. In some cases, gravel contents are determined in the laboratory. This is the case for soils in which gravel contents are low and cannot be estimated in the field with any measure of accuracy. In that case, gravel is separated from the sample material by sieving and is subsequently weighed. The gravel content is then often expressed as a weight fraction (i.e. in $g\ g^{-1}$); profile P9, presented in Table 52, is an example. Weight fractions are easily converted to volume fractions (i.e. in $cm^3\ cm^{-3}$) by multiplying the weight fraction by BD/SD (cf. Equation 91). Clearly the quantity of ultimately available water in the soil (W_a, see Equation 52 in Section 3.2) is affected if the rooted surface soil contains gravel. A soil which contains a fraction 'X' $cm^3 \, cm^{-3}$ of gravel contains only $(1 - X) \cdot W_a$ cm available water. As a consequence, water shortage develops more rapidly on gravelly soils than on similar soils without gravel. This statement is illustrated by a remark on the description form of profile Kpuabu 1, a gravelly soil in the Manowa Series, Sierra Leone (Table 52).

Exercise 68

Calculate SD and SM_0 for the analysed horizons of profile Kpuabu 1, Table 52. Use Equations 93 and 91.

Approximate for each of the analysed horizons a value of γ that applies under conditions of low suction. Use the equation

$$SM_{333} = SM_0 \cdot e^{-\gamma \times (\ln 333)^2}, \text{ or } \gamma = 0.03 \cdot \ln(SM_0/SM_{333})$$

Calculate SM_{100} for each horizon with the low suction γ and SM_0 values established. Use Equation 27.

Calculate the fraction of ultimately available water contained in each horizon ($= SM_{100} - SM_{16000}$).

Convert fraction to cm (multiply each fraction by the depth of the horizon in cm).

Correct for the presence of gravel in each horizon.

Calculate the quantity of ultimately available water in the upper 178 cm of soil (assume that horizon B21 has 2.8 cm available water, before correction for gravel, and a gravel content of 0.6 $cm^3 \, cm^{-3}$).

Calculate the quantity of water that would have been available to the crop if the soil contained no gravel.

Soil morphological indicators

Detailed examination of a soil profile can produce valuable information on soil rootability, homogeneity and water regime. Although it requires much experience to benefit fully from the interpretation of soil morphology, some profile characteristics are rather straightforward reflections of important soil physical conditions. A few morphological clues to maximum and minimum groundwater depths over the year, to past rooting depths and to the intensity of biological soil homogenization will be discussed briefly.

– Iron mottling and water table fluctuation

Most soils contain iron, originating from rock weathering or from influx from outside. Depending on the aeration status of the soil material, iron ions occur in a trivalent (Fe^{3+}, 'ferric') or in a bivalent (Fe^{2+}, 'ferrous') form. Well aerated soil layers contain ferric oxides ('rust') which are stable and immobile as long as oxidized conditions persist. However, under conditions of prolonged water saturation of the soil, ferric ions are transformed to ferrous ions, which are easily transported with moving water. This ferrous iron is reconverted to the immobile ferric form near air – filled pockets in the otherwise reduced soil mass. Thus, accumulation of iron oxides takes place at these places and this shows up as red, brown or yellow rust mottling. The abundance, size, colour, shape and contrast of these mottles are indications of the intensity of groundwater – table fluctuations in the past. For our purposes a few rules suffice:

- In subsoil horizons which are permanently below groundwater, rust mottling is completely absent and matrix colours are commonly greyish.
- Surface horizons that are not saturated commonly have reddish or brownish matrix colours, witnessing evenly distributed ferric oxides. (Organic matter accumulation may cause dark brown colours but there are no signs of iron redistribution).
- Distinct iron mottling indicates the zone of past groundwater – table fluctuations. (A groundwater table shallower than a few metres at the time of the soil examination is recorded in the profile description).

Exercise 69

Check the profile descriptions of profiles P9 and Kpuabu 1 for depth and abundance of reddish mottles. Which one of these two profiles has the deeper groundwater table?

Figure 53 presents the groundwater table level in P9 during the second half of 1968. Are the groundwater table depths recorded in that period exceptional?

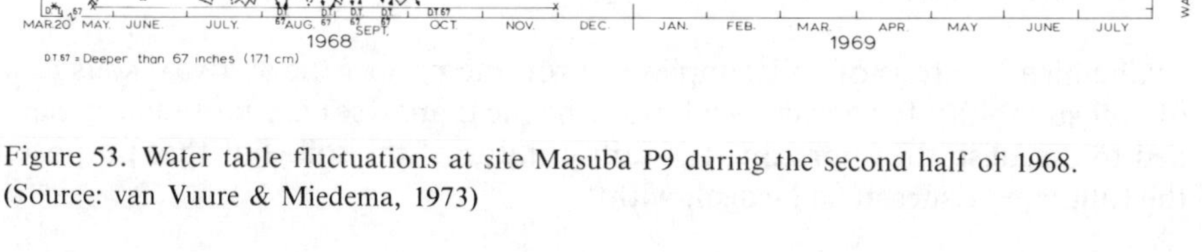

Figure 53. Water table fluctuations at site Masuba P9 during the second half of 1968. (Source: van Vuure & Miedema, 1973)

– Depth of undisturbed subsoil and root prints

Surface soil homogeneity is an important assumption in the crop production model presented here. This condition may result from deep weathering of homogeneous parent rock or from prolonged sedimentation in a stable environment. More commonly, however, soil homogeneity results from biological mixing of the surface soil by the soil fauna and plant roots. The depth and intensity of homogenization processes can be inferred from soil descriptions.

Soils that are intensively homogenized by biological activity lack abrupt horizon boundaries and have a comparatively uniform matrix texture througout the root zone. A well homogenized surface soil is porous with many 'macropores' and 'mesopores', as a result of the activity of small organisms, and is perforated by worm holes, ant holes, etc. Plant roots also contribute to the biological mixing of soil material. When roots die, they desintegrate and leave voids in the soil matrix. Dark coloured surface soil material often leaches to deeper strata along these former root channels. Under wet conditions, the channels act as air ducts bringing atmospheric oxygen to the poorly aerated subsoil. Transformation of soluble ferrous iron to ferric oxides can then produce a reddish lining of rust along the channel walls which become visible as 'root prints'. The number and size distribution of root prints in a soil reflect the intensity of root activity in the past. Where biopores and root prints are absent, the soil is called 'undisturbed'. If the upper boundary of the undisturbed subsoil is deeper than approximately one metre below soil surface and evidence of surface soil homogenization is clear, it is not to be expected that the root activity of crops will meet any mechanical impediment. Absolute criteria for the degree of soil homogenization cannot be given because soil homogeneity depends not only on the intensity of mixing but also on the uniformity and stratification of its parent material, which is a geological datum.

Exercise 70

Check the profile descriptions of profiles P9 and Kpuabu 1 for depth and abundance of roots. Pay also attention to evidence of ant and worm activity. Is rooting impeded in any of the two soils?
In which of the two soils is biological soil homogenization more intense?

5.3.2 Soil chemical data

Chemical analysis of soil samples is a routine part of the study of soils and of soil suitability for agricultural use. Chemical analyses are particularly needed to establish the approximate fertility status of the soil. For that purpose, the following issues must be dealt with:

- what are the total quantities of plant nutrients in the soil?
- to what extent are these nutrients available for uptake by plants?
- what is the capacity of the soil to temporarily store nutrients for later release to the soil solution and then to the crop?
- are there soil chemical anomalies that hamper crop production?

Soil chemical data as such are of very limited value. In association with measured plant performance they gain significance, but reliable and generally applicable correlations between soil chemical data and crop production cannot be expected. Exogenous conditions that differ each year (e.g. the weather) obscure such correlations and allow at best the identification of generally safe ranges in the chemical status of the soil. Outside these ranges, plants are exposed to either nutrient deficiency or to excess levels of elements in the soil solution, which have a toxic effect on crops.

In this subsection those soil chemical data will be discussed that are routinely collected for soil profiles representing the major soils of important land units. The significance of soil chemical data as indicators of nutritional disorders will be explained to identify soil chemical constraints on plant production in a number of situations.

Total element contents

The determination of total element contents helps to indicate structural nutrient deficiency (e.g. in mineralogically poor soils or in exploited and chemically exhausted soils) but supplies no information on the exact quantities of nutrients that crops can actually take up from the soil. Total nutrient contents are normally determined after digestion of a soil sample with a strong agent such as nitric or perchloric acid. Modern analysis techniques, e.g. X–ray spectroscopy, permit non–destructive determination of total element contents. In view of the limited significance of total element data for practical agriculture, this subject will not be treated any further.

Exercise 71

Calculate the total quantity of phosphorus (P, expressed in kg ha^{-1}) contained in the upper 53 cm of soil profile Kpuabu 1, Sierra Leone. For soil chemical data and BD values of this profile see Table 52.

Repeat the calculations for potassium (K, expressed in kg ha^{-1}).

Note: 1 kg K_2O is equivalent to 0.83 kg K.

Contents of 'available' nutrients

Only a limited part of all nutrients in a soil can actually be taken up by crops in the course of one growth cycle. The other part is lost or tied up in minerals and/or in stable organic and organo–mineral compounds and becomes available to plant roots only after weathering or biochemical decomposition of organic structures ('mineralization'). Soil fertility specialists have long tried to find extraction agents that mimic the action of plant roots. Such agents should remove available nutrients from the soil while leaving the unavailable fraction untouched. Their quest has not been entirely unsuccesful. Many 'mild' (i.e. weakly acid or alkaline) agents have been identified, which give, under specific conditions, a reasonable correlation between element extraction from a soil and total contents of the same element in the tissue of plants grown on that soil. An extractant that reflects nutrient availability to any crop under all conditions does not exist, because the demand of a crop for nutrients and its competitive position relative to other consumers is partly dependent on exogenous factors. In practice, preference for the use of a certain extractant is based on local experience. Extractants that are often used are listed in Table 53. As this table suggests, there are no extractants for estimating the availability of nitrogen in soils.

Sections 4.1 and 4.2 contain the implicit information that there is inorganic nitrogen in the plant – soil – atmosphere system, e.g. fertilizer N, as well as organic nitrogen, incorporated in soil organic matter and in soil micro – organisms. Nitrogen losses through leaching, denitrification or volatilization act mainly on nitrogen in its inorganic form. Organic nitrogen is largely responsible for the N status of unfertilized soils; it becomes available for uptake by plants through mineralization during decomposition of the organic matter. The quantity of nitrogen that is mineralized from the organic structures in the course of decomposition is a function of the quantity of carbon used by soil micro – organisms to support their metabolic activities (Bouldin et al., 1980). This is the reason that in some soil survey reports the C/N ratio of the soil organic matter is specified in addition to the organic matter or organic carbon content. A C/N ratio lower than 15 indicates a situation in which decomposition of soil organic matter yields appreciable quantities of mineral nitrogen; a

Table 53. Some common extractants used for the determination of available phosphorus or potassium.

Element extracted	Name of extractant	Chemical composition of extractant
P	Bray No. 1	0.03 N NH_4F + 0.025 N HCl
P	Bray No. 2	0.03 N NH_4F + 0.01 N HCl
P	Truog	0.002 N H_2SO_4 (pH 4.8)
P	Olsen	0.05 N $NaHCO_3$ (pH 8.5)
P	Modified Olsen	0.5 N $NaHCO_3$ + 0.5 N NH_4F (pH 8.5)
K	—	23% $NaNO_3$
K	—	1 N NH_4 — acetate

(Sources: Dabin, 1980; Odell et al., 1974)

high C/N ratio of 30 or more is indicative of low nitrogen levels in unfertilized soils.

In many traditional agricultural systems an equilibrium situation exists in which inputs of nitrogen equal losses by crop removal, denitrification, leaching, etc. In such situations, the quantity of nitrogen furnished by natural soil fertility can be reliably estimated with the procedures presented in Sections 4.1 and 4.2. However, when this equilibrium is disturbed, e.g. by a change in management system, a non – steady state occurs in which the soil organic – nitrogen content changes towards a new equilibrium level. This involves changes in the rates of decomposition and resynthesis of soil organic matter. Both these processes are extremely complicated and depend on quantity and composition of the organic matter in the soil and also on environmental conditions such as soil moisture content, soil pH and soil temperature. Consequently, there are no generally valid guidelines for estimating, by tests on soil samples, the ability of soil organic matter to supply nitrogen to plants.

Available phosphorus in the soil originates from weathering rock, dissolving phosphorus compounds and also from the organic soil component, but the latter plays a less prominent role than in the case of nitrogen. Tropical soils at an advanced degree of development have lost most of their weatherable minerals and have therefore a high probability to be structurally low in phosphorus. However, P deficiency is also widespread in younger soils because dissolved phosphorus may be inactivated by the solid soil phase (phosphorus fixation), precipitate as an insoluble phosphate or be (temporarily) tied up in microbial bodies. Dabin (1980) quotes a study by Roche et al. (1978), indicating that the phosphorus fixation capacity of 140 soils from 20 different countries correlated with the contents of clay and fine silt ($r = 0.51$), organic matter ($r = 0.7$), total aluminium ($r = 0.63$), 'exchangeable' aluminium ($r = 0.86$), and total iron ($r = 0.55$), all significant at the 0.1% level. With so many variables involved (and there are more!), it is to be expected that the correlation between extractable phosphorus and phosphorus taken up by a crop varies among soils and extractants. In other words, different extractants give different results and locally established correlations are not transferable to other areas. In many countries, Olsen – type extractants are preferred for the determination of phosphorus availability; soil fertility specialists in Sierra Leone (Tables 51 and 52) use Bray No. 1 and/or Bray No. 2. Table 54 illustrates the limited significance of 'available P' – figures without thorough local experience. The table presents the phosphorus uptake by rice from a number of important soil series in Thailand, and 'available P' in these soils obtained with different extractants (Puh & Khunathai, 1971).

The difficulty in interpreting such values is increased by the necessity to judge the availability of phosphorus in relation to the contents of other essential elements, because crop response to the presence of a specific nutrient in the soil depends very much on the balance or imbalance of the total nutrient supply (Section 4.1). Dabin (1980) states that the critical values of extractable

Table 54. Phosphorus uptake by rice and available phosphorus (ppm) by four methods

Soil Series Code	P content of control plants (mg P per pot)	Bray No. 1	Bray No. 2	Truog	Olsen
Bn (1)	59	6.6	9.0	2.7	7.3
Bn (2)	40	1.2	6.0	0.9	8.0
Db	65	3.0	6.9	5.4	4.6
Sb (1)	176	6.1	12.8	6.9	9.9
Cn	143	2.3	8.8	3.6	13.1
Ok (1)	4	2.8	7.8	2.2	3.3
Sm	88	0.6	9.0	3.3	5.8
Np (1)	119	3.8	16.0	5.4	12.2
Hd	56	2.3	7.2	5.1	5.5
Re (1)	33	4.2	8.5	2.2	6.2
Bn (3)	222	4.4	21.8	9.1	8.4
Rs	52	7.1	26.0	5.8	2.9
Ok (2)	68	10.7	29.5	9.1	6.6
Sb (2)	49	3.3	8.4	2.9	3.2
Re (2)	135	9.2	11.9	2.9	3.6
Ub	47	8.2	8.8	1.0	1.6
Lp	63	5.3	7.3	2.9	2.3
Np (2)	135	7.5	13.8	3.5	3.2
Pb	93	15.7	53.2	11.3	9.6
Bk	—	4.7	5.9	2.9	2.5

Source: Puh & Khunathai, 1971.

soil P are co – determined by the soil's total nitrogen content. For modified Olsen P he gives the following empirical rule: for soils with average to low fixation capacity of phosphorus (i.e. roughly Class I and the upper half of Class II in Table 47), the critical phosphorus level is estimated at 0.025 times the total nitrogen content. For soils with a moderate to high fixation capacity (lower half of Class II in Table 47) that level can be twice as high and in the case of soils with very high phosphorus immobilization capacity the level can be increased even further.

Phosphorus immobilization is particularly prominent in certain young volcanic soils with a high content of hydrated iron and aluminium oxides in the clay fraction ('allophane' – rich soils) and is also widespread in old reddish and yellowish tropical soils with a high sesquioxide content. As explained in Subsection 5.3.1, the profiles Masuba P9 and Kpuabu 1 from Sierra Leone (Tables 51 and 52) are examples of such soils, which formed as a result of progressive weathering of rock in the absence of rejuvenation through mineral influx from elsewhere.

Immobilization of potassium also occurs, but the process is different from phosphorus fixation. In the case of potassium, the ions are incorporated in the lattice of certain clay minerals. Many crops take up similar quantities of potassium and of nitrogen, particularly root and tuber crops and many vegetables and fruits. Soil potassium originates from weathering clay minerals (biotite, hydrous mica, illite) and – at a much slower rate – from the breakdown of feldspar and mica. Flood water or irrigation water is another source of K. Many weathered tropical soils, such as quartzitic sands and red or yellow residual soils on geologically old parent rock, are structurally deficient in potassium. Immobilization of potassium by clay minerals with a high selectivity for potassium can be of importance to practical agriculture. Mitra et al. (1958) working with pure clays from India found that the clay minerals bentonite and illite fixed some 4 mg K g^{-1}. The most common clay mineral in tropical soils, kaolinite, fixed only some 0.13 mg K g^{-1}. This, and the fact that the K – fixation capacity is lower in acid soils than in neutral or alkaline soils, explains why potassium fixation is less prominent – and less well researched – in tropical soils than in certain soils in temperate regions. Data by Kemmler (1980) suggest that available potassium levels are 'low' where extraction with 1 N ammonium acetate removes less than 6 mg K per 100 g soil material.

'Minor elements' such as calcium and magnesium, and/or 'trace elements', such as copper, zinc and manganese, are often in short supply in old, leached tropical soils. Calcium deficiency is associated with soil acidity ($pH < 5$); application of lime (Ca) or ground magnesium limestone (Ca, Mg) remedies the situation. Sulfur deficiency also occurs, but published information indicates no association between sulfur deficiency and soil or climatic factors that would be useful in predicting areas of likely deficiency (Blair et al., 1980).

Exercise 72

Compare the total quantity of phosphorus contained in the upper 53 cm of profile Kpuabu 1 (Exercise 71 and Table 52) with the quantity of available P extracted with Bray No. 1. Can you explain the difference?

Suppose that 3000 kg of rice grain was to be produced on a soil like Kpuabu 1. Would P fertilization be necessary? Explain your answer. (Recall that the slope of the yield – uptake curve is 50 to 70 kg grain per kg N and the P/N ratio in the tissue is around 0.1 under balanced supply).

Is there a need for K – fertilizers on Kpuabu 1? Explain your answer.

Ion adsorption in soils

Plants take up nutrients from the soil solution which is essentially a low concentration solution of a mixture of ions. This 'soil solution' would be rapidly exhausted if ions, taken up by the crop, were not replenished. Fortunately, most soils have a capacity to maintain a stock of positive ions (cations) adsorbed at the negatively charged surfaces of solid soil particles (both mineral particles, particulary in the clay fraction and organic matter). The most important cations in this respect are Ca, Mg, K, Na and H. These ions enter the soil solution if the existing equilibrium between the concentration of a certain ion in the soil solution and the quantity adsorbed at the 'exchange complex' of the soil is disturbed, e.g. through nutrient removal by plant roots. A new equilibrium will then be established. The reverse also applies: an increase in the concentration of a certain ion in the soil solution induces adsorption of that ion at the exchange complex. This implies that cations added to the soil, for instance with fertilizers, are not necessarily lost if not instantly taken up by a crop, because part of the ions become available over a longer period via a process of exchange between the exchange complex and the soil solution. The capacity of a soil to adsorb cations depends mainly on the surface properties of the solid soil particles, notably their specific surface (i.e. the total surface area per unit weight of soil) and the electric charge per unit surface area. Finely textured (clay) soils have a high specific surface and therefore a higher cation exchange capacity — 'C.E.C.', expressed in milli — equivalents per 100 g soil — than coarser — textured soils. The mineralogical composition of the solid soil phase is also of influence because clay minerals vary in their surface charge. Table 55 presents typical values for the specific surface and C.E.C. of some common solid phase components. It shows that soil organic matter has a comparatively high cation exchange capacity, therefore the overall C.E.C. of soil material is often higher in the organic matter containing surface horizon than lower in the profile.

The cation exchange complex plays an important role in the determination of soil acidity, i.e. the concentration of hydrogen ions in the soil solution. Often in field situations, a substantial part of the exchange complex may be occupied by hydrogen ions, even at relatively low concentrations of that element in the soil solution, because these ions are adsorbed preferentially. This

Table 55. Specific surface area and C.E.C. of selected solid phase components.

	surface area ($m^2\,g^{-1}$)	C.E.C. me $(100\,g)^{-1}$
Clay minerals:		
kaolinite	10-20	1-10
montmorillonite	600-800	80-120
vermiculite	600-800	120-150
mica	70-120	20-40
chlorite	70-150	20-40
allophane	70-300	10-150
Soil organic matter	800-900	100-300

(Source: Bohn et al., 1979).

preferential adsorption is on the one hand associated with replacement of Al and Mg ions in the clay minerals by H ions, and on the other hand with the surface properties of the soil organic matter. The result of these processes is that soils exhibit an extended buffering range, i.e. addition of relatively large amounts of H ions will only slightly increase their concentration in the soil solution. Hence, the pH (defined as $-\log$ (concentration H^+ ions)) will only decrease very little. On the other hand, if the soil has a low pH value, and thus a high concentration of H ions in the soil solution, considerable quantities of a competitive cation, such as calcium, may be necessary to increase the pH. That quantity expressed in kilograms of CaO required to change pH by 0.1 unit is referred to as the 'liming factor'.

To characterize soils with respect to the composition of the adsorbed ions, the sum of adsorbed calcium, magnesium, potassium and sodium – in me per 100 g soil – expressed as a fraction of the total exchange capacity of the soil, is defined as the soil's 'base saturation'. A low base saturation, i.e. less than 0.4 is typical for most cultivated soils in the tropics. The remainder of the exchange complex is occupied by such ions as hydrogen, aluminium, iron, etc.

Exercise 73

Approximate the C.E.C. of the mineral soil material in the surface horizon of Kpuabu 1 (Table 52). Assume that the soil organic matter has a C.E.C. of 100 me per 100 g; and recall that an organic matter content of 0.01 kg kg^{-1} corresponds with an organic carbon content of 0.0055 kg kg^{-1}.

Is the presence of organic matter in the surface soil of Kpuabu 1 important for the supply of nutrients to a crop? Explain your answer.

Element toxicities

Element toxicity occurs if elements have accumulated in the soil in quantities that are noxious to plants. However, toxic effects can also occur at low absolute element concentrations if the soil solution is dominated by one or only a few elements and, consequently, ion supply to the roots is grossly imbalanced.

Accumulation of high levels of salts in (surface) soils may occur in arid and semi-arid regions, where a distinct evaporation surplus is associated with a predominantly upward movement of soil solution and ions. When this soil moisture evaporates, soluble salts accumulate on top of the soil or in the rooted surface horizons and hamper root activity: the soil has become saline. The degree of soil salinity is established by measuring the electrical resistance of a water-saturated soil paste. The inverse value of electrical resistance is the electrical conductivity (EC_e, expressed in S cm^{-1}, S standing for Siemens). Normal plant growth is inhibited at conductivity values in excess of 16×10^{-3}

S cm^{-1}, corresponding to a salt concentration in the soil of approximately 0.01 kg kg^{-1}. Salt damage becomes apparent already at much lower electrical conductivity values ($EC_e \geq 4 \times 10^{-3}$ S cm^{-1}). Salt accumulation can be remedied by leaching of the soil with fresh water. The harmful effects of salts in the root zone gain prominence if the accumulated cations are dominated by sodium ('saline – alkali soils'). Salt accumulation outside arid or semi – arid regions is usually limited to low – lying coastal areas subject to saline seepage or to periodic inundation with saline or brackish water.

Element toxicity at low salt levels in the soil solution is widespread in acid (pH < 5) soils in the humid tropics. Soil acidity, associated with high levels of hydrogen ions in the soil solution, hampers crop growth directly, but it is often more harmful because it results in the formation of toxic levels of aluminium and/or manganese in the soil. Accumulation of aluminium ions is particularly harmful: crop production is affected already at low levels of aluminium and is drastically reduced when aluminium occupies more than two – thirds of the exchange complex. Aluminium poisoning is most prominent on dry land; it rarely occurs where wetland rice is grown, with the exception of certain acidifying soils in coastal areas ('Acid Sulphate Soils'). Wet-

Table 56. Criteria for the recognition of some important constraining soil chemical conditions.

Poor parent material/ leaching of cations	<7 me cations/100 g soil material in 1N ammonium acetate extractant at pH 7.0 (C.E.C. <7 me/100 g)
N deficiency	Widespread agronomic evidence of nitrogen deficiency in principal crops. No quantitative soil definition
P deficiency associated immobilization	Top soil clay fraction dominated by allophane (X-ray amorphous), or >20% of clay fraction as Fe and Al oxides and >30% clay in topsoil, or pH <7.3, or fizzing of soil material with HC1 (= presence of free $CaCO_3$)
K deficiency	<0.6 mg K per 100 g soil in top 50 cm.
Deficiency of other nutrients	Soil test values below critical levels for main crops
Al toxicity	>67% Al saturation of C.E.C. in top 50 cm or pH <5.0
Mn toxicity	<100 ppm acid extractable Mn in top 50 cm.
Fe toxicity	>300 ppm ferrous iron in topsoil.
Salinity	$EC_e > 4 \times 10^{-3}$ mS cm^{-1} in top 50 cm.
Alkalinity	>15% Na-saturation of C.E.C. in top 50 cm.
Acid sulphate soil conditions	pH drops to values <3.5 upon drying of soil material.

Source: Buol et al., 1974; Sanchez and Cochrane, 1980.

land rice is known to suffer from excess organic acids or hydrogen sulphide but such damage is comparatively rare. More common is damage to wetland rice by toxic levels of ferrous iron. The mechanism of iron accumulation in intermittently wet soils has been outlined in Subsection 5.3.1. Iron poisoning occurs when rice on flooded land is exposed to a soil pH lower than 6.5 and a dissolved iron content in excess of 300 ppm (Moormann & van Breemen, 1978). In soils with low nutrient levels, especially P and K, an iron content as low as 30 ppm may even be toxic. Table 56 summarizes indicative criteria for the most common chemical soil disorders.

Popular misconceptions in soil chemical analysis

Although chemical soil analysis is a practice of long standing, it suffers from a number of persistent imperfections that complicate the interpretation of analysis results. Those shortcomings, which pertain to the routinely collected soil chemical data discussed in this subsection will be mentioned here:

- Soil chemical data are traditionally expressed on a weight basis, e.g. as percentage, ppm or me per 100 g. Such data give no information on the quantities of nutrients (in kg ha^{-1}) present in the rooted soil volume. It is therefore essential that bulk density values (which vary by a factor of almost two, viz. between 0.85 and 1.55 g cm^{-3}, in mineral soils) are specified with the element concentrations. Unfortunately, this is rarely the case. An extreme example may demonstrate the possible implications of this omission. Peat soils are generally considered high in nitrogen with a typical nitrogen concentration between 0.01 and 0.02 kg kg^{-1}. However, a normal peat soil with a bulk density of 0.20 g cm^{-3} and a nitrogen concentration of 0.02 kg kg^{-1} has a total nitrogen content of 4000 kg ha^{-1} in its upper 10 cm layer, whereas a mineral surface horizon with a nitrogen concentration of only 0.005 kg kg^{-1}, but a bulk density of 1.5 g cm^{-3}, has a total nitrogen content of 7500 kg ha^{-1}. In spite of the seemingly high N concentration, nitrogen deficiency is quite common on peat soils. For that reason analysis results should be presented on a weight – per – unit – volume basis. (Driessen, 1978).
- The stoniness of a soil must be specified if critical element levels (Table 56), defined to indicate deficiencies or toxicities of elements, are to be useful. This is rarely done in routine soil analysis.
- Cation – exchange capacity is traditionally determined by measuring the ammonium retention of an aliquot of soil in a neutral ammonium acetate solution (buffered to pH 7.0). However, the surface charge of organic matter and of some clay minerals (notably kaolinite, chlorite and allophane, see Table 55) is strongly pH dependent. As field pH is normally lower than 7, the actual exchange capacity of tropical soils is often lower than suggested by chemical analysis data. Only C.E.C. data determined at field pH would allow a meaningful comparison of soils with regard to exchange capacity and base status.

Exercise 74

Deeper than 25 cm below soil surface, Profile Kpuabu 1 (Table 52) has a gravel content of more than 0.7 kg kg^{-1}. What does this information imply with regard to this soil's nutrient status?

Profile P9 (Table 51) contains less than 0.02 kg kg^{-1} of gravel throughout. How does its real capacity to retain cations compare with that of profile Kpuabu 1?

5.4 Plant data

H. van Keulen

In the foregoing chapters it was shown that a quantitative assessment of plant growth and crop productivity can be obtained in different production situations, on the basis of knowledge about basic properties of plants and soils. In this section some of the required plant data will be summarized, methods to obtain them outlined and indicative values provided for first estimates.

5.4.1 Photosynthetic capacity

As outlined in Section 2.1, gross CO_2 assimilation of a canopy, F_{gc}, may be calculated from the leaf area index and measured or estimated values for F_g, the diffusion – limited maximum gross assimilation rate of a single leaf at high light intensity. The latter characteristic may be experimentally determined by measuring the CO_2 absorption by a leaf of known surface area at high light intensities. This is most conveniently done by enclosing the leaf in a chamber through which air is passed at a known flow rate. The concentration of CO_2 is measured at the entrance to the chamber and at the outlet, so that the amount of CO_2 absorbed can be calculated. As assimilation and respiration proceed concurrently, the measured value represents the net assimilation rate, that is the difference between assimilation and respiration. Thus to obtain F_g, as used by Goudriaan & van Laar (1978a), the measured value must be augmented by the value of the dark respiration (Subsection 5.4.2) implicitly assuming that it proceeds at the same rate in the light. Such experiments are normally carried out under controlled conditions, with the disadvantage that it is often difficult to obtain sufficiently high light intensities from artificial light sources to reach light saturation, especially for plants of the C_4 type. Although the value of F_g varies with such characteristics as age of the leaf, its nitrogen content and its position in the canopy (especially its position with respect to the sun), characteristic values of maximum CO_2 assimilation range from 15 – 50 kg ha^{-1} h^{-1} for plants of the C_3 type and 30 – 90 kg ha^{-1} h^{-1} for C_4 species. If the value of F_g is unknown for a species, 40 kg ha^{-1} h^{-1} for a C_3 species and 70 kg ha^{-1} h^{-1} for a C_4 species is generally a reasonable guess.

Lists of C_4 species have been published by Downton (1975) and Raghavendra & Das (1978). Whether a species has the C_3 or the C_4 type photosynthetic pathway can be deduced from anatomical features. In C_4 type plants, mesophyll cells are arranged radially around chlorenchymatic bundle sheaths, whereas in C_3 type plants mesophyll cells are laminar. These differences are clearly visible under a microscope.

An alternative method is based on the fact that the CO_2 compensation point, that is the value of the CO_2 concentration in the outside air, where net CO_2 assimilation becomes zero, is about 10 ppmv for C_4 species and 120 ppmv for C_3 species. When an unknown species is grown in an airtight environment in the presence of a C_4 species, it will cease functioning long before the C_4 species if it happens to belong to the C_3 group. It will hold out approximately equally long if it is a C_4 species. In the closed system, the photosynthetic process will lead to gradually declining concentrations of CO_2. When these have fallen below 120 ppmv, respiration of the C_3 species exceeds assimilation, whereas the C_4 species continue to assimilate at a positive rate untill about 10 ppmv.

5.4.2 Respiratory losses

Measurement of respiration is in principle identical to that of assimilation, i.e. the CO_2 exchange rate is determined in the absence of an energy source ('dark respiration'). When plants that have been in the light for some time are transferred to the dark and CO_2 exchange is measured, the general picture is that at first the rate of CO_2 evolution is high but gradually declines with time and finally reaches a more or less stable value. In analyzing such data it is often assumed that the stable value represents maintenance respiration, and that the additional component in the beginning results from the conversion of assimilates in structural plant material, i.e. growth respiration.

The magnitude of the respiration losses depends primarily on the chemical composition of the structural material of the plant (Section 2.1). Both maintenance respiration and growth respiration are higher, when the protein (nitrogen) concentration of the material is higher.

Values of the maintenance requirements may be estimated from the protein concentration and the ash concentration of the material, by assuming that at 20 °C the proteins require about 0.035 kg CH_2O per kg per day for maintenance and the minerals about 0.07. In a situation where each of these two components makes up about one tenth of the total weight of the standing crop, the maintenance requirement will thus be 0.0105 kg CH_20 per kg dry weight per day.

Exercise 75

Calculate the carbohydrate requirement for maintenance respiration in kg ha^{-1} d^{-1}, for a crop with a dry weight of 10 000 kg ha^{-1}, a nitrogen concentration of 0.025 kg kg^{-1} and an ash concentration of 0.01 kg kg^{-1} (N.B. the nitrogen concentration of proteins is 0.16 kg kg^{-1}).

In most cases, however, such detailed analyses cannot be performed, because insufficient data are available. Moreover, in the hierarchical approach presented here, the nitrogen concentration of the material is not known when the respiratory losses have to be calculated (Sections 2.3 and 3.4). In that situation approximate values may be applied, as illustrated in Table 3.

The effect of growth respiration may be expressed by the conversion efficiency, explained in Section 2.1. Thus, if the conversion efficiency is 0.7, growth respiration amounts to a fraction 0.3 of the weight of the primary photosynthates. Thus again, high proportions of proteins, but also fats, lead to high losses in growth respiration.

Exercise 76

Calculate the rate of increase in dry matter in kg ha^{-1} d^{-1} for a crop which, after maintenance, has a carbohydrate surplus of 400 kg ha^{-1} d^{-1}:

– if cassava tubers are filled (composition: proteins 0.075 kg kg^{-1}; starch 0.925 kg kg^{-1})
– if maize grains are filled (composition: proteins 0.125 kg kg^{-1}; starch 0.875 kg kg^{-1})
– if soybean seeds are filled (composition: proteins 0.4 kg kg^{-1}; starch 0.4 kg kg^{-1}, lipids 0.2 kg kg^{-1})

What are the carbohydrate losses due to growth respiration in these cases?

If no exact data are available for specific situations, the conversion efficiencies as tabulated in Table 3 may be applied for the various groups of crops. Chemical composition of storage organs of various crops are given by Sinclair & de Wit (1975) and Penning de Vries et al. (1983).

5.4.3 Phenology

One of the most important crop characteristics that has to be taken into account in establishing production potentials is phenological development (Section 2.2). It has been explained that various methods exist to describe phenological development as a function of temperature and day length. The most simple method, which, however, yields good results in many cases is referred to as the Thermal Unit approach. In that approach it is assumed that for each phenological period, the rate of development can be described as a linear function of temperature. The rate of development during such a period is then defined as the inverse of the duration of that period, thus:

$$R_v = 1/N_d = b_t(T_a - T_o) \qquad (94)$$

where

R_v	is rate of development (d^{-1})
N_d	is length of a phenological period (d)
b_t	is a constant ($d^{-1}\,°C^{-1}$)
T_a	is average daily air temperature (°C)
T_o	is apparent threshold temperature or base temperature (°C), below which phenological development comes to a standstill

---

Exercise 77
Calculate the rate of development for the wheat crop of the following example. N_{hm} denotes the number of days from heading to maturity.

Location	T_a	N_{hm}
Harrow	21.1	31
Ottawa	20.6	29
Normandin	15.0	54
Swift Current	19.1	41
Lacomoe	14.7	59
Beaverlodge	13.9	54
Fort Vermilion	15.3	48
Fort Simpson	16.4	38

Make a graph of the results, temperature on the horizontal axis and development rate on the vertical. Draw an eye – fitted straight line and estimate the values of T_o and b_t. How many days will it take this wheat variety from heading to maturity at an average air temperature of 12.5 °C?

Application of the Thermal Unit concept leads to the notion that a constant temperature sum, counted above the base temperature, has to be accumulated during a certain phenological period. That period can be between any two phenological events that are important from the point of view of production. For modelling purposes, in general two phases are distinguished, one before the moment that growth of the storage organs starts, one after that. In cereals that is the moment of anthesis, in root and tuber crops it is the moment of the start of bulking. For both periods a scale from 0 to 1 is then assigned. In most cases it is convenient to start the pre – storage organ phase from emergence, because the temperature relations for germination may be quite different.

In practice, the relation between temperature and development rate can be determined experimentally by observing the dates of emergence and that of the start of the storage organ growth and measuring the ambient temperatures during that period. If only data for one year are available, base temperatures as given in Table 7 (Section 2.2) may be used to construct the linear relation. For a more accurate description, it will be necessary, as a rule, to perform at least two experiments under different temperature conditions to be able to derive the base temperature. Such experiments can easily be carried out under controlled conditions, i.e. in climate rooms or greenhouses. For the period after the start of storage organ growth, experimental determination may be more difficult, because the moment of maturity is often not clearly recognizable. For different crops, different criteria have been specified (for maize for instance formation of the black layer at the base of the kernel). On the basis of these criteria, the maturity point may then be estimated and an identical procedure applied. As a rule, the slope of the development rate – temperature relation is different for the period before the storage organ starts growing and the period thereafter.

Exercise 78
Calculate the rate of development for the period emergence to anthesis for two spring wheat cultivars. T_a is average air temperature in °C; N_{ac} is duration of the period in days.

cv. Timgalen		cv. Orca	
T_a	N_{ac}	T_a	N_{ac}
12.5	93	11.6	95
14.1	72	13.2	78
17.0	61	17.0	69

Do the same for the period of anthesis to maturity.

cv. Timgalen		cv. Orca	
T_a	N_{am}	T_a	N_{am}
19.5	49	17.6	43
20.5	48	17.6	40
22.1	41	16.9	39

Plot the calculated data as in Exercise 77 for the two different periods in two separate graphs. Draw eye – fitted curves. What do you notice about the slope of the two lines?

For modelling purposes the relation between temperature and development rate will thus have to be established for at least each group of cultivars (short duration vs. long duration) within a species.

5.4.4 *Dry matter distribution*

The distribution of the dry matter formed between the various plant organs is a major determinant for both total dry matter production and economic yield. If in the early stages of crop development a large proportion of the dry matter formed is invested in the root system, leaf area development is relatively slow, hence light interception will be incomplete for a long period of time and total production will remain low. In the same way, economic yield may be affected if during the period of storage – organ growth other organs of the plant still claim a substantial proportion of the available assimilates or if the green area rapidly declines because of translocation of essential elements to the developing storage organs, as may be the case under nitrogen shortage.

Since the basic processes governing dry matter partitioning are poorly understood, the subject is treated in an empirical way. The simplest method of analyzing distribution patterns would be to describe the increase in dry weight of the various plant organs in relation to the total increase in dry weight of the plant, i.e. to its age. However, as indicated in Section 2.2, the initiation and development of plant organs is governed by phenological development of the plant, which is a function of environmental conditions, such as temperature and day length, rather than of age. The partitioning pattern in plants is therefore related to the phenological stage of the crop. This starting point severely limits the availability of suitable experimental data for the determination of partitioning patterns, because the number of reports in which not only sufficiently detailed crop data are reported, but also environmental conditions, is very low.

If, however, all of these data are available, an analysis may be carried out, yielding information that can be used for extrapolation and prediction. This is illustrated with an experiment on bunded rice, reported by Erdman (1972). The basic data of the experiment are given in Table 57. These data are used to calculate the derived data as given in Table 58: for each harvest date the development stage of the vegetation is determined as the ratio between accumulated temperature sum and the temperature sum required from transplanting to anthesis (1950 d °C for this variety). As the distribution factors are calculated for periods between two consecutive harvests, the value of the development stage halfway between these two dates is entered in Table 58. For each period, the increase in dry weight of the various plant organs is calculated as the difference between the weight at two consecutive harvests. From these partial values, increase in total dry weight is obtained and from these numbers the fraction allocated to each of the organs in each period is calculated.

Table 57. Basic data on dry matter distribution from a field experiment with bunded rice.

Time (days after transplanting)	T_a (°C)	dry weight of plant organs (g plant $^{-1}$) roots	leaf blades	stems sheaths + ear structures	total
	26.0				
11		0.15	0.26	0.24	0.65
	26.3				
18		0.26	0.94	0.84	2.07
	26.3				
25		0.34	2.19	2.23	4.76
	26.3				
32		0.81	3.98	4.48	9.27
	26.3				
39		1.93	6.06	9.37	17.36
	26.2				
46		2.81	7.82	12.76	23.39
	26.2				
53		2.99	9.76	21.46	34.21
	26.2				
60		3.17	10.77	26.83	40.77
	26.2				
67		3.36	11.85	21.48	36.69
	26.3				
74		3.37	9.11	21.94	34.42
	26.3				
81					
	26.3				
88					

The accuracy of the individual observations is not high. Even with a reasonable number of replicates, the variance in dry matter yield is high, so the increase in dry matter of individual organs is even less accurate. In general, therefore, observations of several experiments have to be combined to obtain a useful relation. In Figure 54, the data of Table 58 are summarized by eye-fitted curves, illustrating the variability in the data. However, for the time being it is necessary to rely on this type of analyses.

After anthesis, only the grains increase in dry weight, the other organs either remaining constant or decreasing in weight, when translocation of substances, nitrogenous compounds as well as carbohydrates, to the developing grain takes place.

As explained in Section 3.4, this translocation process is taken into account in the calculations by assuming partitioning of part of the assimilates to the

Table 58. Derived data on dry matter distribution from a field experiment with bunded rice.

Period (days after transplanting)	R_d	Increase in dry weight					
		leaf blades		'stems'		roots	
		g $plant^{-1}$	fraction	g $plant^{-1}$	fraction	g $plant^{-1}$	fraction
0-11	0.075	0.26	0.40	0.24	0.37	0.15	0.23
11-18	0.195	0.68	0.48	0.60	0.42	0.14	0.10
18-25	0.29	1.25	0.46	1.39	0.52	0.05	0.02
25-32	0.385	1.79	0.40	2.25	0.50	0.47	0.10
32-39	0.475	2.08	0.26	4.89	0.60	1.12	0.14
39-46	0.57	1.76	0.29	3.39	0.56	0.88	0.15
46-53	0.665	1.94	0.18	8.70	0.80	0.18	0.02
53-60	0.76	0.99	0.15	5.37	0.82	0.18	0.03
60-67	0.855	1.08	n.a.**	negative	n.a.	0.18	
67-74*	0.955	negative	n.a.	negative	n.a.	0.01	

* Anthesis at Day 74
** Not applicable

grain before anthesis.

The grain weight calculated with the model is on a dry weight basis. For comparison with field data, the air dry moisture content of the grains has to be taken into account. Normally grain yields are reported at 12 – 14% mois-

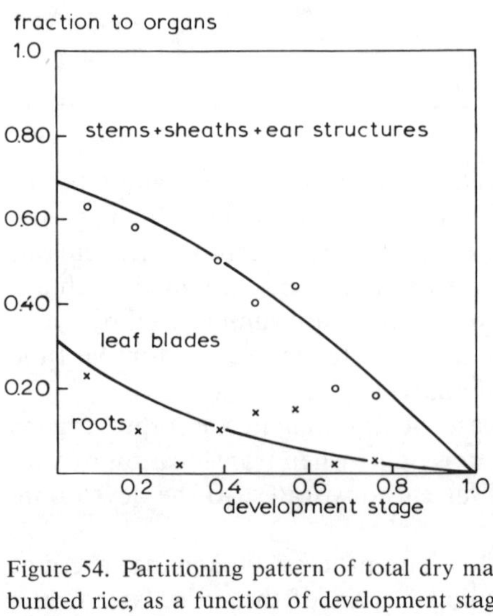

Figure 54. Partitioning pattern of total dry matter increase between various organs of bunded rice, as a function of development stage.

ture content, hence the calculated dry matter must be multiplied by a factor of 1.13. For crops other than grain crops special moisture contents may have to be taken into account. This emphasizes, however, that for experimental work too an accurate definition of the reported data is necessary.

For rice the situation is somewhat more complex: in most cases yield is expressed in rough rice (paddy) at a moisture content of 14%. Rough rice, however, consists of the true fruit (brown rice) and the hull, which consists of leaf – like structures. Brown rice is largely the endosperm and the embryo, but still contains several thin layers of botanically different tissues that are removed during milling, after which white rice, milled rice or polished rice is obtained. In the model, brown rice is the result of the calculations. To convert from white rice to brown rice a conversion factor of 1.25 is generally used, from brown rice to paddy a factor of 1.2.

If the growing conditions are suboptimal, i.e. when shortage of water or nutrients occurs, the plant will generally respond with a change in its distribution pattern. In both cases, the most common response is an increased investment in the root system, because that part of the plant is responsible for its supply with water and nutrients. This phenomenon is often referred to as the functional balance (Brouwer, 1963). A quantitative treatment of this phenomenon is difficult within the scope of the present approach, moreover the data base is narrow.

5.4.5 Stomatal behaviour

Stomatal aperture in plants can change in response to internal or external conditions, so that plants can react to the environment. One such reaction is that of closure of the stomata under conditions of water stress, which effectively reduces transpiration. However, even under optimum moisture supply, stomatal opening may vary. One mechanism governing stomatal aperture is the response to light, i.e. stomata open in the light and close in the dark, the transition zone being rather narrow. However, in certain plants, stomatal aperture is regulated in such a way, that the concentration of CO_2 inside the stomatal cavity is maintained within narrow limits. This may either be a constant value, which is around 120 ppmv for plants with the C_4 type of photosynthesis and around 210 ppmv for plants with the C_3 type. An alternative is that the internal concentration is adjusted in such a way that a constant ratio of about 0.7 for C_3 species and about 0.4 for C_4 species between external and internal CO_2 concentration is maintained (Goudriaan & van Laar, 1978b). Under present conditions there is hardly any difference between the two mechanisms in the field, because at an external value of 340 ppmv about the same internal values result for both situations.

The type of stomatal control is of prime influence on water use efficiency (Section 3.2). The ratio of the amount of CO_2 fixed to the amount of water lost is about twice as high for plants with CO_2 induced stomatal regulation as for plants without that mechanism. It is therefore important to know what type of regulation operates in a given crop. Unfortunately, that is difficult to predict, because not only differences exist between different species, but apparently the same species may react in different ways, depending on external or internal conditions. The only way to determine experimentally the type of regulating mechanism is by measuring concurrently CO_2 assimilation and transpiration at a range of external CO_2 concentrations. Such data permit the construction of the type of graphs, schematically presented in Figure 55. Increased CO_2 concentrations in the external air lead to stomatal closure in regulating plants, hence decreased transpiration and a constant assimilation. In non – regulating plants assimilation increases, but transpiration is hardly

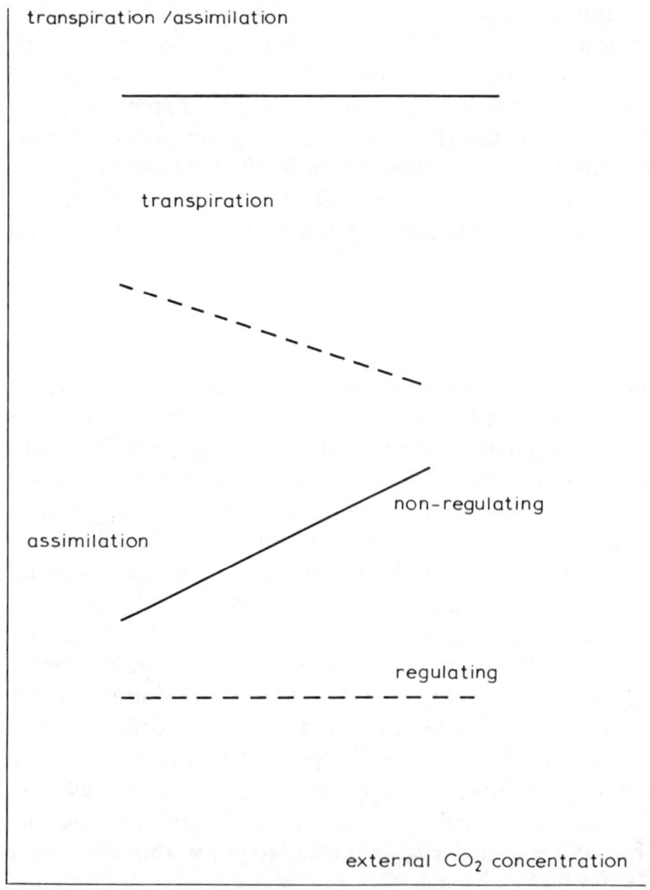

Figure 55. Schematic representation of transpiration and assimilation as a function of external CO_2 concentration for plants with and without CO_2 induced stomatal regulation.

affected as the stomata remain open. The most common type seems to be that in which a constant ratio between external and internal concentration is maintained. The absence of CO_2 induced regulation is most likely associated with maintenance of optimum moisture supply throughout the plant's life cycle. In situations where no information on stomatal behaviour is available, assumption of the constant ratio is most realistic.

5.4.6 Nutrient requirements

In Section 4.1 it was shown for maize and rice that under conditions where yield is determined by availability of a specific nutrient element, the concentration of that element at a given development stage of the crop reaches a species – characteristic minimum value. The determination of total nutrient requirements as proposed in this approach, require the minimum element concentrations at maturity. Such data can be determined for specific crops in pot experiments, because these values are independent of the growing conditions. As a first approximation, values have been collected from the literature pertaining to groups of crops. The determinant factor for each group is the chemical composition of the material formed. The data used in the model structure are summarized in Table 59.

5.4.7 Some additional data

For initialization of the calculations for Production Situation 1 or 2 (Sections 2.3 and 3.4), the values of above – ground dry weight and root dry weight are necessary. It is possible to determine these values experimentally for a specific trial. However a more general approach is to look at the conversion of the substrate contained in the seed into structural plant material. Experiments with a number of species have shown that about half of that energy is lost in respiration, whereas the remainder is equally divided between shoot and root (Penning de Vries et al., 1979). Thus both root and shoot dry weight at emergence may be estimated as about one fourth of the seed rate,

Table 59. Minimum concentrations (kg kg^{-1}) of the macro-elements in economic products and crop residues for a number of crop groups

Crop group	Economic product			Crop residues		
	N	P	K	N	P	K
grains	0.01	0.0011	0.003	0.004	0.0005	0.008
oil seeds	0.0155	0.0045	0.0055	0.0034	0.0007	0.008
root corps	0.008	0.0013	0.012	0.012	0.0011	0.0033
tuber crops	0.0045	0.0005	0.005	0.015	0.0019	0.005

Table 60. Indicative specific leaf area of major crops ($m^2\ kg^{-1}$).

Crop	Specific leaf area (S.L.A.)
Barley	25
Cassava	22
Chick pea	13
Chillie	27
Cotton	20
Cow pea	25
Grass pea	13
Groundnut	28
Jute, capsularis	31
Jute, olitorius	28
Kenaf	25
Lentil	33
Maize	18
Mungbean	30
Mustard (black)	23
Onion	25
Rapeseed	23
Rice	25
Sesame	23
Sorghum	20
Soybean	26
Sugarcane	10
Sweet potato	22
Tobacco	16
Wheat	20

provided that germination is almost complete.

The development of leaf area, important for interception of irradiance, depends on the one hand on the amount of dry matter invested in the leaf blades (Subsection 5.4.4), and on the other hand on the area that each unit of leaf blade dry matter produces. In Sections 2.3 and 3.4 this parameter was introduced as the specific leaf area. It follows directly from simultaneous measurement of dry weight and area of the leaf blades. Special leaf area meters are available that electronically scan the areas of the leaves. A widely used method that requires no sophisticated equipment is to measure leaf width and leaf length and establish a relation between the product of the two and actual leaf area, depending on leaf shape. For situations where experimental data on specific leaf area are not available, indicative values for different crops have been summarized in Table 60.

In general, it may be concluded that for the type of calculations at which the present approach is aiming, and their degree of accuracy, the available plant data are of sufficient quality. It means that these estimations can be made without having to go into detailed plant physiological research.

6 CROPPING SYSTEMS

6.1 Crop calendar, workability and labour requirements

H.D.J. van Heemst

6.1.1 *Introduction*

In the preceding chapters the growth of crops was analyzed in relation to solar radiation, temperature, day length, water and nutrient availability and crop characteristics. It was shown that weather and soil conditions determine crop growth and production. So far, the human factor has not been considered. However, agriculture is distinguished from the natural situation by the intermediary role of man in the process of converting solar energy into edible energy by means of plants and animals. This involves tillage of the soil, cultivation and harvesting of crops, and care and breeding of livestock. Therefore, in addition to the growth factors discussed so far, human activity is an indispensable factor in the agricultural production process. In this section attention will be focussed on aspects of human activity in agriculture. These aspects are:

- availability of human activity
- demand for human activity
- type of human activity required
- choice between different possibilities of applying human activity.

In Chapter 1 it was emphasized that in this monograph only the technical-agricultural aspects of food production would be considered. The aspect of availability of human activity is therefore not treated, because that depends on such factors as the size of the population, its demographic composition, the level of participation, the number of working hours per day and the number of days per year available for agricultural activities. The other aspects will,however, be discussed in this section.

6.1.2 *Crop calendar*

The demand for human activity in agriculture at a particular moment depends in the first place on the growing season, the type of crop and its development stage. The timing of the crop growing season is in most cases determined by temperature and/or water availability. At high latitudes crop cultivation during winter is impossible because of low temperatures. In regions with a distinct dry season, it is impossible to cultivate a crop during that season without irrigation. Even if temperature and soil moisture availability permit growth of certain crops, the conditions may not be suitable for all crops or crop varieties. Some crop species need higher temperatures than others, some are more susceptible to (temporary) water stress than others and day length requirements are different among crops or cultivars. Thus, if an existing day length and temperature regime suits one cultivar of a crop species, not all cultivars of that particular crop may mature within the available growing season. Therefore, in most cases only a limited number of crops or crop varieties is suitable for a specific environment.

The demand for human activity depends further on the time required for land preparation, an activity strongly related to the physical condition of the soil, as will be discussed in Subsection 6.1.3. Before sowing or planting – an absolute prerequisite for crop growth – a suitable seedbed has to be prepared. Once the crop is seeded or planted, the length of the growing period of a given crop variety is fairly predictable, as mean temperatures at a specific site in a particular season are reasonably constant (Section 2.2). Therefore the average development rate, and thus the growth duration of that crop, will be approximately the same each year if the start of the growing period has not been delayed too much because of, say, unusual weather conditions.

During the growing period of a crop, the farmer has to perform a number of field operations to create or maintain the most favourable conditions for crop growth. Among these are the reduction of competition by unwanted plants (by destroying them), optimizing the availability of nutrients and water (by fertilizer application, drainage and irrigation), and protection of the crop (by pest and disease control). At crop maturity, harvesting is necessary. Some on – farm processing may also take place and the products are stored or marketed. Most of these activities have to be carried out in a given order and each of them within a limited time span. Therefore a crop activity calendar can be constructed, indicating the most favourable timing of the various activities and the type of operations required. Such crop activity calendars are elaborated for rice, maize and cassava in Subsection 6.1.5.

6.1.3 Workability

Workability expresses the possibility to perform a certain activity in a given environment. Whether a situation is workable depends on the availability of labour, farm equipment and required materials, the state of the crop, weather conditions and state of the soil, and type and nature of the field operation. In this subsection attention is focussed on the relation between the state of the soil and workability. In general, a soil is considered tractable, that is suitable to be worked, if a tractor or any other required farm machine can move on that soil and satisfactorily perform its function without causing temporary or lasting damage to the soil. The most common form of damage is soil compaction. That is basically a reduction in volume of a given mass of soil. It may be expressed as a change in bulk density or porosity (Section 3.2). Many of the soil physical properties, such as hydraulic conductivity and soil moisture characteristics are affected by bulk density. If bulk density becomes too high, resulting in insufficient air space in the soil, plant – root development may be hampered, which will conceivably lead to a reduction in crop yield. The degree of soil compaction is, among other factors, determined by the pressure applied by the machine and the soil – moisture content at the time of the operation. The pressure applied by the machine is a function of the weight of the farm equipment and the contact area between the equipment and the soil. The soil – moisture content at a given moment depends on the water balance of the root zone (Section 3.2).

The limiting upper soil – moisture content for satisfactorily performing a field operation by hand labour is, rather arbitrarily, set at a soil – moisture suction of 10 cm, i.e. nearly saturated. For operations using draught animals it is set at 100 cm, i.e. near field capacity. For operations using power equipment it is set at 500 cm. These limits do not apply to the cultivation of bunded rice, as special equipment is required for wet – land preparation and an important aim of land preparation for that crop is the creation of a compacted subsurface layer to minimize drainage. Land preparation at the end of a dry season, may, especially on heavy soils, be hampered by a low soil – moisture content. In that case workability is limited, especially if the work has to be carried out completely by hand or with animal traction, because the soil is so hard that cultivation requires too much force. In the present approach, the soil is considered workable if, after a long dry period, at least 75 mm of rain have fallen in the last 20 days before the actual operation has to be carried out.

6.1.4 Labour requirements

Information on the time spent on different agricultural operations is generally more readily available and more accurate for mechanized operations than for systems where manual labour or animal traction are employed. Preferably, labour requirements are defined as the time required to carry out a well – defined operation, under standard conditions, by a skilled healthy labourer, working at normal pace, using standard equipment and at maximum efficiency. This time requirement includes, apart from the actual operation:
- the time required for actions necessary for smooth continuation of the job, e.g. operating the filling mechanism of a sowing machine
- the time required to repair minor breakdowns
- the time required to install the implements, to move them between farm buildings and field, and to perform the necessary maintenance.

The total set of activities required in the agricultural production process can be classified on the basis of timing of the major operations involved:
- preparatory activities, such as land preparation and preparation of plant material
- planting or seeding

– crop management, such as application of fertilizer, irrigation, thinning, topping, weed control and pest and disease control
– harvesting
– on – farm processing.

Most of the operations can be executed in various ways, distinguished mainly according to the source of energy used:
– complete manual labour
– use of draught animals
– light or heavy mechanical equipment using fossil energy.

The time required for any operation is expressed in man – hours per hectare or in some cases in man – hours per unit product weight. In the analysis, this definition may lead to certain difficulties. For some operations, the physical constitution may influence the operation time, so that working hours of children, women or aged people cannot be counted indiscriminately, but have to be weighed for conversion to standard man – hours. In other cases however, it is not physical strength, but acquired skill that is of importance. Farringtons (1975) data suggest that the application of weighting factors depending on age and sex, irrespective of the activities, results in underestimation of labour time, while indiscriminate use of actual working hours results in overestimation. As there is no general agreement on the value of the weighting factors, and assuming that activities requiring more physical effort are mainly carried out by men, preference is given to an indiscriminate use of actual working hours, unadjusted for sex or age.

As the time requirement for manual product handling varies with the quantity of the product, the time required for these activities is expressed in man-hours per unit weight. When machines or draught animals are used, the labour requirement will in general be independent of the quantity of the product, hence for these operations time is expressed in man – hours per unit area.

Labour requirement data for operations carried out manually or with draught animals are scarce. Most of that type of information originates from socio – economic surveys and case studies, in which the time spent on agricultural activities is but one of the aspects treated. In many of these studies, working conditions are described in a rather rudimentary way, the operations are ill – defined from an agro – technical point of view and, consequently, the data are not very accurate. The labour requirement estimates for the various agricultural operations (Tables 61 – 68), are based on a literature survey by Van Heemst et al. (1981). A major part of these data is derived from the type of studies indicated above and they show, therefore, large variations. If actual labour requirements are available for a specific region, use of these data is preferable to the broad averages presented in this section.

Table 61. Labour requirement estimates for land preparation (h ha^{-1}).

Activity	Manual	Draught animal	Power equipment light	Power equipment heavy
ploughing, per operation		28	17	6
hilling up, per operation	85	9	9	7
harrowing, per operation		24	10	3
puddling, per operation		29	13	7
levelling, per operation		34	4	3
bunding plus plastering	20			
digging by hoe	300			
spading, 20-25 cm	500			

Table 62. Labour requirement estimates for the pre-treatment of plant material (h ha^{-1})

Activity	Manual	Mechanical
rice, nursery preparation	80	
preparation of cuttings		
for sweet potato	75	
for cassava	25	7
shelling groundnut	55	1

Table 63. Labour requirement estimates for sowing or planting (h ha^{-1}).

Activity	Manual	Draught animal	Power equipment
rice, transplanting	280		40
rice, seeding	95	15	10
maize	80	15	10
sorghum	80	15	10
millet	80	15	10
cotton	55	15	10
groundnut	85	15	10
soya bean	85	15	10
mung bean	85	15	10
cassava	70		5
sweet potato	70		5
potato	75		
yam	75		
sugarcane	230		5
kenaf	50		
jute	50		
tobacco, transplanting	240		

Table 64. Labour requirement estimates for broadcast fertilizer application (h ha^{-1}).

Activity	Manual	Draught animal	Power equipment
fertilizer application	3	2	1

Table 65. Labour requirement estimates for weed control (h ha^{-1}).

Activity	Manual	Draught animal	Power equipment
first weeding	145		
second weeding	120		
third weeding	65		
fourth weeding	20		
mechanical weeder		7	2-16
sprayer	25	2	1-4
thinning	60		
topping	40		
desuckering	40		

Table 66. Labour requirement estimates for crop protection against seed predation by birds (h ha^{-1}).

Crop	Manual
rice	126
maize	55
sorghum	320
millet	320
groundnut (monkey)	20

Table 67. Labour requirement estimates for harvesting activities

Activity	Manual (h t^{-1})	Draught animal (h ha^{-1})	Power equipment (h ha^{-1})
all harvest activities excluding transport:			
rice	95	40	13
wheat			
cutting	12	5-8	
binding	8-12		
binder		9-16	8-16
making sheaves	3-6		

Table 67. (continued)

Activity	Manual (h t⁻¹)	Draught animal (h ha⁻¹)	Power equipment (h ha⁻¹)
combine			2-10
maize	110		5
sorghum	210		
millet	210		
cotton	620		10
groundnut	195	35	22
soya bean	100		
mung bean (three pickings)	550		
cassava	12	100	30
sweet potato	25		100
potato	5	7	18-68
sugar cane	13		
kenaf	90		
jute	90		
tobacco	800		
transport from field to farm:			
rice	42	30	
wheat		28-30	
maize	42	30	
cassava	16		
sugar cane	8		
stalk disposal:			
cotton	40		2

Table 68. Labour requirement estimates for on-farm processing.

Activity	Manual ($h\ t^{-1}$)	Draught animal ($h\ ha^{-1}$)	Power equipment ($h\ ha^{-1}$)
threshing:			
rice	70 (flair)	190	10
wheat	80		11-25
maize	42		
sorghum	75		
millet	75		
soya bean	200		
mung bean	80		

Table 68. (continued)

Activity	Manual (h t^{-1})	Draught animal (h ha^{-1})	Power equipment (h ha^{-1})
winnowing:			
rice	1-5		
picking:			
groundnut	185		
shelling:			
groundnut	165		
grading, baling:			
cotton	570		
tobacco	1000		
retting:			
kenaf	80		
jute	80		
cleaning, drying:			
kenaf	350		
jute	350		
stringing, airing, curing:			
tobacco	500 (sun curing)		
tobacco	800 (flue curing)		

6.1.5 The choice between alternative applications of human activity

In general, a farmer cultivates more than one crop during the cropping season, to spread risks or to satisfy his subsistence needs. In such a situation, the farmer may have to perform various operations at the same moment, e.g. land preparation for one crop and controlling weeds in another. If insufficient labour is available to carry out both jobs at the same time, a choice has to be made between the two activities: either to abandon cultivation of one crop, with the consequence of no yield at all, or to be content with a lower yield of the other crop because of weed competition. Especially farmers who perform their field operations completely by hand or with animal traction are confronted with this problem of timeliness; they are often forced to minimize the care given to their crops, which results in reduced yields. But also in mechanized systems of agriculture, farmers are confronted with the problem of timeliness. For maximum benefit, all agricultural operations have to be carried out at the very moment they are needed. Working conditions such as the weather, or the state of the soil, or unavailability of labour or equipment, may force a deferment of the required operation, which always results in a yield reduction.

To avoid as much as possible interference between various agricultural activities, a farmer makes his choice among the different crops suited to the specific environment, to obtain a crop mix that meets his labour availability in the course of time. This is done by determining for each crop a crop activity calendar, and calculating the labour demand for each period of the year. Then the cropping pattern is chosen for which the labour demand is distributed as evenly as possible over the year. In no case may labour peaks occur to such an extent that certain operations are impossible. This procedure is explained in more detail in Section 6.2.

6.1.6 Crop activity calendar and labour requirements for transplanted rice, maize and cassava in northeastern Thailand

For this example, a location was chosen in northeastern Thailand, near Udon Thani (17°23′N, 102°48′E), for which average monthly precipitation and potential evapotranspiration are shown in Table 69.

At some time in April, the threshold of 75 mm of rainfall in 20 days is exceeded, hence land preparation for upland crops may start at the end of April or the beginning of May. As the soils in this area are rather sandy, land preparation for cassava may start even somewhat earlier, so that cassava planting can begin early in the wet season. For bunded rice, land preparation can only take place with water standing on the field. Therefore it can start

Table 69. Average monthly precipitation (mm) and potential evapotranspiration (mm) in Udon Thani (17° 23′N, 102° 48′E), northeastern Thailand.

Month	Precipitation	Potential evapotranspiration
January	8	105
February	21	120
March	41	162
April	75	172
May	219	156
June	252	139
July	224	142
August	293	133
September	293	129
October	85	135
November	8	112
December	2	98

only from the moment that precipitation exceeds potential evapotranspiration, towards the end of May.

Land preparation for rice, maize and cassava includes ploughing with a draught animal and an indigenous wooden plough, followed by harrowing once. For maize and cassava the seedbed is ready after these operations. For bunded rice, land preparation is continued by puddling, levelling and preparation or maintenance of the bunds.

Maize is seeded. Cassava is propagated by stem cuttings, which are planted in parallel rows, 1 – 1.5 m apart, with spacing within the row of 0.8 – 1.5 m, depending on the cultivar and local tradition. For bunded rice, seedlings are raised on a nursery bed and transplanted in the field when 5 – 10 weeks old, depending on field conditions. About 0.1 ha nursery bed is required to provide seedlings for one hectare of paddy field. Fertilizers may be applied at seeding or planting as a basal dressing, sometimes followed by a top dressing six weeks later.

Weeds compete with crop plants for essential growth factors, such as energy, nutrients and soil – moisture, which affects crop growth unfavourably and may result in reduced crop yields. The effects of weed competition are most detrimental during the early stages of growth, before a closed crop canopy is formed. At that time essential nutrients are irreversibly incorporated in the weeds, while the crop does not yet intercept sufficient light to prevent weed growth. This critical period lasts for cassava for about three months after planting, and for maize and rice approximately until development stage 0.4 (for more details, see Section 6.3). The duration of the pre – flowering phase is about 85 days for rice, and 75 days for maize for the varieties used at this location, hence the rice crop has to be kept weed – free for the first 34 days after transplanting, and the maize crop for the first 30 days after emergence. For rice, a biocide may be applied at heading to protect the crop against pests and diseases. During the last part of the seed filling period, rice and maize must be protected against seed predation by birds. Bird scaring is generally a child's job.

Finally, at maturity the crops are harvested. The growing season ends for upland crops at the end of September or the beginning of October, the end of the rainy season. For bunded rice, growing on heavier soils with a larger store of available water, the growing season may extend another month.

If it is assumed that all land – preparation operations are performed with animal traction and all other operations with hand labour, crop activity calendars can be constructed for the three relevant crops. For bunded rice, land – preparation activities can start only after the fields are flooded. At the end of May, rainfall exceeds potential evaporation, and water accumulates on the field. Preparations for the nursery bed start in the first ten – day period of June. The soil is ploughed twice, harrowed, puddled and levelled to prepare the seedbed before broadcast seeding. During growth of the seedlings, water supply must be assured, if necessary by lifting water onto the seedbed. After installation of the nursery bed, land – preparation for the paddy field starts in the last ten – day period of June by a first ploughing, followed by a second ploughing plus harrowing in the first ten – day period of July. In the second ten – day period of July the field is puddled, levelled and bunded. Transplanting starts in the last ten – day period of July, the seedlings being about 5 – 6 weeks old. At transplanting, fertilizer is applied broadcast, and the operations are completed during the first ten – day period of August. A top dressing of fertilizer is applied at the maximum tillering stage of the rice crop in the second ten – day period of September. At heading, in the first ten – day period of October, a biocide is applied. Bird scaring is necessary during the last ten – day period of October and the first ten – day period of November. The crop is harvested from the second ten – day period of November onwards. The labour profile of all activities with their labour requirements is shown in Table 70, the appropriate labour times being taken from Tables 61 – 68.

For activities associated with the nursery bed, one overall value is given, including all necessary operations. The harvesting operation is sub – divided into cutting, transport from field to farm compound and threshing. The la-

Table 70. Possible crop activity calendar for transplanted rice, northeastern Thailand.

Date	Activity	Labour ($h\ ha^{-1}$) Field	Nursery
1-10 June	start nursery preparations		80 (1 June – 20 July)
10-20 June	nursery		
20-30 June	nursery; first ploughing of field	28	
1-10 July	nursery; second ploughing and harrowing of field	52	
10-20 July	nursery; puddling, levelling, bunding, plastering field	83	
20-31 July	transplanting	280	
1-10 August	fertilizer application	3	
10-20 August	weed control	145	
20-31 August			
1-10 Sept.	fertilizer topdressing	3	
10-30 Sept.			
1-10 Oct.	biocide application	25	
10-20 Oct.			
20-31 Oct.	bird scaring	63	
1-10 Nov.	bird scaring	63	
10-20 Nov.	cutting (sickle)	238[a]	
	transport	105[a]	
	threshing	175[a]	

[a] Assumed yield 2500 kg ha^{-1}. Land preparation by animal traction, other operations by manual labour, bird scaring by children.

bour requirements for harvesting activities are expressed per unit weight, thus the time requirement per hectare can only be calculated if the yield is known. For the case presented in Table 70, a grain yield of 2500 kg ha^{-1} may be assumed.

Exercise 79

Construct crop – activity calendars for maize and cassava grown at Udon Thani. Land preparation activities are carried out by animal traction. Unless specified otherwise, all other activities are performed by manual labour.
Land preparation for both crops requires ploughing twice and harrowing once. Fertilizer is applied at sowing or planting only.
For maize, start land – preparation in the second ten – day period of May. Hand weeding is done once, one month after sowing. Assume a total growing period of 110 days and a grain yield of 3000 kg ha^{-1}.
For cassava, start land preparation in the first ten – day period of April. Mechanical weeding is done once, 20 days after planting, manual weeding is done once, 20 days later, and hilling up is carried out once, 30 days after the manual weeding. Start harvesting the cassava in the first ten – day period of November, assuming a yield of 15 t ha^{-1} of fresh roots.

Exercise 80
Calculate for each of the three crops, cultivated according to Table 70 and Exercise 79, the total number of labour hours per hectare.

6.2 Low – input farming

H. Schouten

6.2.1 Introduction

Up to this point in the monograph only those production situations have been presented where crop growth was determined by a limited number of well – defined factors that could be at sub – optimal levels. The hierarchical models presented describe methods for a quantitative estimation of possible yield levels under well – defined conditions. In agricultural systems where land and labour are the predominant inputs, the conditions are less well – defined. In such low – input systems it is practically impossible to estimate yield in similar ways, because of lack of knowledge. A general treatment therefore must proceed along different lines.

A main characteristic of low – input farming systems is the emphasis on production of basic needs, in particular food, by a farm family using its own labour, possibly supplemented by animal traction. For a family to subsist, the yield of the agricultural land has to be sufficiently large. Therefore in many low – input agricultural systems, relations can be recognized between the size of a family, the number of draught animals, the soil fertility, the yield and the cultivated acreage. Some of these relations will be discussed in this section.

6.2.2 Basic needs

In most low – input farming systems a family is primarily growing crops to satisfy its basic needs. Today it rarely occurs that all basic needs for food, clothing, shelter, heating and social activities are produced or gathered within the framework of small communities. Subsistence farming refers, therefore, in general to the situation where self – sufficiency in food, fuel and shelter exists, and if possible part of the harvest is marketed to enable the purchase of other necessities: clothing, education, health care, to mention a few. As the emphasis is on food subsistence, the family does not aim at yields that are as high as possible under the prevailing environmental conditions, but rather at maintaining a production level that is high enough to cover food needs and keeps the risk of crop failure to a minimum. In that situation, the first question that arises is how much food is required for the family, and the second one, how it can be produced.

How much food is required depends on the size of the family and its composition in terms of age, sex and body weight, and the activities in which it is engaged (Figure 56). Although minerals, vitamins and proteins are indispensable, they constitute only a minor part of the total food requirements. Most

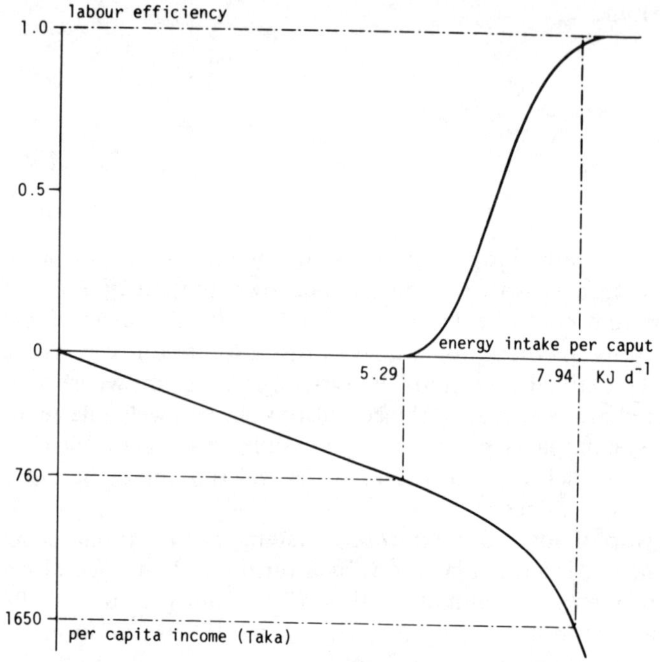

Figure 56. The relation between income, energy intake and labour efficiency. Estimates based on the Bangladesh Household survey; adapted from Stolwijk (1983).

food norms are based therefore on energy demand, assuming that if the energy demand is met shortages of minerals, vitamins, proteins and other nutrients will not occur. The energy requirement is usually expressed in Joules per kilogram of body weight, to account for differences in age and sex. For an average family, consisting of three adults and three children, with an average energy requirement of 8.4 MJ per day, the annual energy requirement is about 18.5 GJ. The energy content of rough rice at a moisture content of 0.1 kg kg^{-1} is 15.5 MJ per kg. If this is the main food, this family of six members needs about 1200 kg yr^{-1} for consumption. Allowing for preparation losses of 10%, pounding losses of 20% and harvest and storage losses of 15%, 1950 kg of paddy has to be produced. If it is assumed that an excess of 30% has to be produced to meet other basic needs and to buffer against the risk of crop failure, the total production requirement will be 2500 kg of paddy per family per year. Whether this is sufficient, depends on market conditions and on what basic needs are considered.

The production target of 2500 kg of paddy per year in a low – input system can be achieved with varying areas of land. On rich volcanic soils or on alluvial soils that receive an annual supply of nutrients through silt in irrigation water, that yield may be harvested from less than one hectare. On most soils, however, yields will be much lower, consequently a farmer has to cultivate more than one hectare to reach that production. It should also be taken into account that the necessary sowing seed, about 60 kg ha^{-1}, has to be withheld for the next year.This comprises only 2.4 % of the yield at a yield level of 2500 kg ha^{-1}, but already 24 % at a level of 250 kg ha^{-1}. Therefore, at low yields a considerable area is necessary to meet the production target. In Figure 57 the relation between yield per area and total area under cultivation is shown for a production target of 2500 kg of paddy.

Exercise 81

The tillering potential of wheat is smaller than that of rice , so that for that crop about 150 kg ha^{-1} sowing seed is needed. Construct a similar graph as the one in Figure 57 for wheat, assuming a production target of 2500 kg per family per year.

If cassava is the staple food, determine the yield that has to be at least maintained to sustain the family (the energy content of the fresh cassava root is 6.0 MJ kg^{-1} and the non – edible fraction is 0.2 kg kg^{-1}).

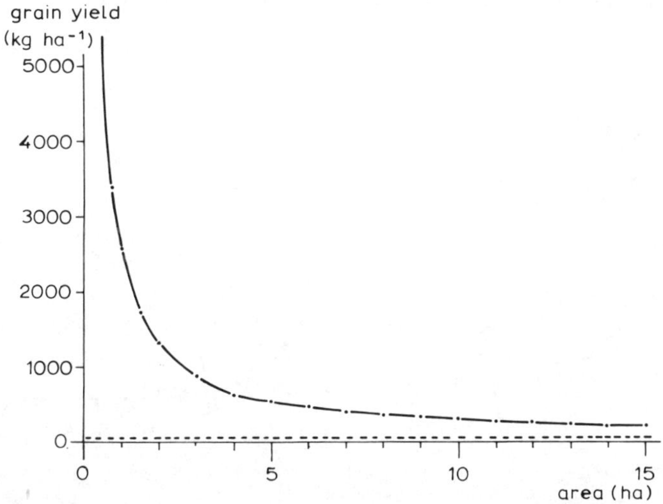

Figure 57. Grain yield required to obtain a production target of 2500 kg paddy from varying areas of crop land. The dotted line represents the amount of sowing seed (60 kg ha^{-1}) that has to be preserved.

A family practicing low – input farming can, however, not extend the cultivated area to an unlimited extent, even if enough land is available, because the maximum area that can be cultivated depends on the availability of labour during periods of peak demand. The required yield level to sustain the family is thus determined by both the cultivable acreage and the total production target. The extent to which actual yields in some low – input agricultural systems differ from these minimum yields will be considered later in this section for some agricultural systems.

6.2.3 Equilibrium yields in agricultural systems

A crop production system in a certain environment submitted to the same cultural practices year after year will eventually reach a state of equilibrium, as is illustrated in Figure 58 for yields of rye in the period from 1880 – 1960. The data are derived from a diluvial, loamy sand soil in Germany receiving an annual precipitation of 500 mm. The year – to – year fluctuations are large, but the trend represented by the eye – fitted curve is evident. That curve could well be described by an exponentially decreasing function of time. However, such a descriptive approach does not explain the dynamic characteristics of equilibrium. What will happen if cultural practices change? How long will it take to reach a new state of equilibrium? What will be the associated yields? Explaining the dynamic properties of equilibrium exhaustively, especially in the context of low – input agriculture, is practically impossible, because too many factors may be involved.

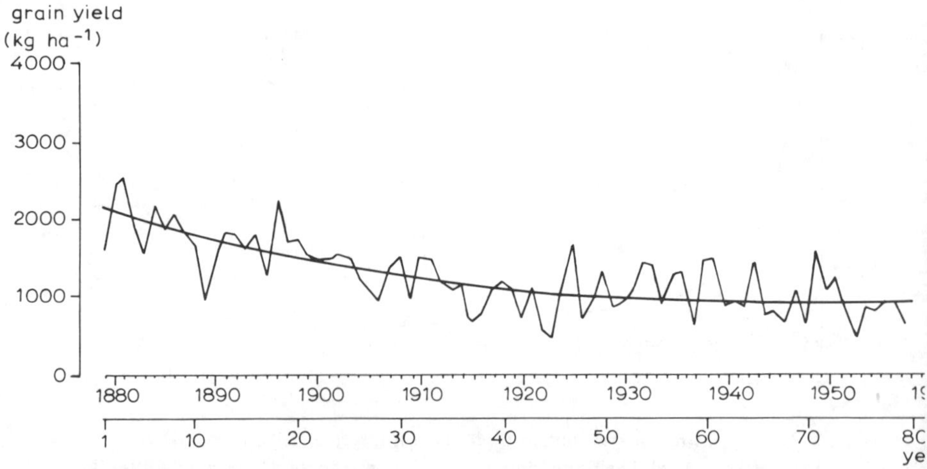

Figure 58. Yield of rye from unfertilized land (adapted from Müller and Reiher, 1966).

Exercise 82
The continuously decreasing line in Figure 58 can be described by an exponential function (Exercise 10) of the form $Y_t = (Y_0 - Y_e)e^{-kt} + Y_e$
What is the meaning of the symbols?
Make rough estimates of Y_0, Y_e and k using the data of Figure 58.

However, simple and general concepts may help to understand the dynamics of the system. One of the earliest concepts was developed by von Wulffen (1823). Although his theory is more than 150 years old it is still worth considering. At that time very little was known about chemical and physical processes involved in plant growth. Therefore the author considered crop yield as a function of the 'Reichtum' (literally: richness, fertility) of the soil which was expressed in the same units as crop yield. A fertility of 20000 kg of rye per hectare means that in total sufficient nutrients are available to produce 20000 kg of this crop from one hectare in the course of time. A single crop cannot extract all fertility in one season. Therefore the 'Tätigkeit' (literally: activity, availability coefficient) was defined as the proportion of the total fertility transferred to crop yield in one cycle. Hypothetically, with an initial fertility, R_n, of 10000 kg of paddy and an activity, T_c, of 0.1, a yield, Y_n, of 1000 kg of paddy would be produced the first year and the fertility next year, R_{n+1}, would be 9000 kg paddy. This can be expressed as:

$$Y_n = R_n \cdot T_c \qquad (95)$$

$$R_{n+1} = R_n - R_n \cdot T_c \qquad (96)$$

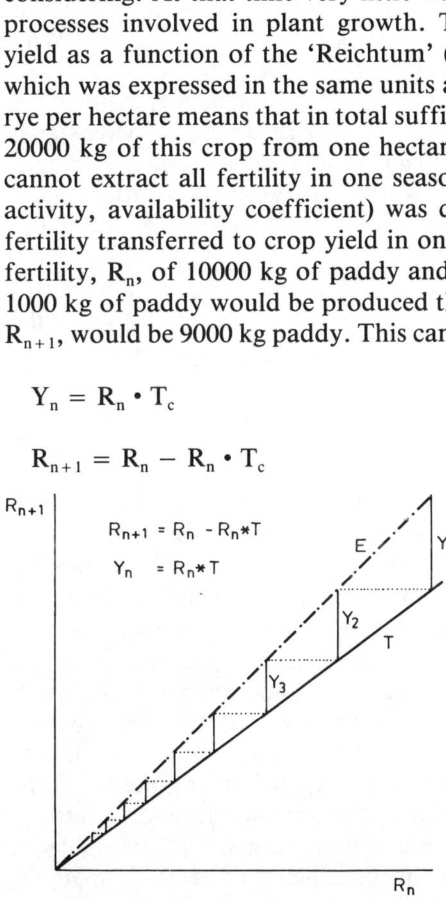

Figure 59. Relation between 'Reichtum' (R_x) in successive years, Tätigkeit (T) and yields (Y_x) in the absence of enrichments.

These equations are graphically presented in Figure 59. The diagonal E represents the points at which R_n equals R_{n+1}. The line T has a slope of $(1 - T_c)$, so that the vertical distance between both lines at any point represents the yield in a given year. Each year, R_n decreases by the fraction $(1 - T_c)$, resulting in a decay to zero in an exponential way.
In practice, however, R_n increases concurrently by weathering, nitrogen fixation, manuring and so on. As a result, R_n and yield will converge towards some non – zero equilibrium value. Fertilizer application serves the purpose of increasing R_n. The activity, T_c, can be manipulated by a variety of cultivation practices, e.g. improving soil aeration, reducing acidity and controlling weeds, pests and diseases.

In the absence of chemical fertilizers, von Wulffen expressed the quantity of manure used in terms of rye equivalents fed to cattle producing the manure and characterized its efficiency by the 'Gattung' (manure coefficient). When the manure obtained by feeding the harvest from a field to man and cattle is returned totally to that field, the manure coefficient equals one, assuming no other additions to the fertility, if the yield of that field neither increases nor decreases. Otherwise, the manure coefficient will be higher than one in some cases (e.g. leguminous crops) and lower when losses occur. The latter is generally the rule. Extended for enrichments, Equation (95) and (96) transform into:

$$Y_n = (R_n + I_r) \cdot T_c \tag{97}$$

$$R_{n+1} = (R_n + I_r) \cdot (1 - T_c) \tag{98}$$

$$I_r = G_a \cdot G_e \tag{99}$$

Here, the newly introduced I_r stands for increase in fertility, G_e for grain equivalents of manure or other sources of fertility such as weathering, and G_a for the manure coefficient. Both yield and R_n now approach equilibrium values greater than zero, as is illustrated in Figure 60. The two lines E′ and T′ are obtained by shifting the original lines E and T in Figure 59 over a distance I_r away from the origin. Now each year the yield is equal to the vertical distance between lines E′ and T′ and the net annual change in R_n to the vertical distance between lines E′ and T′. In the low range this net change is positive; it is negative in the high range. Under otherwise identical conditions, the fertility moves towards the equilibrium point, 'Beharrungspunkt' according to von Wulffen, where yields equal I_r, the annual addition to fertility. Yields could now be controlled by the rate of manuring and by manipulation of T_c and G_a via cultivation practices. In the nineteenth century, many attempts were made to quantify the theory by means of careful book – keeping of yields and manuring. However, the results were disappointing in the long

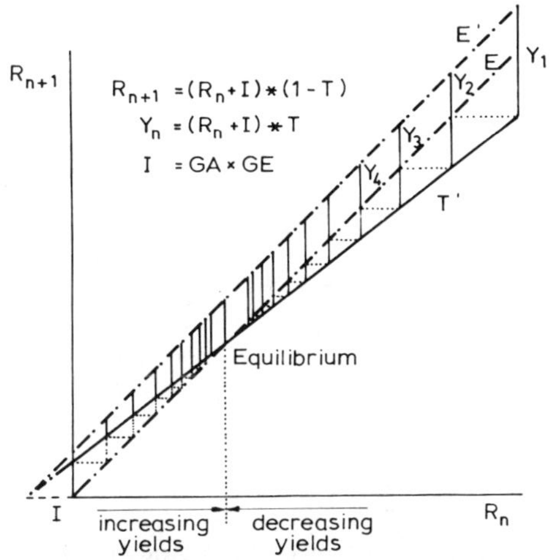

Figure 60. Relation between 'Reichtum' (R_x) in successive years, Tätigkeit (T') and yield (Y_x) with a yearly increase of 'Reichtum' (I).

run, partly because the yields fluctuated too much from year to year, as is illustrated in Figure 58 for rye.

Exercise 83
What is the fertility at the equilibrium point when the activity is 0.09 and the yearly increase in 'Reichtum' amounts to 1200 rye equivalents per hectare?
What will be the annual yield at equilibrium?
What will be the yields in successive years when the activity is increased to 0.16?

6.2.4 Shifting cultivation

Although quantification is difficult, the theory, presented in the preceding subsection, helps to understand agricultural systems. An example in this context concerns shifting cultivation, broadly defined here as any system in which food is produced for a limited period of time from an area of land, after which that area is abandoned temporarily and another piece of land is cultivated (Greenland, 1974). If population density does not impose restrictions on the availability of land, shifting cultivation practices may be succesfully applied to obtain stable food supplies. Under shifting cultivation, the fertility accumulated during many years of rest is depleted by crop yields within a few years. The land is abandoned as soon as yields decline to a level where crop-

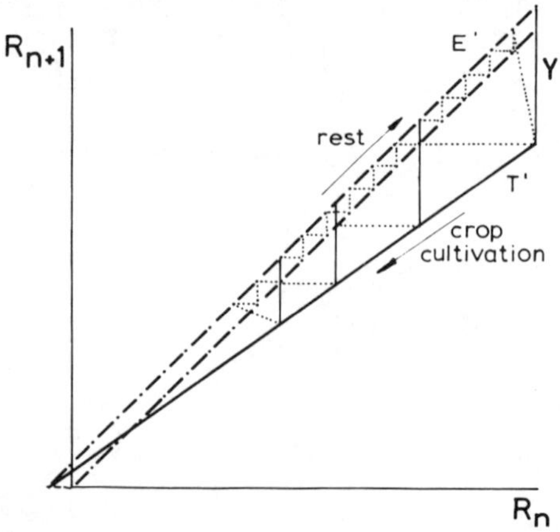

Figure 61. Relation between fertility (R_x) in successive years, activity (T′) and yield (Y) under shifting cultivation presented following von Wulffens theory.

ping is no longer worth the effort. The accumulation of nutrients will then start again by weathering and nitrogen fixation in the semi – natural vegetation.

This course of action is graphically depicted in Figure 61, following the principles outlined in the preceding subsection. There is a small but consistent annual increase in fertility along the line E′ during the years of recovery. After clearing and planting, this accumulated fertility is made available by an assumed activity of 0.1, resulting in its rapid decline along line T. After about four harvests the yield is so low that further cultivation is considered not to be worth the trouble. The land is then abandoned and the cycle starts over again. Figure 62 shows the course of fertility and associated yields with time. In this example, a saw – tooth pattern, with a period of 17 years, is established.

Along these lines it is easy to illustrate what will happen if, for instance, the recovery period, possibly due to increasing need for food, is reduced. The initial fertility at the start of the cropping period will then be lower, resulting in smaller and possibly fewer harvests. Yields may even be too low to justify the effort of any cultivation at all. In principle, it may be possible to overcome an inferior fertility by increasing the activity by better cultivation practices. This assumes, however, that subsistence farmers have not developed good farming systems; in general this assumption is wrong. Moreover, increasing the annual uptake results in a more rapid decrease in fertility, the purpose of which is highly questionable.

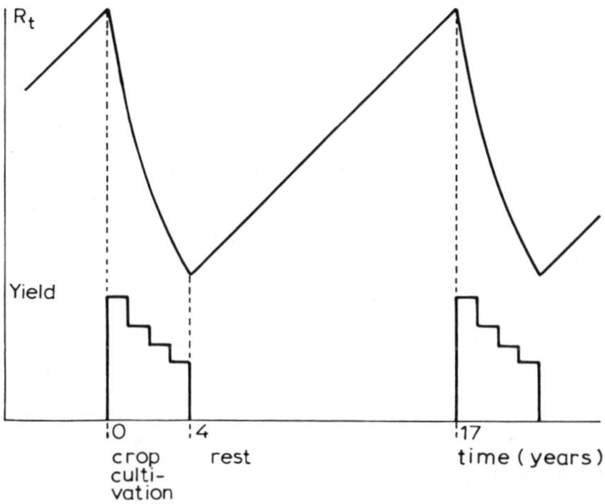

Figure 62. Decline and recovery of fertility and the course of associated crop yields in time under shifting cultivation.

Exercise 84

Calculate the annual levels of fertility and the yield in the case that, due to weathering, the first is increased by 120 kg ha^{-1} each year and for the following situations:

Activity	Fallow period years	Number of harvests
0.08	15	4
0.16	15	4
0.08	10	4
0.16	10	4
0.08	15	3
0.16	15	3
0.08	10	3
0.16	10	3

What are the average yields during the 3 or 4 year cultivation periods; what are they during the whole cycle of 10 or 15 years? In what respect are the 15 years recovery/3 years cultivation schedules superior compared to all others?

Shifting cultivation is a good agricultural system for as long as it lasts. However, with increasing man/land ratio the period of recovery decreases, so that ultimately the yield will decline to a level dictated by the annual inputs of nutrients into the system. These are often so low that they cannot sustain the efforts of farming. Under such conditions other measures should be taken that maintain fertility at a satisfactory level and allow permanent cropping.

One of the possibilities is the exploitation of areas not suitable for arable cropping by keeping animals as part of the agricultural system. From large areas of pasture land the manure can be collected and concentrated on crop land. In the long run, crop yields will be proportional to the amount and quality of manure collected. Because the system is open for nitrogen much more than for the minerals, which remain in circulation, the arable land in such systems tends to become nitrogen deficient. The annual supply of nitrogen to the grassland due to rain and some nitrogen fixation may be about 10 kg ha^{-1}. Even with proper grazing and with proper handling of the manure not more than one – quarter of it is actually available for the arable crop. For example, it may be assumed that 80% of the nitrogen is taken up by the edible portion of the grass, 80% of the grass is grazed off, 20% of the nitrogen is lost by excretion in the field, another 20% during storage in the manure and the recovery in arable crops is 50% or less. Thus at least 4 – 5 hectares of pasture are necessary to collect sufficient manure to increase the nitrogen supply to the arable crop by 10 kg ha^{-1} and to double in this way the grain yield to a level of about 1500 kg ha^{-1} (Section 4.1).

6.2.5 Paddy cultivation in monoculture

In the middle of the nineteenth century the delta of the Irrawady river in lower Burma was almost unpopulated. The few farmers then cultivating the land in this area produced all of the family's food supply, mainly rice, by the efforts of the various members of the family. Birma traded rice with Europe in exchange for textiles and other Western commodities, but the quantities exported were limited. The opening of the Suez Canal in 1859 provided a shorter route to Europe. This made it much more profitable to grow rice for export, so that many more farmers settled here to reclaim the land for paddy cultivation. The land is very suitable for paddy, and because of the uniformly heavy rainfall during a fairly reliable monsoon season, there is little danger of crop failure by lack of water. For six months, rainfall exceeds potential evapotranspiration (Figure 63), so that about three months are available for land – preparation and transplanting, the most labour intensive activities. With bullocks or water buffaloes these activities require about 400 h ha^{-1} (Table 70), depending on the intensity of puddling and levelling. If it is assumed that for a family of six the labour availability is 3 man – days per family per day and that there are 25 eight hour working days per month, 600 working hours per month are available. Such a family could thus cultivate 600/400 or 1.5 ha per month. If, as is the case in lower Burma, the transplanting period may be extended to three months, the family may be able to cultivate 4.5 ha.

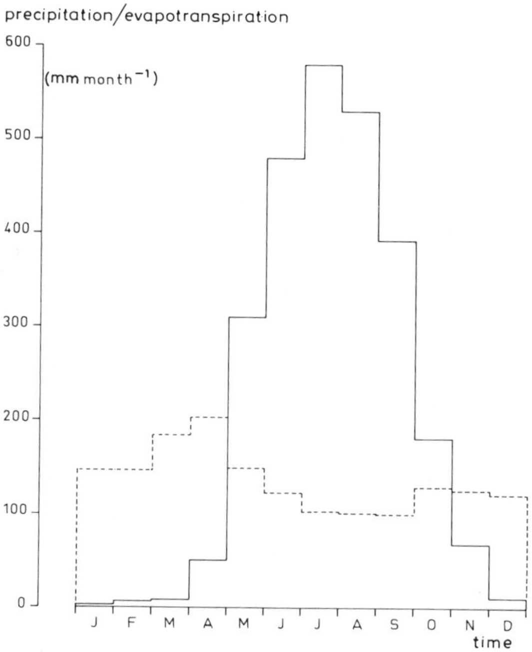

Figure 63. Average monthly precipitation and potential evapotranspiration (dotted line) at Rangoon, Burma.

According to Andrus (1948), this is close to the farm size in lower Burma of a family without hired help. The family of six needs about 2000 kg of paddy per year for consumption (Subsection 6.2.2) and, according to Andrus, it sells about half of the paddy to cover needs other than food. Hence, paddy yields have to be close to 1000 kg ha^{-1}. This appears to be the yield level that may indeed be maintained in many parts of the delta without the use of external inputs (chemical fertilizers). Because in this system a substantial part of the production is marketed, it is not an example of a pure subsistence system. Nevertheless, it illustrates very well how agricultural systems are constrained by labour availability on the one hand, and fertility of the soil on the other.

Further east, the rainy season is shorter and less reliable, resulting in a shorter transplanting period for rice and hence a smaller acreage that can be handled. Northeastern Thailand may be taken as an example. In this region, rainfall exceeds potential evapotranspiration for five months only (Figure 64) and the transplanting period is restricted to one month. This means that the activities related to transplanting, requiring 400 man – hours per hectare, have to be performed in a single month. Assuming a family size of six members, again providing a labour capacity of 600 man – hours per month, at most 1.5 ha of paddy can be grown. Given the subsistence requirement for a six – mem-

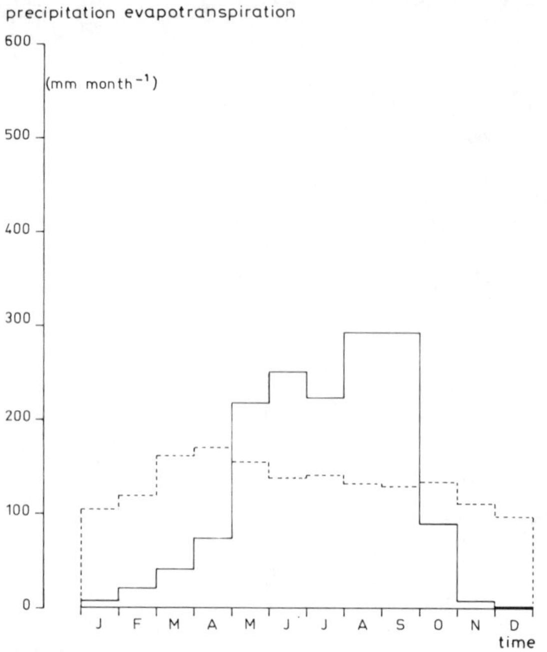

Figure 64. Average monthly precipitation and potential evapotranspiration (dotted line) at Udon Thani, N.E.-Thailand.

ber family of 2500 kg paddy (Subsection 6.2.2), the yield must exceed 1670 kg ha^{-1}. Compared with the actual yields of 1350 kg in northeastern Thailand, this figure is high. It shows that farmers there cannot be self – supporting by paddy cultivation alone under low – input agriculture. On the other hand, much labour remains idle at other times of the year that could be used for other production activities.

6.2.6 A crop mix example for northeastern Thailand

As shown in the preceding subsection, rice monoculture does not provide the subsistence requirements in northeastern Thailand. In that situation the only possibility is the introduction of other production activities with a utilization pattern for the constraining resource that differs from that for rice. Because the only resources used in low – input agriculture are land and labour, the emphasis is placed on making good use of the resource requirements of different crops. The labour demand of a crop in a certain region can be visualized in a labour profile as in Figure 65 for rice in northeastern Thailand. This picture shows that the area under paddy cultivation is predominantly limited by the labour requirements for levelling, puddling and transplanting,

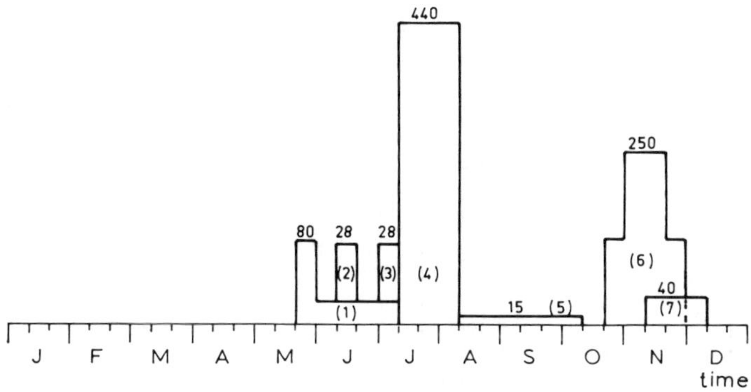

Figure 65. Labour film for paddy cultivation in northeastern Thailand using draught animals; labour requirement in manhours per hectare. (1) seedbed preparation and maintenance. (2) first ploughing. (3) second ploughing. (4) transplanting and related activities. (5) crop maintenance. (6) bird scaring. (7) harvesting.

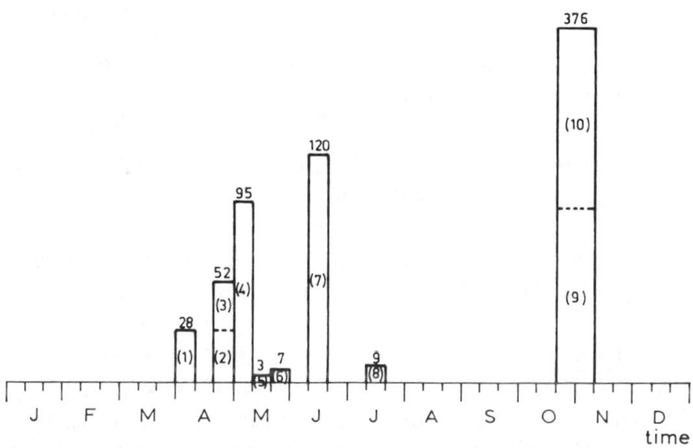

Figure 66. Labour profile for kenaf cultivation in northeastern Thailand using draught animals; labour requirement expressed in manhours per hectare. (1) clearing. (2) first ploughing. (3) second ploughing. (4) harrowing. (5) sowing. (6) fertilizing. (7) cutting (at 1500 kg ha^{-1}). (8) bundling (id.). (9) soaking (id.). (10) handling retted kenaf.

which have to be performed within a single month. Only other crops that do not compete for labour at the same time, can be grown.

Kenaf, a fibre crop, is such a complementary crop, as shown by the labour profile in Figure 66. Comparison of both labour profiles shows that there is practically no concurrent demand on labour at any time. However, kenaf production is severely limited by the heavy labour requirement for retting and associated activities. A yield of 800 kg retted and baled kenaf per hectare

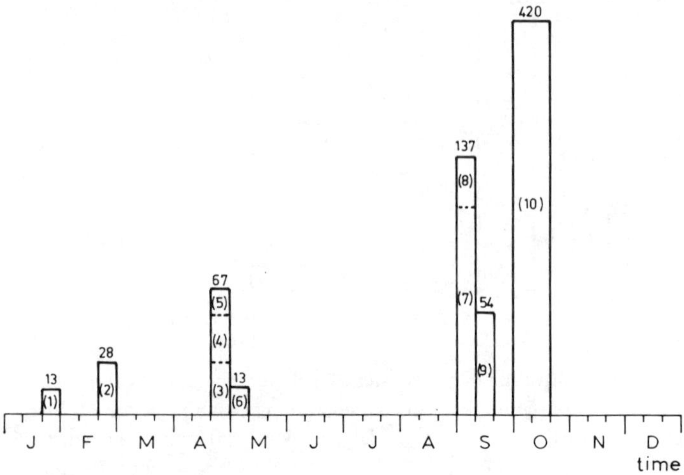

Figure 67. Labour profile for cassava cultivation in northeastern Thailand using draught animals for land preparation, cultivation and hilling-up; all other activities by hand labour; labour requirement expressed in manhours per hectare. (1) first ploughing. (2) second ploughing. (3) harrowing. (4) planting. (5) fertilization. (6) cultivation. (7) weeding. (8) earthing-up. (9) topping, lifting (at 12 t ha^{-1}). (10) transporting.

requires as much as 225 man – hours within twenty days. Because a six – member family is able to produce about 400 man – hours in such a period, at most 1.8 hectare of kenaf can be grown. Since the price of 1 kg kenaf is about equal to the price of 1.5 – 2.5 kg paddy, an equivalent of about 3000 kg paddy can be grown in this way. This is sufficient to fill the gap between actual and required production.

Still another crop could be introduced to optimize labour use. With respect to labour demand, the next most favourable crop is cassava. Only hilling – up (Figure 67) coincides with the transplanting of paddy, but this requires very little labour. The fairly heavy labour demand for transport at the end of the growth cycle refers to commercial cassava growing, where harvesting is concentrated. If cassava is grown for private consumption, harvesting is spread over a much longer period, thus reducing the labour requirement. A six – member family, living on cassava only, would require about 4700 kg fresh cassava roots annually, a yield that can be obtained from one hectare under most circumstances. From the labour profiles it appears that it is possible for one family to grow 1.3 ha of paddy, 1.1 ha of cassava and 1.8 ha of kenaf. All of the latter is then sold and, depending on price and preference, part of the paddy and the cassava could be marketed, leaving sufficient food for the family to cover subsistence needs.

6.3 Weeds, pests and diseases

F.H. Rijsdijk

6.3.1 Introduction

Factors influencing crop production can be divided into three schematic groups: yield – defining factors such as radiation; yield – limiting factors, such as the availability of water and plant nutrients; and yield – reducing factors, such as weeds, pests and diseases. Yield – defining and yield – limiting factors have been treated in previous chapters. In this section emphasis will be placed on an analysis of yield – reducing factors.

Yield reductions caused by weeds, pests and diseases are common in agricultural practice. The actual yield reduction varies with the crop, soil, climate, current weeds, pests and diseases, crop rotation, the level of control and many other factors. The effects of weeds, pests and diseases can be taken into account by multiplying the result of the preceding production estimate by a factor one minus the mean proportion of loss. The result is only a very rough estimate of the effects of weeds, pests and diseases without discriminating between production levels, climatic conditions, etc. Estimates of yield losses obtained from experiments are highly variable, as shown in Figure 68, giving the relation between the relative yield without weed control and the frequency of its occurrence for transplanted, flooded rice (Van Heemst, 1979). The expected mean, m_e, and its standard deviation, q_e, are 0.51 and 0.23, respectively. The expected mean is a crop characteristic and the high value of q_e is an expression of the variability in weed species, weed density and the variability in the crop itself. The variability in loss estimates due to pests and diseases is of the same order of magnitude. Therefore, correcting crop production estimates using this type of information yields only a rough approximation of reality and is not very satisfying. Hence a sounder method of evaluating yield losses should be developed. In this section a methodology is suggested for assessing the effects of weeds, pests, and diseases in a more detailed way by the use of simple explanatory models. On the basis of such models it may be possible to relate the impact of weeds, pests and diseases to the production level that is pursued.

6.3.2 Weed models

Damage to crops through weeds is essentially caused by the competition for radiation, water and nutrients between weeds (unwanted plants) and the crop. However, the degree of weed control in many crops in high – input farming systems, seems to be poorly related to the risk of competition. In such situa-

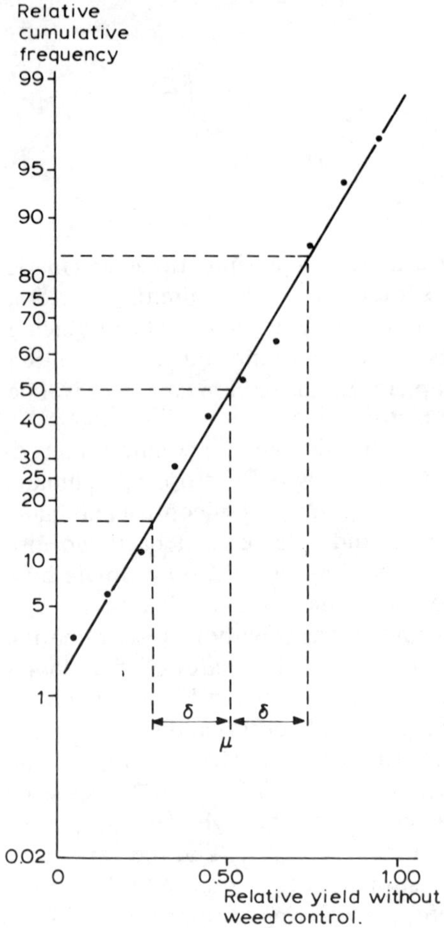

Figure 68. The relative cumulative frequency of the relative yield of transplanted flooded rice without weed control, plotted on normal probability paper.

tions other considerations are of greater importance, such as loss of quality of the harvested product, unfavourable effects during harvest and the need for weed suppression to a level below competition risk in view of crop rotation schemes. Here, only competition aspects will be treated.

Some theoretical aspects

If it is assumed that the physiological characteristics of the weeds and the crop are similar, the growth rates for weeds and crop growing in a mixture can be described by:

$$G_c = (L_c/(L_c + L_w)) \cdot G_t \text{ and } G_w = (L_w/(L_c + L_w)) \cdot G_t \qquad (100)$$

or $G_c/G_w = L_c/L_w$

where

G is growth rate (kg ha^{-1} d^{-1})
L is the leaf area index; the subscripts c, w, t refer to crop, weeds and total, respectively.

If the growth rate for both crop and weeds depends only on the total leaf area index, the ratio between L_c and L_w is maintained during the entire growth cycle. The final total dry matter production is thus distributed over crop and weeds in proportion to that ratio, still under the assumption of identical physiological characteristics. This implies that the damage of weeds to a crop can be derived directly from the ratio of the leaf area indices of weeds and crop at the onset of competition, i.e. at emergence. As the growth of seedlings follows an exponential pattern (Exercise 10), it can be described by:

$$Y_t = Y_0 \cdot e^{rt} = N_s \cdot W_0 \cdot e^{rt} \qquad (101)$$

where

Y_0 is total dry matter yield at time 0, i.e. emergence (kg ha^{-1})
Y_t is total dry matter yield at time t (kg ha^{-1})
N_s is the number of seedlings
W_0 is the average weight of an individual seedling (kg)
r is the relative growth rate (d^{-1})

Hence the relative start position of crop and weeds is defined by the number of seedlings and their weight at the start of the competition. Even under the crude assumption of identical growth characteristics, some general conclusions can be drawn from this description. Planted and transplanted crops will be less susceptible to weed competition than seeded crops because of their relative advantage in leaf development. Small-seeded crops, like sugarbeet, are more susceptible than big-seeded crops because the weight of the seedling is highly correlated with seed weight. Slow germinating species have a disadvantage in comparison to fast germinating species.

Crops and weeds

Clearly, the assumption of identical characteristics for crop and weeds does not hold in many situations. An important difference between a crop and weeds may be their maximum height and the time needed to reach that height. When species differ in height, the tallest species will have an advantage over

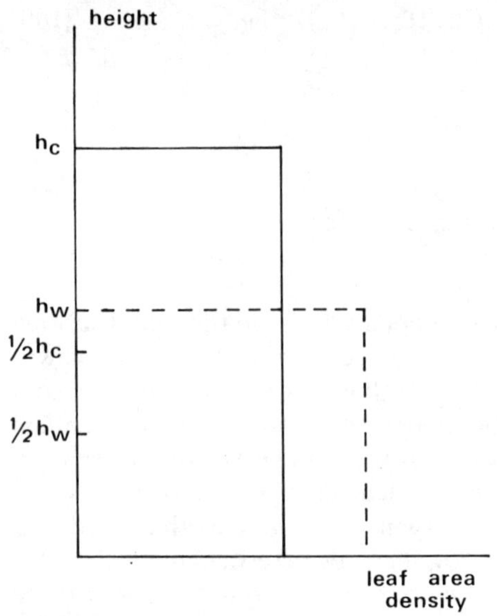

Figure 69. Schematic representation of leaf area density distribution for a mixture of two crops of different height. h_c and h_w represent the height of the crop and of the weeds, respectively.

the shorter one because of shading, even if their leaf area indices are about the same. A quantification of the effects of height differences is given by Spitters & Aerts (1983).

Figure 69 presents an example in which the weeds have reached a height H_w and the crop a height H_c. The leaves of a species are assumed to be evenly distributed with height and its growth rate to be proportional to the leaf area index and the radiation intensity at half the height, H_h, of the crop or the weed. The radiation intensity at H_h is a function of the leaf area above H_h (Section 2.1). The extinction of radiation can be described by:

$$I = I_0 \cdot e^{-k_e \cdot L} \tag{102}$$

with k_e the extinction coefficient and L the total leaf area index above the point of measurement.
The leaf area index above H_{hw}, half the height of the weeds, is:

$$L(H_{hw}) = L_w/2 + ((H_c - H_w/2)/H_c) \cdot L_c \tag{103}$$

If H_w is more than double H_c the last term has to be omitted, because there is no influence of the crop at a height H_{hw}. The leaf area index above H_{hc}, half the height of the crop, is:

$$L(H_{hc}) = L_c/2 + ((H_w - H_c/2)/H_w) \cdot L_w \quad (104)$$

The growth rates can now be described by:

$$G_c/G_W = L_c/L_W \cdot e^{(-k_e \cdot (L(H_{hc}) - L(H_{hw})))} \quad (105)$$

$$G_c + G_W = G_t \quad (106)$$

Exercise 85
Calculate the ratio of the growth rate for weeds and crop, using Equations 105 and 106, for $H_w = 1.2$ and $H_c = 0.8$ and $L_w = L_c = 1.5$ and for $H_w = 0.8$ and $H_c = 1.2$, with $k_e = 0.65$ in both cases.

The ratio between the growth rate of the crop and that of the weeds will now also vary in relation to their heights. A description of the increase in height with time is necessary to calculate the result of the competition process in terms of partitioning of total dry matter between crop and weeds. In Table 71 the equations are given to calculate the growth of crop and weeds over time. The growth conditions are assumed to be constant for the sake of simplicity. The results are given in Tables 72a and b.

6.3.3 Weeding

In almost all agricultural systems, removal of weeds by hand or by the use of herbicides is common practice. Because our main interest is crop production in developing countries, hand weeding will be treated in some detail. Before planting or drilling a new crop, the land is cleaned from weeds as part of the seedbed preparation. The crop is planted and after some time the farmer will judge the need for weeding. As competition for radiation between crop and weeds will only become significant at a total leaf area index above 1.5, weeding is supposed to take place if the total leaf area index, L_t, exceeds 1.5 and the proportion of weeds in L_t is higher than 0.2. Weeding will remove nine – tenths of the weed biomass, reducing at the same time its average height to one tenth. Tables 73a and 73b show the results of the competition when weeding is practiced for a crop with a relatively high competitive ability such as wheat and for a crop such as sugar – beet, which has a much lower competitive ability.

Table 71. Basic data and equations for calculation of competition between crops and weed populations

Basic data		
Potential daily gross CO_2 assimilation	P_{gs}	$= 300$ kg CH_2O $ha^{-1}d^{-1}$
Development stage	DVS	see Table 72
Specific leaf area	C_f	$= 20\ m^2\ kg^{-1}$
Conversion efficiency for dry matter production	E_g	$= 0.7$
Relative maintenance respiration rate	R_m	$= 0.015\ d^{-1}$
Equations to be used sequentially for each time interval		
Reduction factor for assimilation	RA	$= f(L_t)$
Relative rate of leaf dying	d_s	$= 0.\ d^{-1}$;
for DVS > 1	d_s	$= 0.02\ d^{-1}$
Fraction dry matter for leaf growth	FL	$= f$ (DVS)
Potential gross assimilation rate	PGASS	$= P_{gs}.RA$ kg $ha^{-1}d^{-1}$
Maintenance respiration	MRES	$= TDW.R_m$ kg $ha^{-1}d^{-1}$
Assimilates for increase in dry matter	ASAG	= PGASS – MRES kg $ha^{-1}d^{-1}$
Rate of dry matter increase	DMI	$=$ ASAG.E_g kg $ha^{-1}d^{-1}$
Leaf area index above ½ H_c	$L(H_{hc})$	$= L_c/2 + ((H_w - H_c/2)/H_w) \bullet L_w$ for $H_w > H_c/2$.
Leaf area index above ½ H_w	$L(H_{hw})$	$= L_w/2 + ((H_c - H_w/2)/H_c) - L_c$ for $H_c > H_w/2$.
Rate of dry matter increase of the crop	G_c	$DMI \bullet L_c \bullet e^{-0.65 \bullet L(H_{hc})} / (L_c \bullet e^{\bullet L(H_{hc})} + L_w \bullet e^{-0.65 \bullet L(H_{hw})})$ kg $ha^{-1}d^{-1}$
Rate of dry matter increase of the weeds	G_w	$= DMI - G_c$ kg $ha^{-1}d^{-1}$
Height of the crop	H_c	$= H_c + hi_c\ .\ \Delta t$ if $H_c < h_{mc}$
Death rate of the leaves of the crop	$DWLV_c$	$= q.WLV_c$ kg $ha^{-1}d^{-1}$
Weight of the leaves of the crop	WLV_c	$= WLV_c + (FL.G_c - DWLV_c)$ x Δt kg ha^{-1}
Weight other organs of the crop	WGO_c	$= WGO_c + G_c(1 - FL)$ x Δt kg h
Leaf area index of the crop	L_c	$= WLV_c\ .\ C_f\ .\ 10^{-4}$
Total dry weight of the crop	TDW_w	$= WGO_c + WLV_c$ kg ha^{-1}
Height of the weeds	H_w	$= H_w + hi_w\ .\ \Delta t$ if $H_w < h_{mw}$
Death rate of the leaves of the weeds	$DWLV_w$	$= q.WLV_w$ kg $ha^{-1}d^{-1}$
Weight of the leaves of the weeds	WLV_w	$= WLV_w + (FL.G_w - DWLV_w)$ x Δt kg ha^{-1}
Weight of other organs of the weeds	WGO_w	$= WGO_w + G_w(1 - FL)$ x Δt kg
Leaf area index of the weeds	L_w	$= WLV_w\ .\ C_f\ .\ 10^{-4}$
Total dry weight of the weeds	TDW_w	$= WGO_w + WLV_w$ kg ha^{-1}
Total leaf area index	L_t	$= L_w + L_c$

Table 72a. Competition in a 'wheat-like' crop.

	crop	weeds	
Initial height	0.05	0.05	m
Height increase	0.01	0.01	$m\ d^{-1}$
Maximum height	1.0	0.80	m
Initial weight of the leaves	50.0	50.0	$kg\ ha^{-1}$
Initial weight of other organs	50.0	50.0	$kg\ ha^{-1}$

TIME	DVS	RA	FL	PGASS	MRES	ASAG	DMI	LHHC	LHHW	GC	GW	HC	DWLVC
10	0.10	0.18	0.38	54.0	3.0	51.0	35.7	0.12	0.12	17.8	17.8	0.15	0
20	0.20	0.37	0.40	110.5	8.4	102.1	71.5	0.29	0.29	35.7	35.7	0.25	0
30	0.30	0.64	0.43	192.2	19.1	173.1	121.2	0.65	0.65	60.6	60.6	0.35	0
40	0.40	0.87	0.46	260.7	37.3	223.4	156.4	1.29	1.29	78.2	78.2	0.45	0
50	0.50	0.98	0.48	292.7	60.7	232.0	162.4	2.19	2.19	81.2	81.2	0.55	0
60	0.60	1.00	0.48	300.0	85.1	214.9	150.5	3.17	3.17	75.2	75.2	0.65	0
70	0.70	1.00	0.42	300.0	107.6	192.4	134.7	4.06	4.06	67.3	67.3	0.75	0
80	0.80	1.00	0.32	300.0	127.8	172.2	120.5	4.77	4.77	60.3	60.3	0.85	0
90	0.90	1.00	0.21	300.0	145.9	154.1	107.9	5.08	5.40	59.5	48.4	0.95	0
100	1.00	1.00	0.10	300.0	162.1	137.9	96.5	5.01	5.96	63.0	33.5	1.00	0
110	1.20	1.00	0.00	300.0	176.6	123.4	86.4	4.95	6.22	60.5	25.9	1.00	45.7
120	1.50	1.00	0.00	300.0	176.0	124.0	86.8	3.96	4.97	57.7	29.1	1.00	36.6
130	1.80	1.00	0.00	300.0	178.2	121.8	85.3	3.17	3.98	54.1	31.2	1.00	29.2
140	2.00	1.00	0.00	300.0	182.3	117.7	82.4	2.53	3.18	50.2	32.2	1.00	23.4

Table 72a. Competition in a 'wheat-like' crop.

	crop	weeds	
Initial height	0.05	0.05	m
Height increase	0.01	0.01	$m\ d^{-1}$
Maximum height	1.0	0.80	m
Initial weight of the leaves	50.0	50.0	$kg\ ha^{-1}$
Initial weight of other organs	50.0	50.0	$kg\ ha^{-1}$

TIME	WLVC	WGOC	LC	TDWC	HW	DWLVW	WLVW	WGOW	LW	TDWW	LT
10											
20	117	161	0.29	278	0.15	0	117	161	0.29	278	0.58
30	260	376	0.65	636	0.25	0	260	376	0.65	636	1.30
40	518	724	1.29	1242	0.35	0	518	724	1.29	1242	2.59
50	878	1146	2.19	2024	0.45	0	878	1146	2.19	2024	4.39
60	1267	1568	3.17	2835	0.55	0	1267	1568	3.17	2835	6.34
70	1625	1963	4.06	3588	0.65	0	1625	1963	4.06	3588	8.12
80	1907	2354	4.77	4261	0.75	0	1907	2354	4.77	4261	9.54
90	2097	2767	5.24	4864	0.80	0	2097	2767	5.24	4864	10.49
100	2222	3236	5.56	5458	0.80	0	2199	3149	5.50	5348	11.05
110	2285	3803	5.71	6088	0.80	0	2232	3451	5.58	5683	11.29
120	1828	4408	4.57	6236	0.80	44.6	1786	3710	4.46	5496	9.03
130	1462	4985	3.66	6447	0.80	35.7	1429	4001	3.57	5430	7.23
140	1170	5526	2.92	6696	0.80	28.6	1143	4313	2.86	5456	5.78
	936	6029	2.34	6965	0.80	22.9	914	4635	2.29	5549	4.63

Table 72b. Competition in a 'sugar-beet-like' crop.

	crop	weed	
Initial height	0.01	0.05	m
Height increase	0.005	0.01	$m\ d^{-1}$
Maximum height	0.3	0.8	m
Initial weight of the leaves	5.0	50.0	$kg\ ha^{-1}$
Initial weight of other organs	5.0	50.0	$kg\ ha^{-1}$

TIME	DVS	RA	FL	PGASS	MRES	ASAG	DMI	LHHC	LHHW	GC	GW	HC	DWLVC
10	0.10	0.10	0.38	29.7	1.7	28.0	19.6	0.12	0.06	1.7	17.9	0.06	0
20	0.20	0.22	0.40	67.1	4.6	62.5	43.8	0.25	0.15	3.7	40.1	0.11	0
30	0.30	0.44	0.43	133.4	11.2	122.2	85.5	0.57	0.35	6.4	79.1	0.16	0
40	0.40	0.73	0.46	220.2	24.0	196.2	137.3	1.25	0.77	8.21	29.1	0.21	0
50	0.50	0.92	0.48	276.7	44.6	232.1	162.5	2.43	1.51	6.5	156.0	0.26	0
60	0.60	1.00	0.48	300.0	69.0	231.0	161.7	3.89	2.45	3.9	157.9	0.30	0
70	0.70	1.00	0.42	300.0	93.2	206.8	144.7	5.38	3.38	2.0	142.7	0.30	0
80	0.80	1.00	0.32	300.0	114.9	185.1	129.5	6.80	4.13	1.0	128.5	0.30	0
90	0.90	1.00	0.21	300.0	134.4	165.6	115.9	7.73	4.64	0.6	115.3	0.30	0
100	1.00	1.00	0.10	300.0	151.8	148.2	103.8	8.22	4.94	0.5	103.3	0.30	0
110	1.20	1.00	0.00	300.0	167.3	132.7	92.9	8.43	5.07	0.4	92.5	0.30	3.1
120	1.50	1.00	0.00	300.0	168.6	131.4	92.0	6.75	4.06	0.6	91.4	0.30	2.5
130	1.80	1.00	0.00	300.0	172.3	127.7	89.4	5.40	3.25	0.8	88.6	0.30	2.0
140	2.00	1.00	0.00	300.0	177.6	122.4	85.7	4.32	2.60	1.1	84.6	0.30	1.6

Table 72b. Competition in a 'sugar-beet-like' crop.

	crop	weed	
Initial height	0.01	0.05	m
Height increase	0.005	0.01	$m\ d^{-1}$
Maximum height	0.3	0.8	m
Initial weight of the leaves	5.0	50.0	$kg\ ha^{-1}$
Initial weight of other organs	5.0	50.0	$kg\ ha^{-1}$

TIME	WLVC	WGOC	LC	TDWC	HW	DWLVW	WLVW	WGOW	LW	TDWW	LT
10	11	16	0.03	27	0.15	0	118	161	0.29	279	0.32
20	26	38	0.07	64	0.25	0	278	402	0.69	680	0.76
30	53	75	0.13	128	0.35	0	614	857	1.54	1471	1.67
40	91	119	0.23	210	0.45	0	1208	1554	3.02	2762	3.25
50	122	153	0.31	275	0.55	0	1957	2365	4.89	4322	5.20
60	141	173	0.35	314	0.65	0	2707	3194	6.77	5901	7.12
70	149	185	0.37	334	0.75	0	3306	4022	8.27	7328	8.64
80	153	192	0.38	345	0.80	0	3711	4902	9.28	8613	9.66
90	154	197	0.38	351	0.80	0	3953	5813	9.88	9766	10.27
100	155	201	0.39	356	0.80	0	4056	6743	10.14	10799	10.53
110	124	205	0.31	329	0.80	81.1	3246	7667	8.11	10913	8.42
120	99	211	0.25	310	0.80	64.9	2596	8581	6.49	11177	6.74
130	79	220	0.20	299	0.80	51.9	2077	9467	5.19	11544	5.39
140	63	230	0.16	293	0.80	41.5	1661	10313	4.15	11974	4.31

Exercise 86
Calculate the effect of each successive weeding on dry – matter production of the crops in Table 73 and plot the results.
Explain why the dry – matter production of a completely weed – free crop is higher than that of a crop that is weeded several times during its development.

The response of crops and weeds to sub – optimum growing conditions may differ. Crop plants consist, by selection and breeding, of populations with uniform properties, tailored to the needs of mankind. Weeds are plants that are unwanted and a population contains many species that fill the gaps ('niches') not used by the crop. Sub – optimum growing conditions for the crop, such as excess or shortage of water, lack of nutrients, low or extremely high temperatures, favour those species in a weed population that are better adapted to such conditions than the crop itself. So, as a rule, any condition that will interfere with normal crop development not only affects crop production directly, but also increases the risk of crop losses due to weeds. For example, to counteract the effects of weeds in rice cultivation, the crop is, if possible, flooded because the crop is resistant against flooding, but many weeds are not. When flooding fails, an outburst of weed development is the result.

The competition model presented here only demonstrates the principles of competition. Coupling of such models with more elaborate crop growth models can supply more quantitative information, if sufficiently accurate data on growth characteristics of weeds are included. The explanatory value of the competition principle can be tested with data summarized by van Heemst (1985). Table 74 provides facts derived from the literature on the relative yield of a number of crops without weed control and specifies the time, expressed relative to the total crop growth period, that crops should be kept weed free to avoid losses of more than 5%.

Exercise 87
Try to explain differences in crop loss without weed control and in necessary weed – free periods among the crops in Table 74, by applying information given in this section.

Table 73a. Weeding in a 'wheat-like' crop.

	crop	weed	
Initial height	0.05	0.05	m
Height increase	0.01	0.01	m d^{-1}
Maximum height	1.0	0.80	m
Initial weight of the leaves	50.0	50.0	kg ha^{-1}
Initial weight of other organs	50.0	50.0	kg ha^{-}

TIME	DVS	RA	FL	PGASS	MRES	ASAG	DMI	LHHC	LHHW	GC	GW	HC	DWLVC
10	0.10	0.18	0.38	54.0	3.0	51.0	35.7	0.12	0.12	17.8	17.8	0.15	0
20	0.20	0.37	0.40	110.5	8.4	102.1	71.5	0.29	0.29	35.7	35.7	0.25	0
30	0.30	0.64	0.43	192.2	19.1	173.1	121.2	0.65	0.65	60.6	60.6	0.35	0
40	0.40	0.68	0.46	203.2	20.5	182.7	127.9	0.65	1.29	120.0	7.9	0.45	0
50	0.50	0.90	0.48	269.8	39.7	230.2	161.1	1.34	2.38	154.7	6.4	0.55	0
60	0.60	0.99	0.48	297.9	63.8	234.1	163.9	2.27	3.71	159.8	4.1	0.65	0
70	0.70	1.00	0.42	300.0	88.4	211.6	148.1	3.22	4.94	145.5	2.6	0.75	0
80	0.80	1.00	0.32	300.0	110.6	189.4	132.6	4.03	5.84	130.7	1.9	0.85	0
90	0.90	1.00	0.21	300.0	130.5	169.5	118.6	4.57	6.35	117.0	1.6	0.95	0
100	1.00	1.00	0.10	300.0	148.3	151.7	106.2	4.90	6.59	104.7	1.4	1.00	0
110	1.20	1.00	0.00	300.0	164.2	135.8	95.0	5.06	6.44	93.5	1.5	1.00	78.9
120	1.50	1.00	0.00	300.0	166.2	133.8	93.7	4.06	4.89	91.5	2.2	1.00	63.1
130	1.80	1.00	0.00	300.0	170.4	129.6	90.7	3.25	3.91	88.4	2.3	1.00	50.5
140	2.00	1.00	0.00	300.0	176.1	123.9	86.7	2.60	3.13	84.3	2.4	1.00	40.4

Table 73a.(continued)

TIME	WLVC	WGOC	LC	TDWC	HW	DWLVW	WLVW	WGOW	LW	TDWW	LT
10	117	161	0.29	278	0.15	0	117	161	0.29	278	0.59
20	260	376	0.65	636	0.25	0	260	376	0.65	636	1.30
30	518	724	1.29	1242	0.04	0	52	72	0.13	124	1.42
40	1070	1372	2.67	2442	0.14	0	88	115	0.22	203	2.89
50	1812	2176	4.53	3988	0.24	0	119	148	0.30	267	4.83
60	2571	3015	6.43	5586	0.34	0	138	170	0.35	308	6.77
70	3182	3859	7.96	7041	0.44	0	149	185	0.37	334	8.33
80	3594	4754	8.98	8348	0.54	0	155	198	0.39	353	9.37
90	3840	5679	9.60	9519	0.64	0	158	210	0.40	369	10.00
100	3945	6621	9.86	10566	0.74	0	160	223	0.40	383	10.26
110	3156	7556	7.89	10712	0.80	3.2	128	239	0.32	367	8.21
120	2524	8471	6.31	10966	0.80	2.6	102	260	0.26	363	6.57
130	2020	9355	5.05	11375	0.80	2.0	82	284	0.20	366	5.25
140	1616	10198	4.04	11814	0.80	1.6	66	308	0.16	373	4.20

Table 73b. Weeding in a 'sugar-beet-like' crop.

	crop	weed	
Initial height	0.01	0.05	m
Height increase	0.005	0.01	$m\ d^{-1}$
Maximum height	0.3	0.8	m
Initial weight of the leaves	5.0	50.0	$kg\ ha^{-1}$
Initial weight of other organs	5.0	50.0	$kg\ ha^{-1}$

TIME	DVS	RA	FL	PGASS	MRES	ASAG	DMI	LHHC	LHHW	GC	GW	HC	DWLVC
10	0.10	0.10	0.38	29.7	1.7	28.0	19.6	0.12	0.06	1.7	17.9	0.06	0
20	0.20	0.22	0.40	67.1	4.6	62.5	43.8	0.25	0.15	3.7	40.1	0.11	0
30	0.30	0.44	0.43	133.4	11.2	122.2	85.5	0.57	0.35	6.4	79.1	0.16	0
40	0.40	0.20	0.46	60.7	4.1	56.6	39.6	0.07	0.20	19.3	20.3	0.21	0
50	0.50	0.44	0.48	131.1	10.1	121.0	84.7	0.26	0.44	42.9	41.8	0.26	0
60	0.60	0.53	0.48	159.6	12.4	147.2	103.1	0.44	0.88	95.7	7.4	0.30	0
70	0.70	0.82	0.42	246.6	27.9	218.7	153.1	1.00	1.68	144.9	8.2	0.30	0
80	0.80	0.95	0.32	284.8	50.8	234.0	163.8	1.85	2.35	155.4	8.4	0.30	0
90	0.90	1.00	0.21	300.0	75.4	224.6	157.2	2.55	2.35	145.8	11.4	0.30	0
100	1.00	1.00	0.10	300.0	99.0	201.0	140.7	3.01	1.82	122.1	18.6	0.30	0
110	1.20	1.00	0.00	300.0	120.1	179.9	125.9	3.22	0.96	95.0	30.9	0.30	46.6
120	1.50	1.00	0.00	300.0	131.5	168.5	118.0	2.59	0.17	86.8	31.2	0.30	37.3
130	1.80	0.96	0.00	288.1	143.2	144.9	101.4	2.08	0.14	80.2	21.2	0.30	29.8
140	2.00	0.92	0.00	276.1	153.6	122.5	85.7	1.67	0.11	71.1	14.6	0.30	23.9

Table 73b.(continued)

TIME	WLVC	WGOC	LC	TDWC	HW	DWLVW	WLVW	WGOW	LW	TDWW	LT
10	12	15	0.03	27	0.15	0	118	161	0.29	279	0.32
20	26	38	0.07	64	0.25	0	278	402	0.69	680	0.76
30	54	74	0.13	128	0.04	0	61	86	0.15	147	0.29
40	142	179	0.36	321	0.14	0	155	195	0.39	350	0.74
50	348	402	0.87	750	0.02	0	36	41	0.09	77	0.96
60	803	905	2.01	1708	0.12	0	70	80	0.18	150	2.18
70	1412	1745	3.53	3157	0.22	0	105	127	0.26	232	3.79
80	1901	2810	4.75	4711	0.32	0	131	185	0.33	316	5.08
90	2207	3962	5.52	6169	0.42	0	155	275	0.39	430	5.91
100	2330	5060	5.82	7390	0.52	0	174	443	0.43	617	6.26
110	1864	6011	4.66	7875	0.62	3.5	139	751	0.35	890	5.01
120	1491	6879	3.73	8370	0.72	2.8	111	1063	0.28	1174	4.01
130	1193	7681	2.98	8874	0.80	2.2	89	1275	0.22	1364	3.20
140	954	8392	2.39	9346	0.80	1.8	71	1422	0.18	1493	2.56

Table 74. Estimated end of critical period relative to total crop growth period and yield without weed control relative to yield with complete weed control for a number of agricultural crops [a].

Crop	Estimated relative end of critical period	Estimated relative yield without weed control
wheat	0.19	0.75
peas	0.21	0.70
potato	0.22	0.68
sorghum	0.26	0.61
cabbage	0.27	0.59
maize	0.27	0.59
soya bean	0.27	0.58
sweet potato	0.29	0.54
transplanted rice	0.30	0.52
sugar-cane	0.33	0.47
flax	0.35	0.42
groundnut	0.36	0.41
beans	0.36	0.41
red beet	0.36	0.40
tobacco	0.39	0.34
okra	0.41	0.31
sugar-beet	0.43	0.26
upland rice	0.44	0.25
yam	0.47	0.19
cassava	0.47	0.18
cotton	0.49	0.14
garlic	0.50	0.12
mungbean	0.56	0.00
carrots	0.56	0.00
onions	0.56	0.00

[a] Other crop-specific agricultural operations as earthing up (potato, sugar-cane), thinning (cotton, sugar-beet), transplanting (tobacco, rice) are included in determining the yield without weed control, although these treatments have effects on weed competition.

6.3.4 *Pests and diseases*

The effects of pests and diseases on crop yields vary strongly among crops and yield levels. The number of different pests and diseases is so large that a general treatment of the effects of pests and diseases is almost impossible. However, in agricultural practice the number of relevant pests and diseases at one site or in a region is limited. Because the aim is not an exhaustive description of the effects of all possible pests and diseases on crops, the causal agents are classified according to their mode of action on the crop and the susceptibility of the crop to each of these groups is defined. Adopting this approach may result in a methodology that can be used for a simple evaluation of potential and actual crop losses in relation to environment and farming practice.

Pests and diseases may be classified according to population development and according to the way in which they interact with the productivity of the crop. In the first classification, a distinction can be made between 'single interest' and 'multiple interest' pests and diseases. Single interest (monocyclic) pests and diseases are characterized by one infection cycle during the growing period of the crop. For this group of causal agents, the expected damage level depends mainly on the initial level of attack, for example seed and seedling removal by pests and diseases during a very limited period in crop development. Smuts and bunts of cereals and some one – generation insect pests belong to this group. Multiple interest (polycyclic) pests and diseases are characterized by the occurrence of more than one generation during the growing season. The damage level depends not only on the initial level of infection, but also on the ability of the causal agent to develop through repetitive life cycles to a level that affects crop production. Since the development of such pests and diseases depends, at least partly, on the crop characteristics and the course of crop development, the effects of such pests and diseases may vary considerably with production level. Important pests and diseases belonging to this group are cereal aphids, leaf blight, leaf spot diseases, rusts and mildews.

Another criterion for classifying pests and diseases is the mode of interaction with the host. Certain pests and diseases remove green tissue or whole plants without affecting the remaining plant parts or plants, except through canopy density. Examples of these are cereal leaf beetles and various soil pests that remove whole seedlings. Many other pests and diseases not only affect the infested tissue but also influence the physiology of plant parts not yet infested, for example through effects on photosynthesis and leaf ageing, such as caused by cereal aphids and many leaf diseases. Detailed evaluation of the effects of this type of infestation is only possible by taking into account crop physiology and population growth of the causal agent concurrently. Examples of such an approach are given by Rabbinge & Rijsdijk (1982). In this section, the emphasis is on a methodology for evaluating effects of polycyclic pests and diseases on crops at different production levels.

6.3.5 Dynamics of polycyclic population growth

In principle, population growth of a polycyclic organism follows an exponential pattern. The growth rate of that population is, according to differential calculus:

$$\frac{dP}{dt} = r \cdot P \qquad (107)$$

After integration the time course of the population is described by:

$$P_t = P_0 \cdot e^{rt} \qquad (108)$$

where

P_0 is the initial level of the population
t is time (d)
r is the relative growth rate (d^{-1})

As the population cannot expand infinitely, a maximum level of the population or a carrying capacity has to be defined. The actual growth rate of the population is influenced by this maximum level, not only at the moment the maximum level P_m is reached, but long before. This may be taken into account if it is assumed that the growth rate of the population is proportional to the fraction of the host that is not yet infected. This inhibition mechanism is explained by non-effective double infections in the case of fungi and by intra-specific inhibition mechanisms in insect populations. The rate of growth of the population is then:

$$\frac{dP}{dt} = r \cdot P \cdot (1 - P/P_m) \qquad (109)$$

The population size at time t follows from integration of Equation 109:

$$P_t = \frac{P_m}{1 + K \cdot e^{-rt}} \; ; \; K = \frac{P_m - P_0}{P_0} \qquad (110)$$

Such a population growth model is called a logistic model. The logistic growth model describes population growth for insects and pathogenic fungi only approximately, because in reality delays occur such as latent periods for fungi and non-reproductive periods as larvae and pupae stages in insect populations. These delays are not explicitely defined in the equations. Introducing those delays, too, leads to numerical models of a more complex nature. Detailed information on crop, environment and pests and diseases is necessary for such models. Nevertheless, logistic models may be used in evaluating effects of pests and diseases on productivity. For that purpose the relative growth rate, r, of the population should be defined not as a constant throughout the growing cycle, but as a function of a crop characteristic such as development stage, which reflects both crop physiology and past environmental conditions, and the resistance of the host. The calculation of the population dynamics should be carried out for sufficiently small time intervals to take account of the effects of changes in its parameter values. The values for the parameters r and P_m as a function of crop development can be obtained from more complex models or from experiments where pest or disease levels are recorded sequentially in combination with crop characteristics.

Exercise 88

Calculate the r values during crop development from disease readings and crop characteristics as given in Table 75. Calculate the growth rate of the population at the time of disease readings.

Coupling of calculations on pathogen population development and its consequences for crop production with calculations of crop production itself will be demonstrated for a cereal rust on wheat. The calculation procedure used in Section 3.4 for Production Situation 2 will be used with some simplification in parameters to avoid excessive use of calculus. The complete calculation procedure is summarized in Table 76. The epidemic of cereal rust takes place by colonization of the leaf tissue by the fungus. The level of infection is expressed in kilograms of living infected leaves per hectare.

In the model the amount of infected leaf tissue is thus calculated as a separate state variable, Y_i. The rate of change of this variable is:

$$\frac{dY_i}{dt} = r \cdot Y_i \cdot (1 - Y_i/Y_m) - Y_d \qquad (111)$$

where

Y_m is the total weight of living leaf tissue (kg ha^{-1})
r is the relative growth rate of the fungus population as a function of the development stage of the crop (see Table 77) (d^{-1})
Y_d is the death rate of the diseased leaf tissue (kg ha^{-1} d^{-1})

Table 75. Disease readings and crop development of an epidemic of yellow rust on wheat.

Time	DVS	Severity (P_i/P_m)
40	0.1	0.00001
70	0.3	0.0002
90	0.5	0.005
110	0.8	0.08
125	1.0	0.2
135	1.3	0.5

Table 76. Parameters and equations for the combined crop-disease model

Potential daily gross CO_2 assimilation	P_{gs}	= 300 kg CH_2O/ha
Development stage of the crop	DVS	= f (TIME)
Specific leaf area	SLA	= 25 $m^2\ kg^{-1}$
Potential evapotranspiration	ETO	= f (TIME)
Total water supply	IM	= f (TIME)
Soil water depletion factor	P	= f (Tm) see Table 20
Soil moisture content of air dry soil	SMa	= 0.03
Soil moisture content at fieldcapacity	SMfc	= 0.225
Soil moisture content at wilting point	SMw	= 0.09
Potential rooting depth	Drm	= 1500 m
Growth rate of the roots	Rr	= 100 mm/decade
Conversion efficiency for dry matter production	Eg	= 0.7
Relative rate of disease senescence	Qd	= .025
Relative maintenance respiration rate	Rm	= 0.015 day^{-1}
Ratio between dying of diseased and healthy leaves	Fd	= 3.
Equations to be calculated sequentially for each time interval		
1. Reduction factor for assimilation	RA	= f(L) see Table 11
2. Proportionality factor for disease severity	Fs	= f(PROPD)
3. Relative rate of leaf senescence	Ds	= 0.0; for DVS > 1 Ds = 0.02/day
4. Fraction dry matter for leaf growth	FL	= f(DVS) see Table 12
5. Fraction dry matter for root growth	FR	= f(DVS) see Table 12
6. Fraction dry matter for stem growth	FS	= f(DVS) see Table 12
7. Fraction dry matter for grain growth	FG	= f(DVS) see Table 12
8. Potential gross assimilation	PGASS	= Pgs.RA.Δt kg/ha
9. Maximum evaporation from soil surface	Em	= ETo.(1 – RA) mm/decade
10. Maximum transpiration	Tm	= ETo.RA mm/decade
11. Actual evaporation	Ea	= Em.(SMr – SMa)/(SMfc – SMa) mm/d
12. Critical soil moisture content	SMcr	= (1 – P).(SMfc – SMw) + SMw
13. Actual transpiration	T	= Tm; for SMcr > SMr
	T	Tm.(SMr – SMw)/(SMcr – SMw)
14. Rooting depth	RD	= RD + Rr, for RD < Drm
15. Moisture added to rooted zone by root growth	dMr	= Wnr.Rr/(Drm – RD) mm/decade
16. Change of moisture in rooted part of the soil	DWr	= IM + dMr – Ea – T mm/decade
17. Soil moisture in rooted zone	Wr	= Wr + Dwr.Δt mm
18. Soil moisture content of rooted zone	SMr	= Wr/RD
19. Amount of moisture in non-rooted zone	Wnr	= Wnr – dMr.Δt mm
20. Actual gross assimilation	GASS	= PGASS.T/Tm kg/ha
21. Maintenance respiration	MRES	= TDW.Rm.Δt kg/ha
22. Assimilation for increase of dry matter	ASAG	= GASS – MRES kg/ha
23. Total relative rate of dying of leaves	Q	= (Dw.(1 – T/Tm) + Ds) . Fs ; for Q < Qd.PROPD Q = Qd.PROPI
24. Dry matter increase	DMI	= ASAG . Eg . (1 – PROPD . (1 – RA)) k
25. Death rate of the leaves of the crop	DWLV	= WLV . Q . Δt kg/ha
26. Weight of the leaves of the crop	WLV	= WLV + FL.DMI – DWLV kg/ha
27. Leaf area index of the crop	L	= WLV . SLA . 0.0001; for DVS > L ⩾
28. Weight of the roots	WRT	= WRT + DMI.FR kg/ha
29. Weight of the stems	WST	= WST + DMI.FS kg/ha
30. Weight of the grains	WGR	= WGR + DMI.FG kg/ha
31. Total dry weight	TDW	= WLV + WGR + WST + WRT kg/ha
32. Total dry weight of dead leaves	TDWD	= TDWD + DWLV kg/ha
33. Death rate of diseased leaves	Yd	= Fd.Q.WLV.Y.Δt/((Fd – 1).Y + WLV)
34. Relative growth rate of the disease	r	= f(DVS) see Table 77
35. Weight of diseased leaves	Y	= WLV/(1 + (WLV – Y)/Y. EXP(– r.Δ for Y < WLV
36. Disease severity	PROPD	= Y/WLV

Table 77. Parameter values for development of an early and a late disease on wheat.

TIME	DVS	'early' r(rust)	'late' r(leafspot)	Y_i/Y_m	F_{ds}
0	0.01	0.11	0.04	0.0	1.0
	0.1			0.1	1.0
	0.2			0.2	1.5
	0.3			0.3	2.0
	0.4			0.5	2.0
	0.5		0.09	1.0	1.0
	0.6				
	0.7				
	0.8				
	0.9				
100	1.0	0.13	0.15		
	1.2	0.05			
	1.5	0.0			
140	2.0	0.0	0.15		

Diseased leaves die, either from senescence or as a result of the disease. The death rate of diseased leaves is not proportional to the total death rate of leaves, since normally the disease is not homogeneously distributed within the canopy. Epidemics take time to develop; fructifications that can cause new infections appear only after a certain latent period, so older leaves, low in the canopy, have a much higher chance to be infected than young leaves. As old leaves die first, it is assumed that diseased leaves die with a relative death rate that is a factor F_d higher than healthy leaves. If the overall relative death rate of all leaf tissue (Section 3.4) equals q_t, the relative death rate of the diseased leaves, q_i, is calculated as:

$$q_t \cdot Y_m = q_i \cdot Y_i + q_i/F_d \cdot (Y_m - Y_i) \qquad (112)$$

which, after some rearrangement yields:

$$q_i = F_d \cdot q_t \cdot Y_m/((F_d - 1) \cdot Y_i + Y_m) \qquad (113)$$

The death rate of the infected leaves is thus equal to:

$$Y_d = F_d \cdot q_t \cdot Y_m/((F_d - 1) \cdot Y_i + Y_m) \cdot Y_i \qquad (114)$$

Dying of non – infected leaf tissue may be caused by stress through lack of water or from senescence. However, disease may also cause death of non – infected leaves,for example enclosures of healthy leaf tissue within infected leaves. When the infestation is relatively mild, the relative death rate of leaves is assumed to be proportional to the level of infestation. When the infection increases, still – healthy leaf tissue in the surroundings of the disease lesions starts to die. The relation between disease severity and death of healthy leaves is characteristic for the host – pathogen combination. A rough estimate of this effect for cereal rusts is given in Table 77. Now q_t is defined as:

$$q_t = (d_w \cdot (1 - T/T_m) + d_s) \cdot F_{ds} \qquad (115)$$

$$F_{ds} = f(Y_i/Y_m)$$

where d_w and d_s are the maximum relative death rate caused by water stress and senescence, respectively, and F_{ds} the proportionality factor for the disease severity.

Finally, the effect of ageing of the pathogen itself should be taken into account. Disappearance of the disease by ageing proceeds at a more or less constant relative rate that is specific for a pathogen – host combination. For cereal rusts it is between 0.05 and 0.01 per day. It is assumed that if this value is smaller than q_t, all dying infected leaf tissue is taken into account in the previous definition. If this value q_d is higher, it will replace q_t.

The effect of the epidemic on crop production is incorporated as follows. Infected leaf tissue is assumed to take part in assimilation and respiration. The assimilates produced are, however, not available for crop growth but are used for growth and maintenance of the fungus, while the maintenance respiration continues as in healthy leaves. The decline in production due to the disease is proportional to the amount of diseased leaf tissue. As discussed earlier, the disease is not evenly distributed within the canopy. This implies that the effect of the disease will be relatively small in crops with a leaf area index of 4 or more, because most of the radiation is intercepted by the healthy leaves at the top of the canopy, and the infected leaves at the bottom contribute very little to assimilation. This effect can be quantified.

First, the distribution of the disease in the canopy will be treated. For that purpose the crop canopy is divided in an upper and a lower half, each with 0.5 LAI. When the proportion of the diseased leaves is close to zero, all the disease will be concentrated in the lower half of the canopy and it will be absent in the upper half. When all leaves in the canopy are infected, e.g. the proportion of diseased leaves equals one, the disease is evenly distributed over the canopy.
The fraction of disease in the lower half of the canopy is:

$$d_l = 1/(1 + P_d) \qquad (116)$$

The fraction of disease in the upper half of the canopy is:

$$d_u = P_d/(1 + P_d) \qquad (117)$$

where

$$P_d = Y_i/Y_m$$

The effect of the disease on the dry – matter production of the crop depends on the radiation intercepted by diseased leaves. As demonstrated above, the disease is unevenly distributed over the canopy. The fraction of the radiation intercepted in the upper half of the canopy is:

$$1 - e^{-k_e \, . \, LAI/2} \tag{118}$$

The fraction of radiation intercepted in the lower half of the canopy is:

$$e^{-k_e \, . \, LAI/2} - e^{-k_e \, . \, LAI} \tag{119}$$

The proportions of the total radiation intercepted in the upper and the lower halves of the canopy are respectively:

$$P_u = \frac{1 - e^{-k_e \, . \, LAI/2}}{1 - e^{-k_e \, . \, LAI}} \tag{120}$$

$$P_l = \frac{e^{-k_e \, . \, LAI/2} - e^{-k_e \, . \, LAI}}{1 - e^{-k_e \, . \, LAI}} \tag{121}$$

The dry matter increase due to interception of radiation in the upper half of the canopy, corrected for the effect of disease, can now be defined as:

$$DMI_u = ASAG \cdot E_g \cdot P_u \cdot (1 - d_u \cdot 2 \cdot P_d) \tag{122}$$

and the dry matter increase due to interception of radiation in the lower half of the canopy as:

$$DMI_l = ASAG \cdot E_g \cdot P_l \cdot (1 - d_l \cdot 2 \cdot P_d) \tag{123}$$

so

$$DMI = DMI_u + DMI_l \tag{124}$$

Exercise 89

Make a plot of the effect of disease on dry-matter production for values of Y_i/Y_{mi} equal to 0.1, 0.3, 0.5, 0.7, and 0.9 for LAI values of 1, 3, 5, and 7. Assume a value of 0.65 for k_e.

A further adjustment has to be made because in comparison to the original model of Section 3.4, this disease only affects leaf blades and sometimes leaf sheaths. However, the heads and stems that are not affected, contribute to assimilation, even if all leaves are dead. Therefore, a minimum value for LAI after anthesis of 0.5 is maintained. On the basis of these assumptions it is possible to calculate crop production and the effect of the epidemic in combination. Table 76 gives the calculation procedure summarized in a FORTRAN programme. The results are given in Table 78.

The disease treated in this example develops mostly during leaf development before anthesis. After anthesis, its development slows down quickly and midway between anthesis and maturity it comes to a complete stop. Other diseases — Septoria leafspot, for instance — develop slowly during leaf formation, but with increasing temperature they continue to develop until crop maturation. The effect of such a 'late' disease can be calculated using a relative growth rate, r, of the fungus as given in Table 77. The impact of both disease types on crop production in a situation with optimum and sub – optimum water supply is presented in Table 79, which gives the final grain yields. The more severe impact of the 'early' disease can be explained by the fact that it affects the maximum leaf area index, which has an effect on the whole post-anthesis period while the late disease only accelerates leaf death after leaf formation is completed.

6.3.6 Interaction of nutrient status with pests and diseases.

When nutrients limit crop growth, the impact of diseases and pests on crop production may be different from that in the optimum growth situation. For example, when N supply is the limiting factor, the yield estimate is adapted for the amount of N available. The dynamics of N in the crop are, however, not considered. If N supply to the crop is limiting, redistribution of N takes place from vegetative organs to the grains. That process accelerates leaf senescence and causes increased leaf death, partly explaining the lower yield that is obtained under N limiting conditions, as the leaf area index decreases more rapidly and assimilation will be considerably lower. As the leaves are the substrate upon which leaf diseases and many pests rely, interaction is to be expected. The effect of limiting N supply can be expressed in the relative death rate of leaves, which governs the leaf area duration, i.e. the integrated value of leaf area index. Table 79 summarizes the results obtained from the calculation procedure illustrated in Table 76, including the effect of non – optimum N supply expressed as an increase in d_s, for two disease patterns and limited availability of water.

The proportion of loss caused by a disease or pest depends, therefore, also on the impact of other growth limiting factors. It demonstrates why crops with a potentially high production level may suffer more than proportionally from a certain infestation of a pest or disease than crops with a lower produc-

Table 78. Effect of an 'early disease' on crop growth using the programme from Table 76.

Without disease					With disease					
TIME	WLV	LAI	WGR	TDW	WLV	LAI	WGR	TDW	Y_i	P_d
0	50	0.100	0	100	50	0.100	0	100	0.200	0.004
10	133	0.266	0	243	133	0.266	0	243	0.423	0.003
20	221	0.443	0	384	221	0.443	0	384	0.771	0.003
30	283	0.566	0	506	282	0.566	0	505	1.345	0.005
40	619	1.238	0	1155	617	1.235	0	1152	2.872	0.005
50	1099	2.198	0	2283	1095	2.190	0	2275	6.179	0.006
60	1436	2.872	0	3382	1430	2.861	0	3368	12.748	0.009
70	1830	3.662	0	5098	1823	3.646	0	5074	28.406	0.016
80	2033	4.067	0	6786	2024	4.048	0	6749	63.629	0.031
90	2127	4.255	781	8349	2116	4.233	771	8292	142.501	0.067
100	1701	3.404	2528	9671	1693	3.387	2465	9563	239.926	0.142
110	1361	2.723	4006	10808	1284	2.568	3820	10508	207.340	0.161
120	1089	2.178	5194	11724	948	1.897	4854	11207	84.400	0.089
130	871	1.743	6084	12396	758	1.517	5607	11770	41.412	0.055
140	697	1.394	6680	12818	606	1.214	6104	12116	19.011	0.301

Table 79. Calculated grain yields of a crop with an 'early' disease, with a 'late' disease, and without disease, for wet and dry conditions, and for optimum and suboptimum nitrogen conditions.

N supply	Disease	Dry	Wet
Optimum	'early'	3435	6105
	'late'	3521	6123
	'no disease'	4026	6680
Suboptimum	'early'	2204	4345
	'late'	2221	4600
	'no disease'	2375	4700

tion capacity, especially when the pest or disease develops mainly after completion of leaf formation.

6.3.7 *Effects of weather*

Effects of weather conditions on the development of pests and diseases is treated in an indirect way using the relation between the relative growth rate of the pest or disease and the development stage of the crop. As crop and pest or disease do not always react in a similar fashion to different weather conditions, such relations are probably weather – specific. Because it is impossible to establish experimentally the relation between the relative growth rate of the causal agent and the development stage of the crop for each weather type, it is advisable to assess effects of differences in weather conditions on population growth separately. This can be done by using more fundamental models for population growth of pests and diseases applied to various weather conditions (Rijsdijk & Zadoks, 1979).

6.3.8 *Other effects on population growth*

Other effects on population development of pests and diseases, such as the direct effect of the nitrogen status of the canopy on the growth rate of the population, are not treated here. There is evidence that at least some important pests and diseases that rely on living tissue, develop more rapidly on crops optimally supplied with N than on crops with a sub – optimal supply of N (Rabbinge et al., 1981, Rijsdijk, 1980; Darwinkel, 1980a, 1980b). The reverse may be true for fungi that use dead leaf tissue for fructification. However, information about these effects should again be assessed using more complex models. The results of such studies may be included in the simpler approach by redefining the parameter values.

6.3.9 Control of weeds, pests and diseases

Control of weeds, pests and diseases is advisable in many situations. Control measures can be classified in preventive measures, such as growing resistant varieties, using crop rotation schemes in which the causal agents are, at least partly, controlled by reducing their population in fallow periods, flooding of the land, etc., and direct control measures. Direct control measures are mainly weeding and application of herbicides against weeds, and the use of pesticides against pests and diseases. In some cases sophisticated techniques of biological control may be applied. These control measures rely heavily on available resources of labour, cash and management. The capital is invested in spraying equipment and the sprayed product; and management refers to the ability of the farmer to use the resources as efficiently as possible. The labour requirement for weed control differs markedly between subsistence farming and high – input farming in the Western world. Manual weeding of a crop demands 50 – 150 times more labour than the application of herbicides with advanced spraying equipment. This heavy labour demand limits the area of a crop that can be tended. In agricultural practice, generally, control of weeds seems to prevail over control of pest and diseases. One reason may be that control of pests and diseases is expensive, so that it is only worthwile in a more or less weed – free crop. Another reason is the fact that no capital is needed for manual weeding. Even in the most primitive agricultural system, weeds can be removed by hand or with a simple implement, while for control of pests and diseases relatively expensive chemicals and at least some spraying equipment – however simple – is needed.

When chemicals are used, a problem is that their application in most instances does not lead to complete control. The reasons for such incomplete control may be a limited effectiviness of the chemical control to each specific weed, pest or disease, an improperly timed application, unfavourable weather conditions, etc. The ability of the farmer to judge the necessity for application of the appropriate chemicals at the proper time depends on the management skill. In this respect, local expertise, an effective extension service, and the education level of the farmer are of great importance. Even under intensive management, control measures are seldom completely effective due to only partial control of the causal agent.

It is clear that the expected loss through weeds, pests and diseases cannot be the only criterion for estimating whether control measures are economically attractive. An approach is necessary that takes into account differences in efficiency of control as related to management level. The essential question is not the magnitude of loss caused by a certain weed, pest or disease but the yield increment that can be gained by control measures. Even under a high management level, control measures are seldom fully effective because of only partial control of the causal agents. An approach to answer that question, taking into account different management levels, is illustrated in Figure 70. It shows a frequency distribution of the effect of one application of a mixture of broad-spectrum fungicides against 'ripening diseases' on grain yield of winter wheat in the Netherlands. Some five diseases may be involved. The fungicide mixture was applied at the beginning of anthesis, irrespective of the intensity of symptoms of the diseases (Rijsdijk, 1979). The mean costs of the treatment, expressed in kilograms grain per hectare, was slightly higher than the mean of the effect; the median of the effect was even lower. Clearly, on average, the cost of the treatment is higher than the benefits, so a routine application is not profitable. A closer examination of the observations on which Figure 70 is based showed considerable differences in disease incidence among fields and among years. If disease incidence had been used as a criterion for fungicide application, many fields would not have been treated, while other fields would have been treated much earlier in the season, to avoid disease levels that would damage the crop irreversibly before the fungicide was applied. This would have significantly increased the cost – effectiviness of the treatments. A routine treatment with a fixed mixture of chemicals at a time fixed by date or development stage of the crop requires very little of the management abilities of the farmer. The only condition is that the standard application must pay in the long run. The mean expected gain of the treatment should be clearly higher than the mean expected costs of the treatment. If the management level is higher, the effectiveness of chemical control can be increased by careful inspection of the crop and adaptation of the chemicals to specific weeds, pests or diseases.

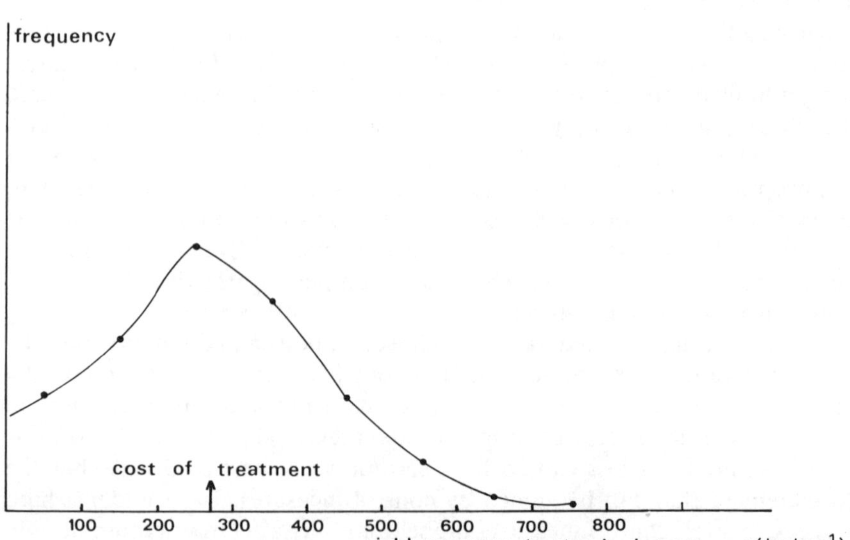

Figure 70. The frequency distribution of yield response to standard treatment with biocides.

Exercise 90

Explain why in a situation of sophisticated management, chemical treatment may be profitable, even if the value of the mean yield increase over the years is less than the mean costs of a treatment.

Considering the effects of the production situation on losses due to weeds, pests and diseases and the prospects of reducing these losses, a hypothesis for their control may be summarized as below.

Production Situation 1:
Production determined by radiation and temperature only.
Chemical control of weeds, pests and diseases. A high effectiviness of biocide application because of sophisticated management.

Production situation 2:
Water supply limiting at times.
Chemical control of weeds, pests and diseases. Effectiviness of biocide application is less due to 'natural' variations in yield. Management form is sophisticated to reasonable.

Productions Situation 3:
Water and nutrients limiting at times.
Chemical and/or manual weed control. Low control of pests and diseases. Relatively low level of management.

Production Situation 4:
Low input farming.
Some manual weeding, no control of pests and diseases.

More specific conclusions for specific situations can,however, only be obtained by defining all the parameters involved and eventually using optimization techniques to find the most profitable combination of input factors for each production situation.

7 LAND IMPROVEMENT

7.1 Reclamation

J. Wolf

7.1.1 Introduction

The productivity of agricultural land is to a large extent determined by the moisture regime of the soil. Potential production can only be attained if the soil-moisture content during the entire crop growth cycle can be maintained within the optimal range. Reclamation can be considered as the total of operations required to lenghten the period of optimal soil-moisture for crop growth.

Reclamation measures consist mainly of excavation and movement of earth. These activitites require a large amount of labour and/or machinery, but they are needed only once or with long intervals in between. This in contrast with many other operations in agricultural production, which mostly have to be repeated for every planted crop (Section 6.1). In this section only reclamation measures carried out on the farm by the farmer and his family are considered. Macro-structures like canals, dams, dikes, etc., are assumed to be constructed by the government and are not taken into consideration. Also, measures like reclamation of saline or very acid soils are not considered here.

Four schematic levels of reclamation can be distinguished:

Reclamation Level I: natural situation; agricultural use limited to extensive grazing, food gathering and fuelwood collection; moisture regime is fully dictated by the weather conditions; this level will not be treated here.

Reclamation Level II: simple clearance; forest clearing or burning of vegetation, more permanent cultivation with or without fallow periods, field-protecting floodwalls, bunds to permit the cultivation of rainfed rice; moisture regime is fully dictated by the weather conditions; range of soil-moisture contents from very low to very high.

Reclamation Level III: control of excess water; completion of clearance; levelled or terraced land; field-protecting floodwalls, drainage by open ditches; range of soil-moisture contents from very low to optimal.

Reclamation Level IV: complete water control; well levelled or terraced land; infrastructure for complete water control; sufficient water available for irrigation; optimal soil-moisture content throughout.

In the following subsections, first the main effects of reclamation are treated. Then the consequences of improving the water regime are further illustrated by discussing the results of some computer simulations for a loamy fine sand in northern Thailand. Last, the physical inputs that are needed on the farm for the transfer of land from one Reclamation Level to the next are considered.

7.1.2 *Effects of reclamation*

Reclamation has various positive effects. In the first place, it makes crop growth possible or improves it, so that higher yields may be achieved. Second, the number of crop species that may be grown is increased, as is the freedom of choice with respect to the time of sowing. In addition, it results in a decrease in fertilizer requirements. These positive effects will be considered in more detail in this subsection.

The gross assimilation rate of a crop is reduced when the soil-moisture content is outside its optimal range, as discussed in Section 3.2 and summarized in Table 80.

At Reclamation Level II, the soil-moisture content may range from very low to very high, at Reclamation Level III from very low to optimal and at Reclamation Level IV the soil-moisture content is in the optimal range throughout, barring exceptional weather conditions. The yield level will be reduced proportionally to the period and the degree of sub-optimum soil-moisture conditions, as discussed in Section 3.4 for Production Situation 2.

It has been shown in Section 6.1 that one of the factors, determining the workability of the land is its moisture content. Depending on the cultivating equipment that is used and the type of soil, there is a lower limit of the soil-moisture suction at which the land can be worked without too much damage. Initially, three technology levels are distinguished for the field work, (1) manual labour, (2) use of animal drawn equipment and (3) light or heavy mechanical equipment. The critical limits for the three levels in terms of soil-moisture suction are assumed to be 10, 100 and 500 cm, respectively. The reclamation level also determines the range of soil-moisture suction in the soil. At

Table 80. The influence of the moisture content of the soil on gross assimilation rate.

Moisture content (SM)*						Reduction factor for gross assimilation rate
very low	SM	$<$	SM_w			0
low	SM_w	$\leq$	SM	$<$	SM_{cr}	$(SM-SM_w)/(SM_{cr}-SM_w)$
optimal	SM_{cr}	$\leq$	SM	$<$	SM_{100}	1
high	SM_{100}	$\leq$	SM	$\leq$	$(SM_0 - 0.05)$	$((SM_0 - 0.05) - SM)$ / $((SM_0 - 0.05) - SM_{100})$**
very high	SM	$>$	$(SM_0 - 0.05)$			0**

* SM_w, SM_{cr}, SM_{100} and SM_0 - 0.05 are the soil-moisture contents at wilting point, at the critical value for optimal growth, at a soil suction of 100 cm and at a soil air content of 0.05 $cm^3\ cm^{-3}$, respectively.
These moisture contents are different for different soils.

** For rice, reduction factor is equal to 1.

Levels III and IV, it is in general not lower than 100 cm, so operations at Technology Levels 1 and 2 are always possible. At Reclamation Level II under natural drainage, the soil-moisture suction may be so low that it is impossible to work the land, even manually, except of course for the cultivation of bunded rice.

At the onset of crop growth, the soil should be sufficiently moist to enable germination and, in the case of a clay soil, to allow the necessary tillage operations. It is assumed that the risk of early crop failure due to lack of water is sufficiently small if sowing is postponed till the moment that the soil-moisture suction is less than 300 cm, or cumulative rainfall over the last 20 days exceeds 7.5 cm. In this way, the beginning of the growing season on soils at Reclamation Levels II and III is fully determined by the rainfall pattern. As a consequence, the season may become too short to grow certain crops that would otherwise be preferred. The possibility of irrigation at Reclamation Level IV makes the farmer, in this respect, independent of rainfall.

At Reclamation Level II shallow groundwater tables may occur. A groundwater table within 10 cm of the rooting depth will result in reduced root activity and eventually lasting damage to the root system due to lack of oxygen in the root zone. The minimum groundwater depth that may occur limits, therefore, the number of crop species that can be grown, because crops differ distinctly in their depth of rooting, as shown in Table 81.

In climates with erratic rainfall on land at Reclamation Levels II and III, deep-rooted crops with a low critical moisture content (Section 3.2, Table 20) must be selected, because these will be less adversely affected by dry periods.

Exercise 91

Which crops are not affected by an average groundwater depth of 1 m below the surface?

Which crops are best adapted to climates with irregular rainfall (in the absence of irrigation)?

The effect of reclamation on the availability of plant nutrients is treated as in Section 4.1. The relation between fertilizer application and yield is laid out in two relations: one between uptake and yield and one between the amount of fertilizer applied and uptake. Both relations are affected by reclamation, as illustrated in Figure 71.

The uptake-yield curve for a particular element is characterized by its initial slope and the maximum yield that may be achieved at an optimal supply of the nutrient. The initial slope is mainly a crop characteristic and is affected only indirectly by environmental conditions, for instance if at Reclamation Levels II or III a sub-optimal water supply during seed filling reduces the seed/straw ratio.

Table 81. Indicative rooting depth of major crops.

crop species	rooting depth (cm)
sugar-cane	160
sorghum	150
cotton	135
maize	135
barley	125
black mustard	125
jute	125
lentil	125
rape seed	125
sesame	125
wheat	125
sweet potato	120
cassava	100
cowpea	95
gram	95
kheshari	95
chilli	90
rice	80
groundnut	75
kenaf	75
mung bean	75
tobacco	75
potato	50
onion	40

Exercise 92

Do you expect an increase or a decrease of the ratio of yield to uptake as a result of a reduced seed/straw ratio. Explain your answer. Draw an uptake-yield curve in Figure 71 that could be the result of the reduced seed/straw ratio.

If the average sub-optimal soil-moisture content during seed filling is known, is it possible to estimate the maximum yield and the ratio between yield and uptake compared to that at optimal water supply during seed filling. How would you do that? Explain your answer.

One of the main purposes of reclamation is to improve crop water supply, which is reflected directly in an increase in the maximum yield that can be

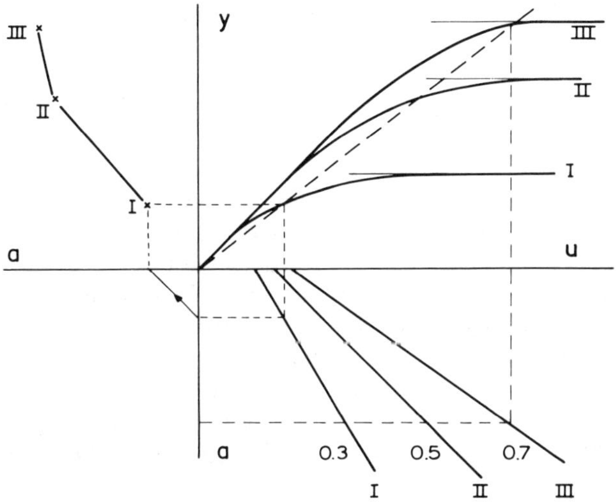

Figure 71. The effect of various reclamation levels (I, II, III) on the relation between fertilizer uptake (u) and yield (y), on the relation between fertilizer application rate (a) and uptake, and on the relation between fertilizer application and yield.

achieved. The nutrient requirement increases linearly with that yield. The amount of fertilizer needed to meet that requirement, depends on the recovery of the fertilizer and the amount of nutrients taken up from unfertilized soil. In general, both are favourably affected by reclamation because better water control stimulates the activity of the root system, contributes to an increased mineralization of the soil organic matter, decreases losses by leaching and leads to lower losses by denitrification because of reduced waterlogging. Although generalizations in quantitative terms cannot be made, it may be concluded that the fertilizer requirement increases less than proportionally with the yield increase due to reclamation. Examples are found in Section 4.1.

The effects of reclamation measures are treated in this section by comparing crop growth at Reclamation Level II with that at higher reclamation levels. Of course these effects are mainly positive, because otherwise the large amount of physical inputs would not be worthwhile. However, when considering clearing of the original vegetation (Reclamation Level I) to allow cultivation of the land, attention should be paid to possible adverse effects.

On slopes of more than 8 to 15% (depending on soil erodibility, slope length, rainfall intensity, etc.) a permanent cover should be maintained by a perennial crop, such as a tree crop, pasture or forest, as indicated for recommended land use in Table 85. Otherwise, soil erosion may result in rapid loss of topsoil, causing a decline in soil fertility and soil structure. Moreover, gullies may be formed accelerating erosion even more. Adverse effects may also appear in coastal areas after reclamation of alluvial soils, deposited in brackish water. Such soils often become very acid after drying and aeration.

Only when the groundwater table can be maintained at a shallow depth (Reclamation Level IV), may cultivation – mainly of rice – be possible on these soils.

Tropical rain forests grow partially on very poor soils. After clearing the original vegetation, the soil organic matter decomposes at a very high rate under the hot, wet tropical climate. This results in a rapid loss of soil fertility built up in the course of time (Section 6.2) and a substantial decrease in the moisture- and nutrient-holding capacity of the soil. Within a short time after clearing the land the production capacity may have declined to such a level that crop production is no longer worthwhile. In addition, fertilizer application is not very effective on these degraded soils because of the low moisture- and nutrient-holding capacity, which results in rapid leaching of nutrients under the high intensity of the tropical rainfall regime. A long bush-fallow period of 10 to 50 years is the best way to restore soil fertility and soil structure. Permanent cultivation on these poor tropical soils appears to be almost impossible.

7.1.3 A simulated example

In this subsection important differences between the Reclamation Levels II, III and IV are illustrated by means of results obtained using a simulation model that calculates the yield levels for Production Situations 1 and 2 (chapter 9). They refer to a slightly undulating, elevated lowland area with a soil of loamy fine sand texture in northern Thailand. The physical characteristics of the soil are given in Table 82.

The climate in the region is of a rather pronounced monsoon type, with the wet season starting in the beginning of April and ending in November. The simulations are for the year 1974, which had a rainfall of 1440 mm and a potential evapotranspiration of 950 mm from April to November. In the calculations, groundnuts are grown both as an early- and as a late-season crop.

The onset of the rains is so distinct that emergence of the early-season crop

Table 82. Physical characteristics of the loamy fine sand, used for the simulation exercise.

Physical characteristic	Value for exercise
Saturated hydraulic conductivity (k_o)	26.5 cm d^{-1}
Total pore space fraction (SM_o)	0.439 cm^3 cm^{-3}
Texture specific geometry factor (γ)	0.0312 cm^{-2}
Field capacity (suction)	ψ = 200 cm (pF = 2.3)
Field capacity (moisture content)	SM = 0.183 cm^3 cm^{-3}

for all three levels of reclamation was during the 14th ten-day period of the year. Under irrigation, emergence could have been shifted to an earlier moment, but this was not done to facilitate comparison. Emergence of the late-season crop was set at the 31st ten-day period of the year, in November. The variety chosen has a growth period of 100 days, so the growth of the late-season crop extends until the beginning of February, well into the dry season. The early-season crop has to be harvested in the middle of the wet season.

It is assumed that during the dry season preceding the early-season crop the land was fallowed and kept free of weeds. Thus after the topsoil has dried out totally by evaporation, no water loss from the profile by evapotranspiration or drainage will take place. This results in a soil at about field capacity with a mulch layer on top at the start of the wet season.

Crop failure will occur when the air content in the root zone is lower than 0.05 $cm^3\ cm^{-3}$ for more than 20 consecutive days. At Reclamation Level II, with no control of excess water, this occurs directly after emergence for both the early- and late-season crop. For the late-season crop this can be avoided by sowing later, but then germination problems may occur because of drying out of the topsoil. In addition, too much water may be lost by direct evaporation from the soil surface. To conclude, the conditions at this low level of reclamation are not suitable for groundnut cultivation. The land should be used for crops that prefer or can stand waterlogged soils, such as bunded rice.

For both the early- and late-season crop of groundnuts grown at Reclamation Level III, the values of the various terms of the water balance, the soil-moisture content and some growth characteristics are given by ten-day period in Table 83, together with the technology level, the rainfall and the potential evaporation. The cumulative values of the terms of the water balance for the whole season and the yield of pods for Reclamation Levels III and IV are summarized in Table 84.

Because of the drainage at Reclamation Level III, the soil-moisture content during growth of the early-season crop is maintained at field capacity, i.e. 0.183 $cm^3\ cm^{-3}$, except between the 18th and 20th ten-day periods, a time of relatively little rain. At Reclamation Level IV, 90 mm of water is given during this period. This results in a slight increase in actual evapotranspiration, but hardly affects yield. As the groundnut crop is supposed to be growing under optimal nutrient supply, the simulated yields at both reclamation levels are those of Production Situation 1.

At the end of the wet season there is considerable drainage during the 30th and 31st ten-day periods, so that at Reclamation Level III the crop does not fail because of waterlogging. The rain stops rather abruptly, so that water stress starts to develop already in the third ten-day period of growth. At maturity, soil-moisture is practically exhausted and due to water stress during the major part of the growing period, the simulated yield is only 1100 kg pods ha^{-1}. At Reclamation Level IV, irrigation starts in the 33rd ten-day period and is continued into the second ten-day period of the following year. A total

Table 83a. The simulated terms of the water balance and some simulated growth charac an early planted crop of groundnut. Emergence at 14th 10 day period of the year.

10 day period	10	11	12	13	14
rain (mm per 10 day period)	0	0	20	160	48
mulch[a]	m	m	m		
potential evaporation (open water), (mm per 10 day period)	60	62	60	56	54
actual evapotranspiration, (mm per 10 day period)	0	0	22	22	20
drainage (mm per 10 day period)	0	0	0	120	30
pF	2.3	2.3	2.3	2.3	2.3
soil moisture content ($cm^3\ cm^{-3}$)	0.183	0.183	0.183	0.183	0.183
leaves, live weight ($kg\ ha^{-1}$)					$\leqslant$100
weight pods ($kg\ ha^{-1}$)					
technology level	3	3	3	2	2

[a]m means present.

Table 83b. The simulated terms of the water balance and some simulated growth characte late planted crop of groundnut. Emergence at 31st 10 day period of the year.

10 day period	27	28	29	30	31
rain (mm per 10 day period)	66	30	8	50	116
mulch					
potential evaporation (open water), (mm per 10 day period)	42	40	40	42	36
actual evapotranspiration, (mm per 10 day period)	16	16	16	16	14
drainage (mm per 10 day period)	50	10	0	30	100
pF	2.3	2.3	2.3	2.3	2.3
soil moisture content ($cm^3\ cm^{-3}$)	0.183	0.183	0.183	0.183	0.183
leaves, live weight ($kg\ ha^{-1}$)					$\leqslant$100
weight pods ($kg\ ha^{-1}$)					
technology level	2	2	2	2	2

of 100 mm of water was given, which more than doubled the yield. Not surprisingly, the growth of such a late-season crop is considerably improved by irrigation.

Only during seedbed preparation and sowing of the early crop, and during harvest of the late crop, is the soil dry enough to permit use of the heavy equipment associated with Technology Level 3. Harvesting of the early crop and seedbed preparation for and sowing of the late crop is carried out well within the wet season, but then only the light equipment of Technology Level

16	17	18	19	20	21	22	23
54	66	12	6	14	136	72	38
50	48	48	48	48	50	46	44
36	40	40	40	40	42	30	22
20	30	0	0	10	40	120	0
2.3	2.3	2.5	3.0	3.4	2.3	2.3	2.3
0.183	0.183	0.156	0.099	0.065	0.183	0.183	0.183
1300	2000	2500	2700	1600	700	200	≤100
		200	500	1000	1600	2100	2300
2	2	2	3	3	2	2	2

33	34	35	36	1	2	3	4
0	0	16	0	0	0	0	0
32	30	28	32	30	32	38	38
16	22	22	26	22	12	4	4
0	0	0	0	0	0	0	0
2.4	2.7	2.8	3.1	3.5	3.8	3.9	3.9
0.169	0.131	0.120	0.090	0.058	0.040	0.035	0.035
800	1400	1900	2200	1500	600	≤100	≤100
		200	500	900	1100	1100	1100
2	3	3	3	3	3	3	3

2 can be used. Fully mechanized agriculture is therefore not possible on this soil in this region.

Exercise 93

Yield of the late-season crop of groundnut at Reclamation Level III is seriously reduced compared to the yield of irrigated groundnut (Table 84). Calculate the critical soil-moisture content at a maximum transpiration rate of 2.5 mm

Table 84. Summarized data for the whole growth period

	Reclamation level III		Reclamation level IV	
	early crop	late crop	early crop	late crop
Rain (mm)	838	304	838	304
Irrigation (mm)	—	—	130	100
Actual evapotranspiration (mm)	378	220	404	252
Drainage (mm)	460	190	560	190
Soil moisture increase (mm)	0	-106	+4	-38
Yield of pods (kg ha^{-1})	2300	1100	2300	2400

d^{-1} on loamy fine sand (SM_0 = 0.439 cm^3 cm^{-3}, γ = 0.0312). Use Table 20 and Equation 53.
Estimate the number of days of unrestricted assimilation starting from the 35th ten-day period (end of rainfall), using a rooting depth of 75 cm and a maximum transpiration rate of 2.5 mm d^{-1}.

Exercise 94
Calculate the reduction in gross assimilation rate (see Table 80) at Reclamation Level III in ten-day periods 36, 2 and 4, compared to the assimilation rate of irrigated groundnuts.

Exercise 95
Repeat the calculations of Exercise 94 for sorghum, wheat and onion using the same data, except for the soil-water depletion factor p and the rooting depth. Which crop will perform best in this period?

7.1.4 *Physical inputs for reclamation*

Physiography, slope and soil type are the main determinants of the type and amount of physical inputs needed for reclamation. For example, erosion control is hardly needed on level lowland, flood control is superfluous in the uplands, the steepness of the slope determines the amount of earth movement for ridging and terracing, and it is more difficult to move clay than sand. In a

Table 85. Geographical classification of Thailand.

Land unit	Characteristics	N_r	h_r, h_b	N_t	Slope (%)	Naturel drainage	Recommended landuse
U1	lowland, level	0	0.5	3	0-3	moderate	bunded rice
U2	elevated lowland, level to undulating	3	0.5	4	0-8	moderate	bunded rice, field crops
U3	upland, level to undulating	3	0.5	4	0-8	well	upland rice, field crops, tree crops
U4	upland, undulating to rolling	6	0.8	7	8-15	well	field crops, tree crops, grazing
U5	upland, rolling to steep	9	1.0	10	15-30	well to excessive	tree crops, grazing, forest
U6	lowland, flooded	3	0.5	4	0-8	various	floating rice, swamp, wet forest
U7	upland, steep	9	1.0	10	>30	excessive	forest, waste land

N_r is number of ridges required for erosion control for a plot of 100 × 100 m.
h_r is height of ridges for erosion control (m).
N_t is number of required terraces for a plot of 100 × 100 m.
h_b is height of bunds, equal to height of ridges (m).

schematized setup, seven land units have been distinguished in Thailand, ranging from level lowland to steep upland. These land units are characterized in Table 85 in terms of slope, the degree of natural drainage and recommended land use. The example of the previous subsection referred to Land Unit U2, level to undulating elevated lowland with a slope between 0 and 8 percent and moderate natural drainage. The soil is a loamy fine sand.

For a proper assessment of reclamation possibilities, reasonable estimates of labour requirements for upgrading land from one reclamation level to the next are necessary, because the labour requirements for reclamation are high if carried out in hand labour. Because most soils can only be worked in the growing season, when labour requirements for other activities are also high, labour is often the limiting resource. That is the reason why in most instances heavy earth-moving equipment is used.

To upgrade an hectare of Land Unit U2 from its more or less natural situation of Reclamation Level I to Reclamation Level II, the vegetation has to be destroyed by slashing and burning. In this activity heavy trees, tree stumps and termite hills are not removed. Manually this would take about 300 man-hours per hectare, but if heavy equipment were to be used the labour requirement is reduced to about 7 man-hours per hectare. This machinery can also be used to build flood walls, needed to protect against high water levels. Manually this would take 100 man-hours per hectare, but with heavy equipment it is reduced to about 8 man-hours. Because the land is undulating, some erosion control by contour ridging is necessary. On this land unit, 3 ridges per 100 metre of field length will on the average suffice. Assuming ridges with a width and height of 0.5 metre and a length of 300 m ha^{-1}, the necessary earth movement amounts to 37.5 m^3 ha^{-1}. If done by hand, that will take for a loamy fine sand about 20 man-hours per hectare.

The calculations presented in the previous subsection show that the chances of crop failure due to waterlogging are high, so that the farmer will probably choose for bunded rice. In that case bunds must be constructed that have the same dimensions as the ridges. To increase the surface storage capacity, the number of bunds has to be higher than needed for erosion control. The bunds are placed at an average distance of about 6 m, so about 1600 m of bund is needed per hectare. The labour requirement for their construction is about 100 man-hours per hectare.

Assuming that a main road system is provided, the land has to be opened up by simple dirt paths. At this reclamation level, 50 m of path, one m wide, per hectare will suffice. This requires an earth movement of 10 m^3 ha^{-1} or 5 man-hours of manual work. The total labour requirement for construction of roads and bunds is about 100 man-hours per hectare. This is still very small compared with the 1000 man-hours of work needed for the cultivation of 1 hectare of bunded rice. It is therefore likely that the time for this work can be found, provided the land is cleared and protected against floods.

To upgrade land from Reclamation Level II to Level III, the land has to be

Table 86. Productivity for excavation and movement of earth.

		Moving distance (m)									units
		0	10	20	30	40	50	60	70	100	
bulldozer	clay	100	85	71	55	46	40	34	31	23	m^3h^{-1}
bulldozer	sandy soil	140	120	100	78	65	56	48	43	32	m^3h^{-1}
manual, digging + loading	clay	1.2	1.2	1.2	1.2	1.2	1.2	1.2	1.2	1.2	m^3h^{-1}
manual, digging + loading	sandy soil	2.3	2.3	2.3	2.3	2.3	2.3	2.3	2.3	2.3	m^3h^{-1}
manual, wheel barrow transport	clay	4	4	2.8	2.1	1.8	1.5	1.4	1.3		m^3h^{-1}
manual, wheel barrow transport	sandy soil	4	4	2.8	2.1	1.8	1.5	1.4	1.3		m^3h^{-1}
manual total[a]	clay	1.08	1.08	1.19	1.31	1.39	1.50	1.54	1.60		hm^{-3}
manual total[a]	sandy soil	0.68	0.68	0.79	0.91	0.99	1.10	1.14	1.20		hm^{-3}
manual total[a]	clay	0.93	0.93	0.84	0.76	0.72	0.67	0.65	0.63		m^3h^{-1}
manual total[a]	sandy soil	1.47	1.47	1.27	1.10	1.01	0.91	0.88	0.83		m^3h^{-1}

[a] Calculation procedure, for example manual, clay, moving distance 30 m:

$\frac{1}{1.2} + \frac{1}{2.1} = 1.31\ h\ m^{-3}$. So the average productivity for digging and transport is $1/1.31 = 0.76\ m^3h^{-1}$.

Table 87. Reclamation activities, coresponding earth movement ($m^3\ ha^{-1}$) and manual and mechanical labour requirements ($h\ ha^{-1}$) on land unit U2, for a plot of 100 × 100 m.[b]

Reclamation level II	Floodwalls[3]	Bunds[4a]	Clearance[5] A	Road construction[6] A	Erosion control[7]
Earth movement:	200	200[8)]		10	38[9]
Labour requirements:					
manual[1)]	135	135	225-430	7	25
mechanical[2)]	8[10)]	2	5-8	—	0.4

Reclamation level III	Clearance[11] B	Road construction[12] B	Drainage ditches[13]	Levelling/Terracing[14]
Earth movement:		40	500	1250
Labour requirements:				
manual[1)]	400-800	27	500	1440
mechanical[2)]	10-30	0.4	20[10)]	16

Reclamation level IV	Irrigation ditches [15]
Earth movement:	500
Labour requirements:	
manual	500
mechanical	20[10)]

[a] Only bunded rice, no erosion control.
[b] Clay is dominant soil texture on this land unit.

1) Labour productivity assumed to be 1.5 $m^3 h^{-1}$ for clay, 3 $m^3 h^{-1}$ for sand, and 1.0 $m^3 h^{-1}$ (clay) or 1.5 $m^3 h^{-1}$ (sand) for excavation of ditches.
2) Labour productivity assumed at 100 $m^3 h^{-1}$ for clay, 140 $m^3 h^{-1}$ for sand, and 25 $m^3 h^{-1}$ for construction of walls and ditches with dragline.
3) Only required in lowlands, length 200 m, 1 m^3 earth movement per m of flood wall
4) Number of required bunds calculated as: 2 × (slope × field length) / (bund height) (Table 85)
5) Slash-and-burn, large obstacles left in the field.
6) Levelling and excavating for dirt path, width 1 m, to a depth of 0.2 m
7) Crops, other than bunded rice, no bunds.
8) Earth movement for bund construction: 0.5 × $(h_b)^2$ × field width × number of required bunds
9) Earth movement for erosion control: 0.5 × $(h_r)^2$ × field width × number of required ridges
10) Dragline
11) Removing all obstacles
12) Levelling and excavating for dirt road to a depth of 0.2 m, length 50 m ha^{-1}, width 4 m
13) Total length of drainage ditches is assumed 100 (N_t + 1), 1 m^3 earth excavated per m of ditch (Table 85)
14) Excavation depth calculated as: (slope × field length) / 2 N_t (Table 85)
Excavation width calculated as: (field length) / 2 N_t (Table 85)
Earth movement for terracing: 0.5 × excavation depth × excavation width × field width × N_t
Labour productivity of excavation and earth movement for terracing is given in Table 86
Moving distance calculated as: 2/3 × field length / N_t
15) Length of irrigation ditches is assumed 100 (N_t + 1), 1 m^3 of earth excavated per m of ditch (Table 85)

Table 88. Total manual and mechanical labour requirement (h ha^{-1}), to convert the original situation into one higher reclamation level on the different landunits[a].

Land unit	Reclamation level			
	I to II	I to II (with bunds)	II to III[b]	III to IV[b]
U1 manual	370-570	420-620	1600-2000	400
U1 mechanical	13-16	14-17	36-56	16
U2 manual	390-600	500-710	2370-2770	500
U2 mechanical	13-16	15-18	46-66	20
U3 manual	240-450	300-500	1680-2080	330
U3 mechanical	5-8	6-9	42-62	20
U4 manual	290-500	530-730	2390-2790	530
U4 mechanical	6-9	11-14	60-80	32
U5 manual	380-580	960-1170	3020-3420	730
U5 mechanical	8-11	21-24	76-96	44
U6 manual	390-600	500-710	-	-
U6 mechanical	13-16	15-18	-	-
U7 manual	380-580	-	-	-
U7 mechanical	8-11	-	-	-

\- not relevant

a It is assumed, that clay is the dominant soil texture in the lowlands (landunits U1, U2, U6) and that sandy soils dominate in the uplands (landunits U3, U4, U5, U7).

b Mechanical excavations of ditches are supposed to be done with a dragline.

cleared of tree stumps, termite hills and big trees. Manually that requires another 600 man-hours per hectare, but with the proper machinery the work can be done in about 20 man-hours per hectare. As it is the intention to grow also other crops than bunded rice at Reclamation Level III, the soil has to be properly drained. It is assumed that this is the case when the land is terraced, with 4 terraces per 100 m along the slope. At an average slope of 4% that requires an earth movement of 1250 m^3 ha^{-1} over an average distance of nearly 17 m. The average productivity for digging, loading and transport per wheelbarrow is 1.0 m^3 per man-hour, so the labour requirement is 1250 man-hours per hectare, if done manually. If a bulldozer is used, only 14 man-hours per hectare are needed (Table 86). In addition, it is necessary to construct 500 m of drainage ditch per hectare, one hundred metres along the slope and four times one hunderd metres along the ridges of the terraces. This requires the movement of 500 m^3 of soil, which takes 400 man-hours if done by hand and 20 man-hours with a dragline. At Reclamation Level III, dirt paths need to be replaced by dirt roads with a width of 4 m to allow the use of animal-drawn and mechanical equipment. That requires the movement of only 30 m^3 ha^{-1}, which takes 15 man-hours per hectare, if done by hand and less than 1 man-hour, if a bulldozer is available.

Hence, the upgrading of one hectare of land from Reclamation Level II to Reclamation Level III takes about 2400 man-hours per hectare, if done manually. As most of the work can only be carried out during the growing season, it can hardly be expected that this time can be made available. Therefore unless machinery is available, it is most likely that the farmer has no option but the cultivation of bunded rice at Reclamation Level II.

To upgrade the terraced land at Reclamation Level III to Reclamation Level IV it suffices to add 500 m of irrigation ditch per hectare, which requires the same amount of work as the construction of drainage ditches. This is negligible, compared to the costs of the total infrastructure for irrigation.

A summary of all reclamation activities described in this section, the corresponding earth movement and manual and mechanical labour requirements on Land Unit U2 are given in Table 87. Because clay is assumed to be the dominant soil texture class on this land unit (lowlands), the labour requirements given in Table 87 refer to clay soils only. A summary of the total manual and mechanical labour requirement, for converting the original situation (Reclamation Levels I, II, III) into one reclamation level higher is given in Table 88.

Exercise 96

Calculate earth movement and labour requirement (both manual and mechanical) for the various activities required to obtain Reclamation Level II, III and IV on Land Unit U4 (sandy soil). The information required is summarized in Tables 87 and 88.

---

8 APPLICATION OF AGRONOMIC INFORMATION

8.1 The use of agronomic information in the socio-economic models of the Centre for World Food Studies

D.C. Faber

8.1.1 Introduction

Although the world produces sufficient food for the total population, and although it has been estimated that agricultural output can be increased at least twenty-fold (Buringh & van Heemst, 1979; see also Chapter 1), more than 400 million people suffer from less than adequate nutrition, and many of them from starvation. To get to the root of this hunger and poverty problem an agro-technical and socio-economic analysis must be made in a global context. In particular, the mechanisms that determine the distribution of resources and purchasing power, through productive employment, have been found to cause an unbalanced distribution of available food (Linneman et al., 1979). To improve the food situation, efforts should therefore be directed towards reduction of income inequalities between rich and poor countries, as well as between different income groups within countries. Policies to influence that distribution could be initiated at the international level, while interdependence of national policies should also be taken into account (see Chapter 1). Global policy analyses have indicated the general direction of solutions but have not been able to quantify the effects of such measures. A necessary starting point of an investigation into the nature and causes of the hunger problem therefore lies in the development and construction of a set of individual country models, that interact through international markets and are linked through international agreements.

The focus on relationships between agro-technical factors and economic processes, and on national policies and linkage of nations, which together determine the characteristics of the food market, requires a multi-disciplinary approach. This approach necessitates starting such an analysis with a detailed description of the physical environment determining actual and potential agricultural production, and its limitations (as outlined in this monograph), as well as an inquiry into the socio-economic structure of the agricultural sector and its interrelationships with the rest of the economy. Such economy-wide models will not be used to forecast or to make predictions (Meadows et al., 1972), but rather to enable the analysis of possible alternative futures. Quantitative modelling of these relationships and interactions can contribute to increased insight into the causes of the hunger problem, which in turn may lead to the formulation of useful policy proposals. At the national level, it is particularly important that such models are applicable for medium- and long-term planning and that they contain all elements that may influence the national food market. At the international level, the model system should provide explicit and reliable suggestions for international policy coordination. Before the explicit linkage between the physical factors and socio-economic structure is discussed, the general structure of the full model is outlined.

8.1.2 Characteristics and structure of the study: a modelling approach

The pursued research objectives are decisive for the choice of the methods to be used. First, the complex system of interacting economies in a world market must be described. Second, at the national level, economy-wide models are needed to describe the agricultural and food sectors in their appropriate context. Third, the formulation of the intricate interrelationships between physical factors that determine agricultural production and the socio-economic factors that govern behaviour require an in-depth study.

Three main components, the main characteristics of the model structure that are considered in line with these objectives can be distinguished:

- the general structure, describing the interaction among country models
- the agricultural supply model
- the physical crop and livestock growth models.

The general structure

The international system – or linkage system – comprises a number of country models connected through the world market. The system contains all countries, some of them explicitly modelled, others aggregated to a 'rest of the world' group. It thus describes world trade. Linkage of models requires a set of stringent conditions and agreements to be followed in the formulation and development of each individual model. Each national model must distinguish four levels of analysis:

- an international market
- a national government and a national market
- classification of the population in producers and/or consumers of agricultural and non-agricultural commodities
- agronomic information at a subregional level.

The unique feature of this approach is that it attempts to model and analyse the mutual dependence between each and all of these levels (Figure 72).

The country model

In the national model, the government, the agricultural and the non-agricultural population groups are the main actors, i.e. the government and each of the socio-economic groups are decision makers and can act and react independently in the model. It is essential, therefore, that a country model describes realistically the behaviour of the actors and their reactions to a changing

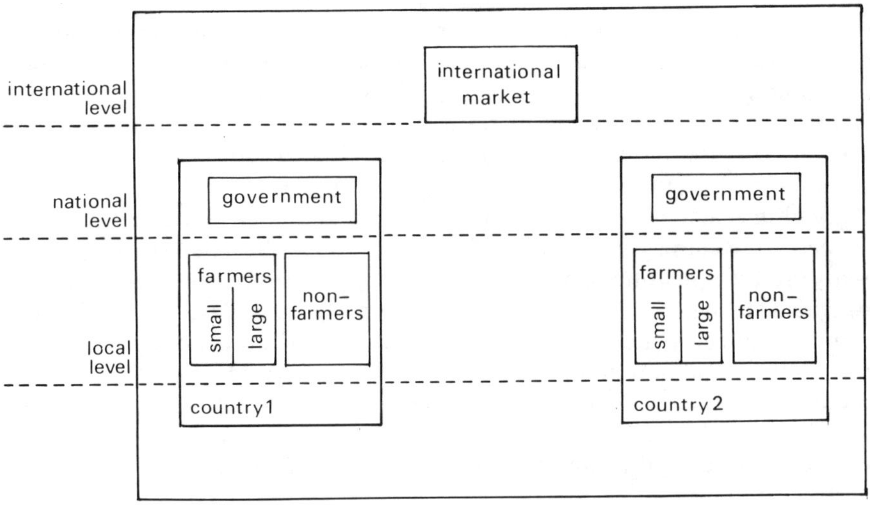

Figure 72. The international linkage system; actors in the economic model.

and fluctuating environment, and a pattern of relationships between the actors. A main feature of the national model is the role played by each actor – including the government – given his endowments (resource structure) and the constraints governing their use. Thus the government must use its instruments – taxes, tariffs, quotas, public investment – to influence the behaviour of the other actors to achieve pre-stated objectives, such as growth, stability and distribution. The production and consumption behaviour of the other actors is governed by the physical and economic environment to which they are exposed.

The agricultural supply model

Within the national model, the agricultural production model describes the farmer's behaviour that is relevant for the determination of agricultural output. Income maximization and/or satisfaction of subsistence requirements are pre-defined objectives and act, therefore, as driving forces in this model. The output of the crop growth model (the production of both marketable yield and crop residues of a given crop) and the necessary farm and non-farm inputs to achieve that yield on the specified land type are then transferred to the input file of a linear programming tableau. The linear programme* is an optimization technique used for maximizing revenues of the agricultural sector (over regions and farm sizes) subject to a number of physical and socio-economic constraints. Part of the information required for the LP tableau is

* See for instance R.R. Benneke & R. Winter (1973) for a popular description of the linear programming technique.

shown in Table 89 (this information and other exemplary material given in this subsection are derived from ongoing research to develop suitable socio-economic models for Bangladesh). The output of the linear programming model consists of the level of production, production pattern and input requirements, where the production pattern is determined by the relative prices, resource structure, agro-technical possibilities and the socio-economic environment.

Agricultural activities – i.e. feasible enterprises that can be undertaken by a farmer, for example the planting of a paddy crop in the *aman* season, is one activity, and the planting of a paddy crop in the *boro* season another – can be distinguished by type. The two main ones, closely interrelated in traditional agriculture, are crop and livestock production, to which processing of agricultural products and household activities can be added (trade, non-agricultural production, renting and hiring of production factors, etc.). Within the agricultural sector distinctions can be made between (agro-ecological) regions, farm sizes and applied technology level. For each region, each farm size and

Table 89. Example of output of the crop growth model and required farm and non-farm inputs for a specific crop on a specified land type under three types of management.

1310

R1/Aman paddy local /transplanted /start: 21/177 days
18 3

	C7IWL2	C7IIWL2	C7IIIWL2
	1306	1307	1308
YLDPADDY1373	–1931.00000	–2006.00000	–5236.00000
PADSTRAW1374	–4754.00000	–4936.00000	–12678.00000
FERTEQVS1322	0.00000	16.00000	192.00000
FRTCPT011323	0.00000	1.00000	1.00000
FRTCPT021324	0.00000	0.00000	1.00000
FRTCPT031325	0.00000	0.00000	1.00000
WEEDCT011328	1.00000	1.00000	1.00000
WEEDCT031330	1.00000	1.00000	1.00000
DRAINCAP1369	0.00000	0.00000	0.00000
IRRCAPDS1371	0.00000	0.00000	0.00000
LODGPREV1372	0.00000	0.00000	1.00000
PDCTRCAP1346	0.00000	0.00000	1.00000
PRRCCT151361	1.00000	1.00000	1.00000
PRRCCT161362	1.00000	1.00000	1.00000
SGSDCAP 1306	1.00000	1.00000	1.00000
HCARCT171431	1.00000	1.00000	1.00000
HCWRCT171493	1931.00000	2006.00000	5236.00000
HCSRCT171555	4754.00000	4936.00000	12678.00000

Table 89. (continued)

1336			
R1/Cotton	/sown/loc	var/start:	26/233 days
	C16IOL2	C16IIOL2	C16IIIL2
	1476	1477	1478
COTTSEED1394	−770.00000	−1015.00000	−3213.00000
COTSTALK1395	−3217.00000	−4241.00000	−13419.00000
YLDCOTTN1393	−415.00000	−547.00000	−1730.00000
FERTEQVS1322	0.00000	108.00000	451.00000
FRTCPT011323	0.00000	1.00000	1.00000
FRTCPT021324	0.00000	1.00000	1.00000
FRTCPT031325	0.00000	0.00000	1.00000
WEEDCT011328	1.00000	1.00000	1.00000
WEEDCT031330	1.00000	1.00000	1.00000
WEEDCT051332	1.00000	1.00000	1.00000
DRAINCAP1369	1.00000	1.00000	1.00000
PDCTRCAP1346	0.00000	1.00000	1.00000
CTSDCAP 1315	1.00000	1.00000	1.00000
THINCT021337	1.00000	1.00000	1.00000
THINCT041339	1.00000	1.00000	1.00000
HCACOT221460	1.00000	1.00000	1.00000
HCWCOT221522	415.00000	547.00000	1730.00000
HCSCOT221584	3217.00000	4241.00000	13419.00000

each level of technology a set of activities, as mentioned above, is constructed. Thus as the degree of detail increases, the number of activities within the agricultural supply model increases at an exponential rate. For example, two paddy growing activities (Aus – HYV and *aman* broadcast paddy) in two agro-ecological zones (Northwest and Sylhet), on small-owner operated and large-owner operated farms, with hand and animal traction technology under two levels of management (no fertilizer and 100 kg fertilizer per ha) would represent 32 individual activities in the model (2^n, n = 5). However, the many combinations may for the larger part be generated in a systematic way, as shown in the treatment of the physical crop-growth model.

The physical crop-growth model

In the physical crop-growth model, environmental characteristics such as soil type and climate are taken as points of departure for a step-wise calculation of possible outputs by crops (already outlined in previous chapters of this monograph). This, in combination with information on options for land development and production technologies, provides the basic information needed to model actual and potential agricultural production activities. The physical crop-growth model, providing the inputs used in the country model is characterized by the hierarchical structure as outlined in Chapters 1-4 of this monograph.The distinguished levels account for all physical factors that influence crop production and yields. The relationship between crop growth, nutrient uptake by the crop and nutrient application by the farmer is described in the fieldwork and management component and is accounted for, together with the other required farm inputs, in the supply model.

The type of information generated by the physical crop-growth model and the field work and management module is presented in Table 89. The heading of the first set of activities in the table indicates the region (R_1), the crop and variety (a local variety of *aman* paddy), the date the crop has to be transplanted (the 21st ten-day period) and the length of the growing period (177 days). The column codes, e.g. C7IWL2, indicate the crop species (C7), Activity Level (I) (i.e. lowest production level), water supply limited (W), and Land Unit (L2). The third column, C7IIIWL2, indicates Activity Level III, the highest production level, where production is only limited by radiation and temperature. In the second set of activities the same information is generated for cotton under optimal water conditions, as indicated by the O in the column headings, e.g. C16IOL2. The three columns give information on yields that can be obtained under specified conditions, indicated by a negative (−) sign on a hectare basis and on the required inputs per ha to achieve that yield, indicated by a positive sign. Most inputs assume the value 1, indicating that for that particular input the physical resources, such as man-hours of labour and a certain type of equipment are being requested from another data file called OPERAT for various levels of technology as specified in Section 6.1.

In this particular case, *aman* paddy will yield under the specified conditions 1931, 2006 and 5236 kg ha^{-1} of paddy and 4754, 4936 and 12678 kg ha^{-1} of paddy straw, respectively, if supplied with the specified required inputs, e.g.:

FERTEQVS 0.0000 16.0000 192.000

This means that at Activity Level I no fertilizer is required, while at Activity Level II and III 16.0 and 192.0 kg, respectively of pure N must be applied. The line

WEEDCF01 1.0 , 1.0 1.0

indicates that at all three activity levels weeding is required, where the coefficient (1.0) is a transfer coefficient calling for the appropriate technical coefficients of weeding 1 ha of *aman* paddy.

The livestock model

The farming systems in, for example, South and South-East Asia have been developed over a very long period in balance with the environment. Under the present circumstances, in which these regions are subject to a number of destabilizing influences such as, for example, high rates of population growth and economic recession, the sustainability of agriculture and, therefore, the environmental balance is being endangered. To be able to recommend better management options to redress this situation, the interrelationships that exist between the two sub-sectors of arable cropping and livestock and the socio-economic environment must be analyzed and, if possible, quantified because they are of considerable importance for the analysis of the problems and potentials of such farming systems.

Therefore livestock is modelled in close relation to the arable farming sector, both in terms of energy provided to cropping as draught power and manure and in terms of consumption of crop by-products, residues and waste. Moreover, grazing of range, forest and waste lands are considered in terms of area and yield (Figure 73). The livestock simulation model itself specifies the individual factors that determine feed intake and the conversion of feed into animal growth, production, and offspring. The model describes the individual animal as a system that converts energy into traction, weight gain, livestock products (milk, wool, meat), offspring or heat loss. These functional relationships provide an insight into the growth characteristics of an animal or herd and yield the estimated quantities of concentrate feed and roughage necessary to achieve the specified growth rates.

The type of data used in the supply model's linear programme consists of feed requirements for maintenance, growth, etc. for different types and age of

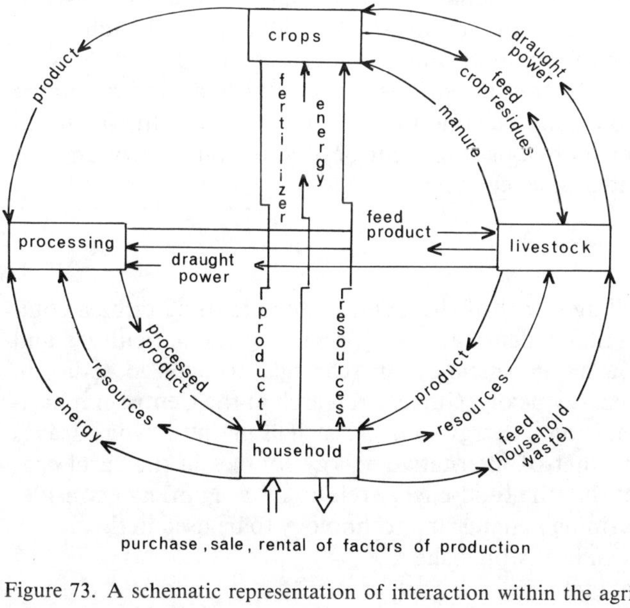

Figure 73. A schematic representation of interaction within the agricultural sector.

livestock. For example, Table 90 shows that the activity COWYMNGA – a young cow, male or female, on a maintenance and growth ration – requires COWYOUNG, i.e. one newborn animal, and will yield an animal one-year old of 55 kg. The FEEDDAYS indicate the number of days the animal needs to be kept on the indicated ration. DMINTAKE, CPINTAKE and MEINTAKE indicate the required dry matter (kg d^{-1}), digestable protein (g d^{-1}) and metabolizable energy (MJ d^{-1}) intake, respectively, on a daily basis. Finally, MANUREPR and NIMANURE indicate the daily manure production in kg of dry matter per day, and the pure nitrogen content in the produced daily manure in g d^{-1}.

The other activities, COWGMSGM, COWGMNGM, etc. define production and requirements for a growing animal (older than one year) that reaches mature weight at different rates of growth. COWGMSGM indicates slow growth of males (1870 days to reach mature weight) and COWGMNGM shows the requirements for normal growth of a male animal. The last activity refers to female animals. The energy and protein content of the various feeds and by-products, as well as waste, are also defined. These data are then linked with the feed availability for the domestic livestock. In this fashion the model will eventually find an equilibrium between the total required feed and roughage stock and the size of the livestock herd.

The integration of the crop and livestock model components into a comprehensive analytical system that is governed by the behavioural relationships of the farm households, is a unique characteristic of this modelling approach. Such a synthetic modelling approach that can subsequently be validated by field experiments, is particularly valuable as it permits quantitative description of alternative methods of crop and livestock production and enables the analysis of alternative and feasible farming systems. The integrated research on crop and livestock production in various farming systems under defined environmental and socio-economic conditions provides a basis for quantitative land evaluation, which obviously goes far beyond the pure agronomic or technical assessments undertaken elsewhere.

8.1.3 Farm energy

The agronomic modelling effort of the Centre attempts to describe a complex system of mutually dependent factors representing the agriculture and food sector. One important dependency, increasingly recognized today in agriculture, is the food-energy nexus. Current research at the Centre emphasizes the role played by traditional energy sources, such as manure, twigs, leaves and firewood, as well as that of alternative energy sources in the rural economy. There is no doubt that the food-energy relation is a significant consideration in the choice of farming systems and technology to be used in them.
Three energy functions can be distinguished:
– energy as agricultural input

Table 90. Activities for livestock growth in Bangladesh.

410
Feed requirements for Bangladesh cattle
9 5

		COWYMNGA	COWGMSGM	COWGMNGM	COWGMFGM	COWGMSGF
		450	451	452	453	454
COWYOUNG	430	1.00000	0.00000	0.00000	0.00000	0.00000
COWGROWG	431	−1.00000	1.00000	1.00000	1.00000	1.00000
COWADULT	432	0.00000	−1.00000	−1.00000	−1.00000	−1.00000
FEEDDAYS	500	365.00000	1870.00000	611.00000	266.00000	1661.00000
DMINTAKE	433	0.51000	3.61000	3.98000	4.74000	3.16000
CPINTAKE	434	27.60000	110.80000	151.20000	382.10000	98.10000
MEINTAKE	435	3.90000	22.63000	27.36000	40.79000	20.14000
MANUREPR	436	−0.26000	−1.99000	−1.99000	−1.90000	−1.74000
NIMANURE	437	−0.00290	−0.02190	−0.02190	−0.02090	−0.01910

- energy as agricultural output
- energy as domestic fuel.

Their integration in the agricultural modelling system permits a number of innovative analytical linkages. For example, the model can trace the effects of energy prices on production, income and the distribution of income in agriculture, or the implications of agricultural energy-extensive technology on energy and food supply. Furthermore, it can give insight into the production possibilities of 'green' energy, as embodied in wood, biogas and alcohol, and into the effects of relative energy scarcity on the productivity of land and livestock.

8.1.4 The socio-economic factors

The socio-economic factors that play such an important role in the process of resource allocation to various production activities, and in achieving income-maximization, are pervasive throughout the agricultural and rural development process. Land tenure and ownership patterns, farm size, family size, allocation of labour within the family, custom and religious factors, on-farm and off-farm employment opportunities, migration and market and credit facilities are all factors that must be accounted for when estimating and analyzing alternative production possibilities. Thus the socio-economic factors govern and constrain to a large extent the actual agricultural production process. The economic analyses based on these alternatives are necessary to supply sufficient knowledge and information to the government as a basis for its decisions to implement certain policies. Hence the rigorous use of quantitative models and methods can accomplish two things: increased insight in the planning situation and a qualitative improvement of decision making on the basis of existing knowledge.

8.1.5 Regional planning

Planning takes many forms and may serve a multitude of purposes and objectives. It comes about because national leaders are convinced that only part of the government's objectives will be realized if left to the forces of the free market and private enterprise. Regional planning may take place because a government pursues a balanced growth within or among regions. Therefore, regional development plans are formulated because a particular region is considered to have physical or socio-economic development potential, or because that region is poorly endowed physically and therefore lags behind in development relative to other regions. Planning models thus make it possible to analyse certain development opportunities through improved existing technologies or existing technologies, that are unused in the region, or completely new technologies derived from experiment stations. Thus the effect of introducing new technologies on, for example, regional income and employment can be estimated.

Development plans must thus describe the actual situation and formulate alternative courses of action that may lead to alternative development paths. To successfully plan and to be useful for decision making the information supplied must be of relevance to the planning situation of the policy makers and must relate to the instruments that the policy maker is able to apply. For example, if the government wants to increase food production it does not have the tools to physically force farmers to increase agricultural production (at least in market economies), but it can manipulate relative prices of agricultural commodities, or it may subsidize fertilizers or other inputs, or invest in land improvement or infrastructural works and thus stimulate agricultural production. The model should then be able to indicate which of these measures would be more effective in terms of satisfying or meeting the government's objectives. In another situation, if the government wants to increase government revenue by raising direct taxes, it might want to know whether such a measure would adversely affect the poor groups more than the higher-income groups; in other words, would it skew the income distribution in the right direction? Would such a measure affect the terms of trade of agriculture? Planning models must be able to accomodate such questions to be of use to decision makers.

8.1.6 International trade and aid

As C.T. de Wit stated in his introduction to this monograph, the food problem is not only a local or national problem, it is a problem with international political and economic dimensions, because food deficits and hunger do not stop at borders but transcend national boundaries in such forms as food aid and capital flows to achieve a more equal distribution of income between and within countries*.

One of the important features of the modelling exercise presented here is that all such country models can be linked to an international linkage system. The output of that system lends itself for an evaluation and analysis of trade policies. In particular it allows an analysis of the effects of international trade (and commodity) agreements and their impact on trade and balance of payments of developing countries and, even more importantly, on the income distribution effects within developing countries.

At the national level one can analyse the effects of international trade

* An alternative view would be that food aid and capital transfers to chronically food deficient countries are the other side of the same coin, namely international migration of labour to eventually achieve an optimal international division of labour. In other words, development aid can be seen to be some sort of pay-off to keep people within their boundaries.

measures and the resulting costs on developing countries. Thus, modelling the economies of individual countries, where each individual country model has been supplied with the policy instruments that each particular government has at its disposal, has the advantage that the system permits analysis of the effects of one country's or a group of countries economic policies on other countries, and vice versa. Thus, capital flows to developing countries resulting from a negative balance of payment add to the resources for development. The model then can be used to analyse the effects of alternative levels, compositions and allocations of these flows. Upon completion of the linkage system (with all its satellite models), the results may indicate those countries that can be expected to have continued food deficits and those population groups that may continue to be subject to hunger and malnutrition. The advantage of this modelling system over a more fragmented modelling approach is that the effects of any set of policies, both through international agreements and emanating from individual countries, can be analyzed.

8.1.7 Summary

The present national model may be used as a tool for analyzing the food situation and for determining policies conducive to combat hunger. Because national model building is done in close cooperation with local researchers and government institutions, the model should be suitable for transfer to local institutions. The presentation of the methodology in this monograph is an example of the way in which the models can be transferred. One of the Centre's objectives is to make its models fully available to the government of the country in question for use as a tool to make its own policy analyses. Policy analysis with the help of the model does provide government decision makers with a number of choices focussing on increasing food production and the implied growth of income and purchasing power of the lowest income groups. For any developing country such information is of crucial importance. For foreign donors it may also provide guidance in cases where they want to channel more of their resources to rural development and in particular to the basic needs of the lowest income groups.

9 COMPUTER MODELS OF CROP PRODUCTION

9.1 A FORTRAN model of crop production

J. Wolf, F.H. Rijsdijk and H. van Keulen

9.1.1 Introduction

In the preceding chapters of this monograph the method to calculate crop production for the various production situations has been illustrated using a desk – top calculator. However, when the method has to be applied in a more practical situation, where the number of alternatives in terms of soil types, weather conditions and crop species is large, the time required for calculations with a desk – top calculator may be prohibitively long. In these situations a microcomputer, that can perform repetitive calculations rapidly is indispensable. In this section a program is presented, written in FORTRAN 66, which is available on most microcomputers, that permits the execution of calculations for various site – specific basic data sets.

The terminology used in this section corresponds to that used by de Wit & Goudriaan (1972) and Penning de Vries & van Laar (1982) in earlier volumes of the Simulation Monographs series. A listing of the simulation model is given in Subsection 9.1.4. Because the acronyms used in this model are not always identical to those used in the preceding chapters, an alphabetical list of acronyms is provided (Subsection 9.1.5).

The model calculates crop growth and production in a dynamic way both for the situation of potential crop production (Production Situation 1) and for that of water – limited production (Production Situation 2). Production for the third production situation, i.e. unfertilized soil, is subsequently determined from the supply of nutrients by natural sources. This supply is introduced in the model as an exogenous variable, derived from fertilizer experiments (Section 5.3). In addition, the fertilizer application required to arrive at both potential and water – limited crop production is calculated from a exogenously derived fertilizer recovery fraction and from the calculated yields of marketable products and crop residues with their crop specific minimum nutrient concentrations. Finally, a yield level is determined as influenced by the effects of weeds, pests and diseases.

In Subsection 9.1.2, a short description is given of the functional relations and the data used in the model. Because they are mostly identical to the ones treated in the preceding chapters, it is superfluous to go into great detail; reference will be made to the equations or subsections, where more extensive information can be found. In Subsection 9.1.3, an outline is given of the data that have to be changed to use the simulation model in the reader's situation.

9.1.2 Description of the simulation model

The lines starting with a C in the first column are in FORTRAN convention comment lines; they are ignored during execution.

Lines

RANDOM

1 – 20 This routine is used to transform sequential files with data on soil characteristics, capillary rise and climate (Tables 92 to 94) to random access files in which the required data can be read per record number. Thus this program is not used during the crop growth simulation. For example, when new data on climate are available, the sequential file CLIMEAN.DAT has to be changed. Next by running the program RANDOM, CLIMEAN.DAT is transformed to a random access file. After that the calculations of crop growth can be executed.

RANDOMP

1 – 9 This routine is used to transform sequential files with data on crop characteristics (Table 95) to random access files in which the required data can be read per record number. It acts in the same way as RANDOM but is only used for plant data.

WOFAUTN

4 – 32 At compilation, the COMMON blocks of the data base called WOFOST.CMN are inserted.

33 – 35 Dimension declaration of data base. In the present example the data base is restricted to six land units, but this number can be expanded without any problems.

36 – 43 For each of the six land units the relevant climatic data set, soil texture classes, surface roughness, slope angle of the land, minimum and maximum depth of the groundwater table and relative area with particular soil texture class are specified. Names of some land units in Bangladesh are given in Table 91.

45 – 46 These lines specify for 24 crop species or cultivars the factors to convert the calculated dry matter yield of the economic plant parts to the marketable product.

47 Harvest losses as a fraction of the economic plant part are specified per crop species or cultivar.

Table 91. Soil texture classes, crop species, land units.

1. =	Coarse sand
2. =	Loamy sand
3. =	Fine sand
4. =	Fine sandy loam
5. =	Silt loam
6. =	Loam
7. =	Loess loam
8. =	Sandy clay loam
9. =	Silty clay loam
10. =	Clay loam
11. =	Light clay
12. =	Silty clay
13. =	Heavy clay
14. =	Peat
1. =	Spring Wheat
2. =	Rice HYV IR8
3. =	Rice BR 3 boro/aus
4. =	Rice BR 7 aus
5. =	Rice BR 9 boro
6. =	Rice BR 4 t. aman
7. =	Rice Nizersail t. aman
8. =	Jute *(C. capsilarius)*
9. =	Jute *(C. olitorius)*
10. =	Groundnut
11. =	Cassava
12. =	Maize
13. =	Lentil
14. =	Chickpea
15. =	Tobacco
16. =	Cotton
17. =	Sugar cane
18. =	Chillies
19. =	Rapeseed
20. =	Barley
21. =	Sweet potato
22. =	Onion
23. =	Sesame
24. =	Potato
1. =	N.E. Barind tract
2. =	N. Tista flood plain
3. =	Gangus tidal flood plain
4. =	Chittagong hill tract
5. =	Dacca se
6. =	Sylhet basin

48 – 50 Crops are divided into two groups that have a different susceptibility to pests and diseases (IDISR = 1 (less susceptible) or 2 (susceptible)). The relation between relative yield (see lines 132 – 147) and the reduction due to pests and diseases is specified in data sets TAB1 and TAB2 for group 1 and 2, respectively.

51 Specification of crop part that constitutes the marketable product: 1 (leaves), 2 (stems) or 3 (storage organs).

54 – 59 To open input files for data on specific crop growth calculations, on soil and crop characteristics and on capillary rise and climate, and to open the output file in which the results of the simulation will be written. The contents of these files are shown in Tables 92 to 96 and they are specified in subroutines PLANTR, CRDAT, CLIMIN, SOILR.

61 – 69 Read specification of land unit, crop species and starting date of crop growth for which the calculation has to be carried out. The dates are counted according to the Julian calendar for a standard year of 365 days. For the specified land unit, the climatic data set, surface roughness, slope angle of the land and minimum and maximum groundwater depth are found.

70 – 74 For the specified land unit, the calculations are carried out for five soil texture classes, unless the relative area for a given soil texture class equals 0. If IOPT is larger than zero, the crop growth simulation without water balance is omitted.

76 – 84 Call subroutines, which are described further on.

86 – 91 A variable IPS controlling output is specified, the variable specifying day number is set equal to the starting date, the base uptake of nitrogen and phosphorus and the recovery fraction of applied nitrogen and phosphorus fertilizer are given.

94 – 100 Call subroutines, which are described later. Crop production is first calculated without water balance (IWB = 0 ; potential production) and subsequently with water balance (IWB = 1).

102 – 113 The output of the crop – growth simulation model is specified. The calculated production of leaves, stems and storage organs is given for potential production (1 – 3), water – limited production (4 – 6), nutrient – limited production (7 – 9) and water-– plus – nutrient – limited production (10 – 12), respectively.

114 – 120 The calculated production of leaves, stems and storage organs, expressed in dry matter, is corrected for harvest losses, and is multiplied by 1.15 to arrive at air – dry material. Depending on the marketable product (leaves, stems or storage organs, i.e. IHP = 1, 2 or 3), the harvested product is converted into the form in which it is generally expressed in the agricultural statistics. Hence, grains are expressed on an air – dry basis, brown rice is converted to paddy; jute is expressed in fibre; groundnut,

sesame, lentil, rape-seed and chickpea are expressed in air-dry seed; cassave in dry matter; tobacco and sugar-cane in air-dry leaves and stems, respectively; onions, sweet potato and potato in fresh tubers; cotton in seed plus lint and chillies in dry matter. The sequence of the 24 crop species or cultivars is given in Table 91.

122 – 131 For the marketable plant organ, the additional yield per unit applied fertilizer nutrient, both for the potential and the water-limited production situation, is calculated.

132 – 147 For Production Situation 4, the effects of weeds, pests and diseases on production are taken into account. The yields in the absence of control measures are calculated from the water-limited and water-plus-nutrient-limited production through application of a reduction factor. The value of this reduction factor is a function of the relative yield, i.e. the yield compared to the potential yield. The relation is such that a greater reduction is assumed at higher yield levels. Two crop groups are distinguished comprising less and more susceptible crop species (lines 48 – 50).

149 – 160 Write in the output file of a particular crop growth simulation the relevant land unit, the texture class with its relative area, crop species, starting date of crop growth, total transpiration for potential and water — limited production and total surface runoff during the crop growth period. Write further the yield increment per unit of applied nitrogen or phosphorus (legumes only) fertilizer, and the production of leaves, stems and storage organs for the potential and the water-limited production situation, for the nutrient-limited and the water- plus-nutrient-limited production situation, both with and without reduction for the effects of weeds, pests and diseases. The limiting nutrient is normally nitrogen, but for nitrogen-fixing crop species, such as legumes, it is supposed to be phosphorus. At the end of a complete sequence of calculations the files are closed and the calculations stopped.

BLOCK DATA

164 – 192 At compilation the COMMON blocks, called WOFOST.CMN, are inserted.

194 – 200 A data block is given here with initial or default values for the variables used in the calculations of crop growth and the water balance. The soil default values refer mainly to Texture Class 7, i.e. loess loam. For other texture classes, data are read from an input file and replace the default values.

202 – 207 The relation between air temperature and a reduction factor for gross CO_2 assimilation rate is given for C_4 and C_3 crop species. The C_3 species are subdivided into those normally growing under temperate and tropical conditions, respectively. The base uptake of nutrients is defined for crops with a growth period of 120 days. For crops with shorter or longer growth periods the base uptake is corrected on the basis of the relation given.

FUNCTION AFGEN

209 – 232 This function allows linear interpolation in a table for a given value of the independent variable X. The table consists of 15 values for the independent variable (odd numbers, i.e. TABLE(1), (3) ...) and of 15 values for the corresponding dependent variable (even numbers, i.e. TABLE(2), (4) ...) The value of the independent variable X is compared with the values of the odd numbers. When the value of TABLE(I) exceeds the value of X for the first time, the slope of the function is calculated from the values of the independent variables nearest to X. This slope is equal to the difference between two adjacent even numbers divided by the difference between the two corresponding odd numbers. The value that results from the interpolation in the tables for the given value X is called AFGEN. For example, TABLE(5) is the first value larger than X. Then,

SLOPE = (TABLE(4) – TABLE(6)) / (TABLE(3) – TABLE(5))

AFGEN = TABLE(6) + SLOPE • (X — TABLE(5))

If the independent variable X is larger than or equal to the largest independent TABLE value or is smaller than or equal to the smallest independent TABLE value, the largest or the smallest dependent TABLE value is chosen.

SUBROUTINE SOILR

236 – 264 At compilation the COMMON blocks, defined in WOFOST.CMN, are inserted.

266 – 271 Read from the input file on unit 7 starting on record 1 for soil texture class, 1 record 2 for texture class 2, etc. In total, data for 14 soil texture classes (Table 91) are included, which are given in Table 92. The data provided per texture class are: saturated hydraulic conductivity, the suction limit (Equation 31), the transmission – zone conductivity, empirical constants 1 and 2, respectively (Equations 30, 31), geometry factor (Equation 27), slope angle of the land, surface roughness, clod/furrow angle,

initial groundwater depth, standard sorptivity and total pore space. In the present set – up the slope angle is associated with a texture class. That is not always necessary, however. The maximum rooting depth is assumed equal to 150 cm.

272 – 275 In these lines a message is defined that appears if soil data for a texture class cannot be read.

SUBROUTINE PLANTR

279 – 307 At compilation the COMMON blocks, defined in WOFOST.CMN, are inserted.

310 – 328 Read from the input file on unit 8 starting on record 1 for crop 1, record 25 for crop 2, record 49 for crop 3, etc. Data for 24 crops or cultivars (Table 91) are included, which are given in Tables 95 and 96. For each crop, the following data are supplied:

– name of crop

– crop type (1 = C_4 crop; 2 = C_3 crop, adapted to temperate regions; 3 = C_3 crop, adapted to (sub)tropical regions; 4 = leguminous crop); with or without air ducts (1 or 0); increase in maintenance respiration rate per 10 °C temperature increase; extinction coefficient for total radiation; efficiency of converting assimilates into leaf, storage organ, root and stem structural dry matter, respectively; life span of leaves; relative maintenance respiration rate of leaves.

– relative maintenance respiration rate of storage organs, roots and stems, respectively; extension rate of roots; relative death rate of leaves due to water stress; relative death rate due to ageing of roots and stems, respectively; specific pod area; specific stem area; minimum nitrogen concentration in storage organs.

– development rate dependent or not on day length; initial rooting depth; initial total dry weight (living); maximum development rate for the pre – anthesis period; critical day length for maximum and for zero development; maximum development rate for the post – anthesis period; minimum nitrogen concentration in crop residues; minimum phosphorus concentration in storage organs and in crop residues, respectively.

– tables of the partitioning factors of assimilates to roots, leaves, stems and storage organs, respectively and the specific leaf area, all as a function of the development stage.

– table of the relative development rate as a function of temperature.

Table 92. The sequential file SOILDATA.DAT

```
1120.        80.      119.23    0.08    .224    .1000   1.5    10.    30.    200.
      50.16      .395      30.         .25      3.       .1
  26.5      200.      17.80     16.4    .0398   .0330   1.5    10.    30.    200.
      19.20      .439      30.         .25      3.       .1
  50.0      175.      30.33     10.9    .050    .0288   1.5    10.    30.    200.
      21.44      .364      30.         .25      3.       .1
  12.0      300.       9.36     26.5    .0248   .0207   1.5    10.    30.    200.
      17.57      .504      30.         .25      3.       .1
   6.5      300.       5.32     47.3    .020    .0185   1.5    10.    30.    200.
      14.46      .509      30.         .25      3.       .1
   5.0      300.       3.97     14.4    .0231   .0180   1.5    10.    30.    200.
      11.73      .503      30.         .25      3.       .1
  14.5      130.       8.88     22.6    .049    .0169   1.5    10.    30.    200.
      13.05      .455      30.         .25      3.       .1
  23.5      200.      16.51     33.6    .0353   .0096   1.5    10.    30.    200.
      19.05      .432      30.         .25      3.       .1
   1.5      300.       1.18     36.0    .0237   .0105   1.5    10.    30.    200.
       6.15      .475      30.         .25      3.       .1
   0.98     300.       0.76      1.69   .0248   .0058   1.5    10.    30.    200.
       4.70      .445      30.         .25      3.       .1
   3.5      300.       2.66      2.77   .0274   .0085   1.5    10.    30.    200.
      10.74      .453      30.         .25      3.       .1
   1.3       50.       0.80     28.2    .048    .0065   1.5    10.    30.    200.
       3.98      .507      30.         .25      3.       .1
   0.22      80.       0.15      4.86   .038    .0042   1.5    10.    30.    200.
       1.93      .540      30.         .25      3.       .1
   5.3       50.       1.86      6.82   .1045   .0112   1.5    10.    30.    200.
       7.44      .863      30.         .25      3.       .1
```

– table of the soil – water depletion fraction as a function of potential transpiration.

329 – 332 In these lines a message is defined that appears if a set of crop data cannot be read

SUBROUTINE CRDAT

336 – 364 At compilation the COMMON blocks, defined in WOFOST.CMN, are inserted.

366 – 371 Read from the output file on unit 9 starting on record 1 for texture class 1, record 11 for texture class 2 etc. In total, data are provided for 14 texture classes. These data are derived from Rijtema (1969) and relate the rate of capillary rise (8 values) and the soil suction in the rooted zone (10 values) to the distance between root zone and groundwater level (RIJTEM (10,8)). An example of this input file derived from Tables 21 to 34 is shown in Table 93. These data are used in subroutine RIJTEMA to calculate the rate of capillary rise.

372 – 375 In these lines a message is defined that appears if a set of data cannot be read.

SUBROUTINE CLIMIN

379 – 407 At compilation the COMMON blocks, defined in WOFOST.CMN, are inserted.

410 – 432 Read from the input file on unit 10 starting on record 1 for climatic data – set 1, record 14 for climatic data – set 2 etc. In total 6 data sets are included. From each set the following data are read:

– name of the location for which the climate applies and its geographical latitude

– values for the 12 monthly averages of daily air temperature, monthly rainfall, daily potential evapotranspiration rate, daily potential evaporation rate, daily gross CO_2 assimilation rate of a closed C_3 or C_4 crop canopy, respectively, monthly effective irrigation and the number of rainy days per month. An example of this input file is shown in Table 94. The climatic data are entered in tables derived from the data statement DAYNUM that contain initially only the day numbers (the middle of each month in the Julian calendar) and zeros for the corresponding climatic data (lines 408 – 411).

433 – 447 For the potential evaporation rate and evapotranspiration rate the monthly average value is calculated and entered in the tables. For each of the climatic variables the values for the middle of the

first and the last month of the year are averaged to obtain the values for the first and the last day of the year.

Table 93. Part of the sequential file RIJTEMA.DAT, containing data for 4 soil texture classes.

20.0	20.0	20.0	20.0	20.0	20.0	20.0	20.0
43.3	44.1	45.0	46.1	46.8	47.6	48.4	49.4
44.4	45.4	46.7	48.5	49.8	51.6	53.9	58.8
44.5	45.5	46.7	48.6	49.9	51.7	54.0	59.2
44.5	45.5	46.8	48.6	49.9	51.8	54.1	59.5
44.5	45.5	46.8	48.6	49.9	51.8	54.2	59.7
44.5	45.5	46.8	48.6	49.9	51.8	54.2	59.8
44.5	45.5	46.8	48.6	50.0	51.8	54.3	59.9
44.5	45.5	46.8	48.6	50.0	51.9	54.3	60.0
44.5	45.5	46.8	48.6	50.0	51.9	54.3	60.0
3.7	4.3	5.0	6.0	6.6	7.5	8.3	9.4
11.6	13.6	16.5	20.9	24.2	28.9	34.4	43.1
13.9	16.4	20.1	26.2	31.0	38.3	48.1	68.8
14.5	17.2	21.1	27.6	32.2	41.1	52.7	81.5
14.7	17.5	21.5	28.2	33.7	42.4	54.8	87.6
14.9	17.7	21.8	28.7	34.3	43.3	56.4	92.3
15.1	17.9	22.1	29.1	34.9	44.1	57.7	96.4
15.2	18.0	22.3	29.4	35.2	44.6	58.5	98.7
15.3	18.1	22.4	29.5	35.5	45.0	59.1	100.6
15.3	18.2	22.5	29.6	35.6	45.2	59.5	101.6
19.7	19.7	19.8	19.9	19.9	19.9	20.0	20.0
47.9	48.3	48.7	49.1	49.3	49.6	49.7	49.9
82.0	84.5	87.4	90.8	92.7	94.8	96.7	98.9
92.8	97.3	103.3	111.7	117.9	126.7	138.3	165.8
94.1	99.0	105.4	115.0	122.1	133.1	148.7	194.6
95.1	100.2	107.1	117.4	125.4	138.0	156.8	218.1
95.9	101.3	108.5	119.5	128.2	142.2	163.8	238.9
96.4	101.9	109.3	120.7	129.9	144.6	167.9	251.0
96.8	102.4	109.9	121.7	131.2	146.2	171.2	260.9
97.0	102.6	110.3	122.3	131.9	147.6	172.9	266.1
19.1	19.3	19.5	19.7	19.7	19.8	19.9	20.0
47.0	47.5	48.1	48.7	49.0	49.4	49.6	49.9
88.0	90.1	92.3	94.7	95.5	97.2	98.3	99.4
137.0	145.4	155.5	169.8	179.5	192.3	206.9	230.0
140.9	150.0	162.0	179.4	192.0	210.6	235.5	297.1
143.3	153.1	166.0	185.3	199.4	222.4	254.7	350.6
145.4	155.6	169.4	190.4	206.7	232.5	271.6	399.9
146.6	157.1	171.4	193.4	210.7	238.4	281.5	429.3
147.5	158.3	173.0	195.8	213.9	243.2	289.5	453.2
148.0	158.9	173.9	197.1	215.6	245.8	293.7	465.9
18.3	18.6	18.9	19.3	19.4	19.6	19.8	19.9
44.2	45.3	46.3	47.5	48.1	48.7	49.2	49.7
81.2	84.2	87.6	91.3	93.3	95.4	97.2	99.0

Table 93. (continued)

127.7	137.2	149.2	165.7	176.8	191.3	207.3	231.3
134.4	145.4	160.0	181.4	197.2	220.5	252.1	326.9
138.6	150.7	167.0	191.9	211.1	241.0	285.4	414.7
142.3	155.3	173.0	201.0	223.2	259.1	315.2	500.3
144.4	157.9	176.6	206.3	230.2	269.6	322.8	552.2
146.1	160.1	179.4	210.6	236.0	278.2	347.0	594.7
147.0	161.2	180.9	212.8	239.0	282.8	354.6	617.4
17.7	18.2	18.6	19.0	19.3	19.5	19.7	19.9
42.2	43.5	45.0	46.5	47.3	48.2	48.9	49.6
74.0	77.7	82.1	87.0	89.8	92.9	95.6	98.5
102.6	111.0	122.1	137.8	148.8	164.0	182.0	214.3
104.6	113.7	125.7	143.1	155.7	174.2	198.4	256.5
105.9	115.3	127.8	146.3	160.1	180.6	209.1	287.2
107.0	116.7	129.7	149.1	163.8	186.2	218.3	314.4
107.7	117.5	130.7	150.7	165.9	189.4	223.7	330.5
108.2	118.2	131.6	152.0	167.7	192.0	228.0	343.5
108.5	118.5	132.1	152.7	168.6	193.4	230.3	350.5
18.9	19.1	19.3	19.5	19.7	19.8	19.9	20.0
43.8	44.9	46.0	47.3	47.9	48.6	49.1	49.7
65.4	69.0	73.3	78.9	82.4	86.8	91.1	96.6
71.5	76.4	83.0	92.7	100.1	111.2	126.6	165.5
74.2	79.8	87.4	99.3	108.8	124.0	147.3	218.3
76.2	83.3	90.8	104.4	115.6	134.1	163.8	264.7
78.0	84.5	93.7	108.8	121.3	142.7	178.7	307.0
79.0	85.8	95.4	111.3	124.7	147.8	168.7	332.1
79.8	86.8	96.8	113.4	127.5	151.9	193.5	352.5
80.3	87.3	97.5	114.4	128.9	154.1	197.1	363.4
19.4	19.5	19.6	19.7	19.8	19.9	19.9	20.0
47.3	47.8	48.3	48.9	49.1	49.4	49.7	49.9
85.1	87.5	90.1	93.0	94.6	96.3	97.7	99.2
110.0	116.3	124.4	136.0	144.3	156.0	170.9	201.7
113.9	121.2	130.9	145.7	157.0	174.6	200.3	272.7
117.0	125.0	136.0	153.2	167.0	189.4	224.5	338.6
119.6	128.2	140.3	159.6	175.6	202.3	245.8	400.5
121.1	130.1	142.8	163.4	180.6	209.8	258.3	437.6
122.3	131.6	144.8	166.5	184.6	215.9	268.4	467.9
123.0	132.4	145.9	168.1	186.8	219.1	273.8	484.0
14.0	14.9	15.9	17.1	17.7	18.4	19.0	19.7
31.0	33.5	36.5	40.0	42.1	44.4	46.4	48.8
48.1	53.1	59.4	67.9	73.3	80.0	86.6	94.9
58.2	65.3	75.0	89.5	100.1	115.3	134.7	175.0
61.4	69.4	80.5	97.4	110.6	130.8	159.4	235.7
64.7	73.5	85.9	105.6	121.3	146.7	185.2	305.6
67.5	76.9	90.5	112.5	130.5	160.4	208.0	371.7
69.1	79.0	93.2	116.5	135.9	168.5	221.4	411.5
70.4	80.6	95.3	119.8	140.2	175.0	232.2	443.9

Table 93. (continued)

71.1	81.5	96.5	121.5	142.6	178.5	238.0	461.1
12.1	13.1	14.3	15.8	16.7	17.7	18.5	19.5
25.6	28.3	31.7	36.0	38.6	41.7	44.7	48.1
37.6	42.4	48.8	57.7	63.8	71.8	80.3	92.0
43.6	49.7	58.2	71.2	80.9	95.2	113.7	153.9
43.9	50.1	58.7	71.8	81.8	96.4	115.8	160.2
44.0	50.3	58.9	72.2	82.3	97.2	117.1	164.0
44.1	50.4	59.1	72.5	82.7	97.9	118.2	167.2
44.2	50.5	59.3	72.7	83.0	98.2	118.8	169.1
44.3	50.6	59.4	72.9	93.2	98.5	119.4	170.7
44.3	50.6	59.4	73.0	83.3	98.7	119.6	171.5
17.1	17.6	18.1	18.7	19.0	19.3	19.6	19.9
40.8	42.4	44.0	45.8	46.8	47.8	48.6	49.5
73.4	77.4	81.9	87.0	89.9	93.0	95.6	98.5
114.5	124.7	137.9	156.0	168.3	184.5	202.4	229.3
122.0	134.0	150.1	173.8	191.4	217.3	252.0	332.9
127.1	140.3	158.3	186.1	207.7	241.3	290.7	432.8
131.3	145.6	165.4	196.7	221.8	262.4	325.5	532.3
133.8	148.7	169.6	203.0	230.1	274.9	346.1	593.1
135.8	151.2	173.0	208.0	236.8	284.9	362.9	642.9
136.9	152.6	174.7	210.7	240.4	290.3	371.8	669.6
12.3	13.3	14.5	15.9	16.8	17.2	18.6	19.5
22.3	24.8	28.0	32.3	35.2	38.8	42.4	47.1
28.0	31.7	36.7	44.4	50.0	58.1	67.9	84.8
33.8	38.9	46.2	58.1	67.7	83.0	105.0	158.0
37.1	43.1	51.7	66.3	78.5	98.8	130.2	220.6
39.7	46.2	56.0	72.6	86.9	111.3	150.7	277.1
41.9	49.0	59.6	78.0	94.1	122.1	168.6	329.4
43.1	50.5	61.7	81.2	98.4	128.4	179.1	360.7
44.2	51.8	63.4	83.8	101.8	133.5	187.6	386.1
44.7	52.5	64.3	85.1	103.6	136.3	192.1	399.7
4.7	5.5	6.7	8.6	10.0	12.0	14.3	17.6
7.9	9.5	11.7	15.5	18.5	23.1	29.0	39.8
9.4	11.3	14.1	19.0	23.0	29.5	38.7	60.1
10.4	12.6	15.8	21.6	26.5	34.6	47.0	82.2
11.0	13.3	16.8	23.0	28.4	37.5	51.7	95.9
11.4	13.8	17.6	24.1	29.9	39.7	55.4	106.8
11.8	14.3	18.2	25.1	31.1	41.5	58.5	116.1
12.0	14.6	18.5	25.6	31.8	42.6	60.3	121.5
12.2	14.8	18.8	26.0	32.4	43.5	61.8	125.9
12.3	14.9	19.0	26.3	32.7	44.0	62.6	128.3
15.4	16.1	16.9	17.8	18.3	18.8	19.3	19.7
22.9	24.8	27.1	30.4	32.7	35.8	39.3	45.0
24.4	26.6	29.6	34.0	37.4	42.5	49.5	66.5
25.9	28.5	32.0	37.6	42.2	49.5	60.8	95.6
26.7	29.5	33.4	39.7	44.9	53.5	67.4	114.4
27.3	30.3	34.4	41.2	46.9	56.6	72.5	129.5
27.9	30.9	35.3	42.5	48.7	59.2	76.9	142.5
28.2	31.3	32.8	43.3	49.7	60.8	79.5	150.1
28.4	31.6	36.2	43.9	50.5	62.0	81.5	156.3
28.5	31.8	36.3	44.3	51.0	62.7	82.6	159.6

Table 94. The sequential file CLIMEAN.DAT for six sites in Bangladesh.

N.E. Barind tract					25.5	
17.6	1.01	.159	.213	595.	468.	1.
19.8	1.23	.249	.311	678.	530.	1.
24.5	2.01	.386	.459	799.	619.	2.
28.3	5.34	.524	.614	879.	679.	4.
28.7	16.6	.532	.640	869.	678.	9.
28.7	34.1	.464	.564	736.	593.	19.
28.9	39.8	.408	.493	611.	510.	16.
28.8	32.7	.411	.504	654.	533.	16.
28.4	30.5	.364	.445	584.	478.	13.
26.7	13.3	.337	.422	666.	525.	6.
22.4	.950	.220	.285	626.	491.	1.
18.8	.080	.149	.202	587.	462.	1.
N. Tista flood plain					25.75	
17.3	1.13	.156	.210	571.	451.	1.
19.5	1.48	.240	.311	704.	548.	1.
23.5	2.82	.366	.448	798.	618.	2.
27.7	8.46	.488	.584	842.	655.	4.
27.8	29.8	.487	.594	831.	653.	9.
28.3	49.2	.436	.532	697.	568.	19.
28.7	41.1	.429	.526	690.	562.	16.
29.2	33.3	.431	.527	690.	558.	16.
28.9	31.7	.382	.467	617.	500.	13.
26.5	16.6	.347	.431	637.	505.	6.
22.4	1.00	.195	.294	576.	457.	1.
18.9	.140	.149	.198	520.	415.	1.
Gangus tidal flood plain					22.5	
19.6	1.59	.215	.271	581.	460.	1.
22.2	1.90	.305	.371	654.	516.	1.
26.6	4.81	.435	.513	709.	559.	3.
29.0	10.4	.520	.618	779.	613.	6.
29.5	19.6	.524	.631	793.	627.	11.
28.7	42.3	.412	.496	585.	492.	16.
28.2	45.1	.396	.482	588.	493.	12.
28.3	39.5	.375	.451	532.	452.	16.
28.6	28.5	.400	.488	630.	510.	12.
27.7	19.0	.353	.434	622.	497.	7.
23.9	4.23	.260	.327	605.	478.	1.
20.5	.990	.189	.250	601.	473.	1.

Chittagong hill tract					23.	
19.6	1.74	.217	.276	588.	465.	1.
21.6	2.88	.308	.380	710.	553.	1.
25.4	5.86	.443	.532	788.	612.	2.
27.8	13.4	.549	.654	848.	659.	5.
28.4	25.1	.548	.662	863.	674.	10.
27.8	49.3	.463	.559	688.	561.	15.
27.4	54.3	.426	.518	644.	531.	19.
27.3	46.9	.430	.529	691.	558.	15.
27.7	30.8	.396	.484	629.	509.	11.
27.1	15.8	.350	.423	573.	464.	7.
24.0	6.36	.273	.351	666.	520.	2.
20.7	3.81	.198	.258	562.	446.	1.
Dacca se					24.	
18.6	.560	.214	.268	587.	464.	1.
21.2	.530	.312	.373	664.	522.	1.
26.0	4.65	.482	.555	738.	578.	3.
28.8	8.99	.597	.690	797.	625.	6.
29.3	24.7	.608	.719	847.	663.	11.
28.6	32.8	.469	.561	657.	540.	16.
28.1	33.4	.454	.550	675.	551.	12.
28.5	33.9	.454	.547	665.	541.	16.
28.5	23.5	.435	.523	630.	510.	12.
27.1	15.2	.386	.467	624.	497.	7.
23.2	10.8	.273	.340	625.	492.	1.
19.3	3.44	.163	.240	576.	455.	1.
Sylhet basin					24.75	
18.5	1.96	.178	.231	574.	454.	2.
20.6	4.74	.261	.325	663.	520.	2.
24.1	8.18	.383	.454	720.	566.	3.
26.4	31.5	.442	.531	746.	591.	11.
26.8	49.4	.435	.534	744.	596.	18.
27.2	97.3	.374	.456	577.	488.	22.
28.0	73.5	.391	.477	606.	506.	25.
28.0	60.0	.382	.467	600.	498.	24.
27.8	42.5	.349	.426	560.	462.	16.
26.2	17.6	.320	.402	625.	497.	8.
22.9	3.53	.232	.298	604.	477.	1.
20.0	1.80	.168	.214	511.	409.	1.

SUBROUTINE APPLE

459 – 487 At compilation the COMMON blocks, defined in WOFOST.CMN, are inserted.

493 – 535 At initialization most variables are set to zero. In the default option the initial soil – moisture content is set equal to a moisture content at a matric suction of 100 cm. The soil – moisture contents at a matric suction of 100 cm and 16000 cm (wilting point) are calculated with Equation 27 and the maximum surface storage capacity with Equation 41. For the latter calculation, clod angle and slope angle are converted from degrees to radians.The calculation proceeds with time steps of 1 day (DELT).

537 – 549 The total amounts of roots, leaves, stems and storage organs at initialization are determined from the total amount of dry matter at emergence or at transplanting, and the partitioning factors for the various plant organs at development stage equal to zero. The total amount of dry matter at initialization is supplied as a crop – specific variable in the plant data. If different sowing or planting densities are to be examined, this variable can be changed. Both the specific leaf area and the partitioning factors depend on the development stage. The leaf area index is calculated from the specific leaf area and the leaf weight. The contribution of photosynthetically active stems and/or pods is also taken into account, if it is of importance.

551 – 556 Write in the output file the soil texture class, the starting day of crop growth, the name of the crop species and that of the climatic data set. Subsequently, write in the output file the day number, the amount of roots, leaves stems and storage organs, the leaf area index, the groundwater depth, the soil – moisture content in the root zone and the development stage, all referring to the date of emergence or transplanting. An example of this output file is shown in Table 97.

558 – 569 The dynamic part of the calculations that is repeated every time step DELT starts here. If the development stage equals two, the crop growth cycle is completed and the calculations are terminated. The duration of the crop growth period is calculated here. Data on temperature, rainfall, gross CO_2 assimilation rate, potential evapotranspiration rate, potential evaporation rate and effective irrigation rate for day IDAY are derived from subroutine CLIMAT. With subroutine DAYLEN the day length is calculated as function of date and geographical latitude. The fraction of intercepted radiation is calculated according to Equation 5. If control parameter IWB equals zero, the simulation run is executed without taking into account the water balance.

573 – 581 The volumetric soil - moisture content at field capacity is calculated with Equation 27 for a matric suction equal to the groundwater depth. The matric suction is calculated with the same equation, solved in the opposite direction for the actual volumetric soil – moisture content in the rooting zone (Equation 68). The hydraulic conductivity results from Equation 30, solved for this matric suction. If the matric suction is higher than a texture – specific suction limit, Equation 31 is used. The maximum infiltration rate is determined by the actual sorptivity plus the transmission zone permeability (Equations 33 and 34).

585 – 590 The actual evaporation rate is calculated from the potential rate, taking into account the reduction due to soil – moisture content and shading by the vegetation (Equations 37, 40 and 71). If crop growth is calculated without water balance, the maximum instead of the actual evaporation rate is calculated. In the case that a water layer covers the field, evaporation takes place out of this water layer instead of from the soil.

594 – 599 The change in surface storage depends on rainfall, irrigation and evaporation. The actual infiltration rate is the maximum of either the calculated change in surface storage or the maximum infiltration rate. The maximum infiltration rate could be corrected for rainfall distribution and slope angle. However, these functions are not yet defined in the model and do not affect the infiltration capacity. Subtracting the actual infiltration rate and the maximum surface storage capacity from the calculated surface storage results in the runoff, which is integrated for the whole crop growth period. These relations are given as Equations 42, 43 and 44.

603 – 614 Below the root zone, the direction of water flow depends on the distance between the bottom of the root zone and the groundwater level and on the matric suction in the root zone. If the suction exceeds this distance, capillary rise will occur, the rate of which is calculated in subroutine RIJTMA. This upward water flow can, however, not result in a moisture content in the root zone above field capacity. In the reverse case, i.e. gravity potential exceeds matric suction, natural drainage will take place (Equation 48), but only the amount of water in the root zone in excess of field capacity can be removed.

618 – 637 The potential and the maximum transpiration rate are calculated according to Equations 51 and 52. The critical soil – moisture content is calculated in accordance with Equation 53, with a soil – water depletion fraction depending on potential transpiration rate and crop species. The actual transpiration rate depends on the soil – moisture content. The reduction in transpiration

rate applied here differs to some extent from that used in Section 3.2 (Equations 54 and 55), especially at high soil – moisture contents. Here the decrease in transpiration rate is directly proportional to the soil – air content, between 0.10 and 0.05 $cm^3\ cm^{-3}$. Reduction in the transpiration rate does not occur in wet soil if the crop has air ducts (AIRDUC = 1).

641 – 642 The change in volumetric soil – moisture content in the root zone is determined by the balance between the actual infiltration rate, the rate of capillary rise, or drainage below the root zone, and the actual rate of evaporation and transpiration, taking into account the depth of the rooting zone (Equation 24). In the case that the simulation is executed without water balance, the actual rates of evaporation and transpiration are not calculated and the change in soil – moisture content is set to zero.

646 – 648 In situations where artificial drainage systems are available, the resulting drainage can be calculated with Equation 59, assuming that the depth of the groundwater table and the drains, the hydraulic conductivity of the soil, and the drain spacing and drain radius are known.

653 – 655 The depth of the groundwater table may change either by natural drainage or by capillary rise and artificial drainage (Equation 60). The pore space emptied or filled with water at a change in the groundwater depth is supposed to be equal to the air – filled pore fraction in the root zone. When groundwater enters the root zone, the change in groundwater depth and the change in soil – moisture content are calculated in a different way. The groundwater depth (ZTF) and the soil – moisture content (SMF) obtain fixed values at the moment that the groundwater enters the root zone (lines 775 – 777, 783 – 785). The change in groundwater depth is then calculated from the rates of infiltration, evaporation and transpiration using the fixed air – filled pore fraction. The relative change in groundwater depth, in comparison to ZTF, is multiplied with the fixed air – filled pore fraction to obtain the change in soil – moisture content.

665 – 679 The development rate before flowering depends on day length and temperature. After flowering, the development rate only depends on temperature and differs from that before flowering at the same temperature. The specific leaf area of the growing material and the partitioning factors for dry – matter allocation to leaves, stems, etc., depend on the development stage and are obtained by linear interpolation in crop – specific tables (Section 2.2).

683 – 689 The gross rate of CO_2 assimilation depends on the potential gross assimilation rate, the fraction of intercepted radiation and

the actual transpiration rate (Equation 5 and Section 3.3). The potential gross CO_2 assimilation rate is calculated before and added to the climatic data. The maintenance respiration rate depends on the amount of dry matter in the various plant organs, the relative maintenance respiration rate per plant organ and the temperature. It cannot exceed the gross assimilation rate, i.e. it is assumed that the vegetation will not be 'self – destructive' in terms of carbohydrates. Gross assimilation rate minus maintenance respiration rate results in the amount of assimilates available for conversion into structural plant material (Equation 6). The conversion efficiency of carbohydrates to structural plant material is calculated as the weighted average of the efficiencies for the various plant organs.

693 – 748 The growth rate of roots, leaves, stems and storage organs depends on the total dry matter increase of the crop, the partitioning factors and the death rate of each of the plant organs. Roots and stems start to die after development stage 1.5 has been reached. Dying of the leaves often starts earlier, because leaves not only die when the oldest leaves exceed their lifespan (DALV), they also die at an increasing rate with increasing water stress, i.e. a decrease in the ratio of actual transpiration rate to the maximum rate (DSLV1). In addition it is assumed that due to self – shading, the leaves in excess of a leaf area index of 7 will die (DSLV2). The growth rate of the leaves is calculated separately per time step, i.e. per day, via LV(1) = GRLV x DELT (line 727); at the same time the corresponding leaf area is calculated. Each day, leaf weight and leaf area are shifted by one class via lines 725 and 726. To the oldest class of leaves, the amount of leaves present at emergence or at transplanting is added (line 729). Leaves that have exceeded their life span will die (DALV = LV(ISPAN); line 721). However, if those leaves died because of stress (DSLV), dying due to senescence is suppressed (via line 709). If the amount of leaves from the earliest date, (LV(IL)), is insufficient to cover leaf death from stress, leaves that are 1 day, 2 days, etc., younger (LV(IL – 1), LV(IL – 2) etc.) will die (line 715). Thus, dying of leaves due to water stress takes place at the cost of the oldest leaves. The total leaf area is the result of combining the leaf areas of all leaf age classes (line 734).

752 – 759 The development stage and the dry weights of living roots, leaves, stems and storage organs are calculated from the amount at time t and the development or growth rate during time step DELT. Combining the dry weights of the living plant organs results in total above – ground dry weight and total dry weight (living) of the crop.

761 – 769 The dry weight of dead roots, leaves, stems and storage organs are calculated from the amount at time t and the death rate during time step DELT. The total dry weights of roots, leaves, stems and storage organs result from combining the weights of the living and the dead plant organs.

774 – 785 The soil – moisture content in the root zone follows from the rate of change in soil – moisture content, but it is confined between the moisture content of air – dry soil and that of almost saturated soil, respectively. When groundwater enters the root zone, fixed values for soil – moisture content (SMF) and for groundwater depth (ZTF) are established (lines 775 – 777 and 783 – 785). Subtracting the actual infiltration and runoff from the total surface storage, which is the total of the original surface storage plus the balance of rainfall, irrigation and evaporation (line 594), results in the new surface storage. The groundwater depth is determined by the change in groundwater depth, but it is confined between a minimum and a maximum value, which are specified for each land unit (lines 40 and 41), as a function of physiography, river level, etc. Root extension ceases when rooting depth has attained its maximum value.

786 – 796 For crops with air ducts this part is not relevant. When the rooting depth is less than 10 cm above the groundwater level, root extension ceases because oxygen shortage prevents it. When groundwater enters the root zone, roots are dying and the rooting depth decreases. When the amount of assimilates partitioned to the roots becomes zero, root extension will be zero too. The number of days the rooting depth is less than 10 cm is calculated via RTDF. After five days of waterlogging (RTDF = 5), a message of crop failure is printed instead of dry weights of plant organs, etc., and the calculation is terminated.

798 – 800 The green area index (leaf, pod and stem) is calculated from the specific areas and the weights of leaves, stems and pods. The rooting depth results from integration of the root extension rate. Water use by transpiration during the whole crop growth period is obtained by integrating actual transpiration.

TEMPORARY OUTPUT

804 – 817 Every 10 days, the date (Julian calendar day), the dry weights of roots, leaves, stems and storage organs, the leaf area index, the groundwater depth, the soil – moisture content and the development stage are written in the output file (Table 97). The calculations are then continued for the next day. When the development stage reaches two or the leaf weight becomes zero, crop

Table 95. Part of the sequential file PLANTDAT.DAT with data for wheat and rice.

WHEAT	2.	0.	2.	.6	.72	.73	.72	.69	60.	.03
	.01	.01	.015	1.2	.03	.02	.02	0.	0.	.01
	.0	10.	100.	0.0292	1.	0.	.0389	.004	.0011	.0005
	0.	.5	.1	.42	.2	.33	.3	.26	.4	.2
	.5	.15	.6	.1	.7	.07	.8	.04	.9	.02
	1.	0.	2.	0.	0.	0.	0.	0.	0.	0.
	0.	1.	.1	1.	.2	1.	.3	.81	.4	.65
	.5	.51	.6	.38	.7	.25	.8	.12	.9	.06
	1.	0.	2.	0.	0.	0.	0.	0.	0.	0.
	0.	0.	.2	0.	.3	.19	.4	.35	.5	.49
	.6	.62	.7	.75	.8	.88	.9	.43	1.	0.
	2.	0.	0.	0.	0.	0.	0.	0.	0.	0.
	0.	0.	.8	0.	.9	.51	1.	1.	2.	1.
	0.	0.	0.	0.	0.	0.	0.	0.	0.	0.
	0.	0.	0.	0.	0.	0.	0.	0.	0.	0.
	0.	.002	2.	.002	.0	.0	.0	.0	.0	.0
	0.	0.	0.	0.	0.	0.	0.	0.	0.	0.
	0.	0.	0.	0.	0.	0.	0.	0.	0.	0.
	0.	0.	35.	1.	45.	1.	0.	0.	0.	0.
	0.	0.	0.	0.	0.	0.	0.	0.	0.	0.
	0.	0.	0.	0.	0.	0.	0.	0.	0.	0.

Table 95 (continued)

RICE HYV-IR8	.2	.885	.3	.8	.4	.7	.5	.6	.6	.55
	.7	.515	.8	.465	.9	.43	1.	.4	0.	0.
	3.	1.	2.	.6	.72	.73	.72	.69	60.	.03
	.01	.01	.015	1.2	.03	.02	.02	0.	0.	.0083
	.0	10.	200.	0.0233	1.	0.	.0435	.004	.0011	.0005
	0.	.6	.1	.325	.2	.225	.3	.14	.4	.075
	.5	.075	.6	.07	.7	.07	.8	.055	.9	.04
	1.	0.	2.	0.	0.	0.	0.	0.	0.	0.
	0.	.94	.1	.59	.2	.55	.3	.53	.4	.52
	.5	.51	.6	.45	.7	.34	.8	.22	.9	.10
	1.	0.	2.	0.	0.	0.	0.	0.	0.	0.
	0.	.06	.1	.41	.2	.45	.3	.47	.4	.48
	.5	.49	.6	.55	.7	.66	.8	.78	.9	.38
	1.	0.	2.	0.	0.	0.	0.	0.	0.	0.
	0.	0.	.8	0.	.9	.52	1.	1.	2.	1.
	0.	0.	0.	0.	0.	0.	0.	0.	0.	0.
	0.	0.	0.	0.	0.	0.	0.	0.	0.	0.
	0.	.0025	2.	.0025	.0	.0	0.	0.	.0	.0

Table 95 (continued)

0.	0.	0.	0.	0.	0.	0.	0.	0.	0.
0.	0.	0.	0.	0.	0.	0.	0.	0.	0.
0.	0.	35.	1.	45.	1.	0.	0.	0.	0.
0.	0.	0.	0.	0.	0.	0.	0.	0.	0.
0.	0.	0.	0.	0.	0.	0.	0.	0.	0.
.2	.885	.3	.8	.4	.7	.5	.6	.6	.55
.7	.515	.8	.465	.9	.43	1.	.4	0.	0.

Table 96. Crop characteristics for wheat

CRPNAM									
WHEAT									
CROPT	AIRDUC	Q10	EXC	CVL	CVO	CVR	CVS	SPAN	RML
2.	0.	2.	.6	.72	.73	.72	.69	.60	.03
RMO	RMR	RMS	RR	PERDL	PERRT	PERST	SPA	SSA	NPERSO
0.1	.01	.015	1.2	.03	.02	.02	0.	0.	.01
DSL	RD	TDW	DVRC1	DLO	DLC	DVRC2	NPERVE	PPERSO	PPERVE
0.	10.	100.	.0318	1.	0.	.0389	.004	.0011	.0005
FRTB (1,30)									
0.	.5	.1	.42	.2	.33	.3	.26	.4	.2
.5	.15	.6	.1	.7	.07	.8	.04	.9	.02
1.	0.	2.	0.	0.	0.	0.	0.	0.	0.
FLTB (1,30)									
0.	1.	.1	1.	.2	1.	.3	.81	.4	.65
.5	.51	.6	.38	.7	.25	.8	.12	.9	.06
1.	0.	2.	0.	0.	0.	0.	0.	0.	0.
FSTB (1,30)									
0.	0.	.2	0.	.3	.19	.4	.35	.5	.49
.6	.62	.7	.75	.8	.88	.9	.43	1.	0.
2.	0.	0.	0.	0.	0.	0.	0.	0.	0.
FOTB (1,30)									
0.	0.	.8	0.	.9	.51	1.	1.	2.	1.
0.	0.	0.	0.	0.	0.	0.	0.	0.	0.
0.	0.	0.	0.	0.	0.	0.	0.	0.	0.

Table 96. (continued)

SLATB (1,30)									
0.	.002	2.	.002	0.	0.	0.	0.	0.	0.
0.	0.	0.	0.	0.	0.	0.	0.	0.	0.
0.	0.	0.	0.	0.	0.	0.	0.	0.	0.
DVRETB (1,30)									
0.	0.	35.	1.	45.	1.	0.	0.	0.	0.
0.	0.	0.	0.	0.	0.	0.	0.	0.	0.
0.	0.	0.	0.	0.	0.	0.	0.	0.	0.
SWDPTB (1,20)									
.2	.885	.3	.8	.4	.7	.5	.6	.6	.55
.7	.515	.8	.465	.9	.43	1.	.4	0.	0.

growth is terminated. In that case the same variables that are written every 10 days are saved in the output file for the last time.

818 – 825 The calculated weights of leaves, stems and storage organs, the growth duration and the total water use by transpiration for the potential production situation are renamed to be distinguished from those of the water – limited production situation.

SUBROUTINE CLIMAT

833 – 861 At compilation the COMMON blocks, defined in WOFOST.CMN, are inserted.

862 – 871 Air temperature, potential evaporation rate, potential evapotranspiration rate and effective irrigation are read from the climatic tables by linear interpolation for a given day number, designated RI40. With the exception of temperature, these data have to be converted from monthly to daily values. Depending on crop type, the gross assimilation rate on a given day is reduced in the case of sub – optimum air temperature. In addition, the assimilation rate expressed in carbon dioxide is converted to carbohydrates. Daily rainfall is obtained from subroutine RAINDA, using the data on monthly rainfall and the number of rainy days per month.

SUBROUTINE RAINDA

874 – 885 Rainfall on a given day is calculated in this subroutine from the total monthly rainfall and the number of rainy days per month. A data statement is given with 30 random numbers, i.e. the random day numbers in a month. From the given yearly day number (IDAG), the monthly day number (ICD) and the month number (ICM) are derived by using the data statement IDGN that contains the numbers of the first day per month.

887 – 902 Random day numbers (IRAND) are compared with the present day number (ICD) for an equal number of repetitions as the number of rainy days (RDA). If these day numbers are equal, rainfall (RN) on the given date is calculated from the monthly rainfall (RTB) for the given number of rainy days. Otherwise the rainfall on the given date is zero.

SUBROUTINE RIJTMA

909 – 944 The capillary rise is derived here from the tables of Rijtema (Tables 21 – 34). In these tables, the rate of capillary rise

(CCR(8)) is given as a function of the distance between the bottom of the rooted zone and the groundwater level (R(8, 10)) and the matric suction in the root zone (CMH(10)) for each texture class. If the matric suction is less than 20 cm, the tables cannot be used. In that low suction range, capillary rise is calculated with Equation 44. If the matric suction equals 20 cm, the values of R(1,1) to R(8,1) are used. If the actual distance is smaller than R(1,1) or larger than R(8,1), the capillary rise is 0.5 or 0.0 cm d^{-1}, respectively. Otherwise, the actual distance is compared with the values of R(2,1) to R(8,1). If, for example, the actual distance, D_i, is less than or equal to R(5,1), the slope of the function is first calculated as:

$$SLOPE = (CCR(4) - CCR(5)) / (R(4,1) - R(5,1))$$

and the resulting rate of capillary rise is:

$$C = CCR(5) - SLOPE \times (R(5,1) - D_i)$$

If the matric suction is equal to or higher than 16000 cm, the values of R(1,10) to R(8,10) are used. The same method is used as for a soil suction of 20 cm. Thus the capillary rise is 0.5 or 0.0 cm d^{-1}, if the actual distance is smaller than R(1,10) or more than R(8,10), respectively. Otherwise, for example:

$$SLOPE = (CCR(4) - CCR(5)) / (R(4,10) - R(5,10))$$

$$C = CCR(5) - SLOPE \times (R(5,10) - D_i)$$

If the actual matric suction is between 20 and 16000 cm, this suction has to be compared with the soil suction data given in the array CMH. If, for example, the actual suction is less than or equal to CMH(4), the slope of the relation between matric suction and distance is obtained from :

$$SLOPE = (R(I,3) - R(I,4)) / (CMH(3) - CMH(4))$$

and the resulting 8 values (I = 1,8) for the distances VMH obtained from interpolation for the actual matric suction values are:

$$VMH(I) = R(I,4) - SLOPE \times (CMH(4) - RMH)$$

In this case, too, the capillary rise is 0.0 or 0.5 cm d^{-1}, respectively, if the actual distance is larger than the highest (VMH(8)) or smaller than the lowest value (VMH(1)) of the distances found by the interpolation. Otherwise the actual distance is compared with VMH(2) to VMH(8). If, for example, the actual distance is less than or equal to VMH(5), the slope of the relation between capillary rise and distance in the table is obtained from:

$$SLOPE = (CCR(4) - CCR(5)) / (VMH(4) - VMH(5))$$

and the resulting rate of capillary rise is:

$$C = CCR(5) - SLOPE \times (VMH(5) - D_i)$$

SUBROUTINE DAYLEN

950 – 955 Day length is calculated here as a function of geographical latitude and declination of the sun. The declination is a function of the date.

FUNCTION ASIN

956 – 965 The arcsine function is, for example, used in the calculation of the day length. It is approximated here with the equations given because the microcomputer used does not have an arcsine function in its FORTRAN library.

SUBROUTINE NUTRI

972 – 1000 At compilation the COMMON blocks, defined in WOFOST.CMN, are inserted.

1002 – 1008 The base uptake of phosphorus and nitrogen obtained from independent data (Section 5.3) is given for a crop with a growth period of 120 days. For crops with shorter or longer growth duration, the base uptake is reduced or increased. The data for this relation are given in BLOCK DATA. The recovery fraction of applied fertilizers is also defined independently. The country to which this example applies (Bangladesh) showed only small variations in base uptake and recovery fraction, so only one value for all soil types is used. In a different situation, tables may be included relating the base uptake and the recovery fraction to soil texture class, land unit, region or soil type. The variables MANF1 and MANF2 are one at this moment, but they offer the possibility to express differences in capability of farmers to increase the recovery fraction of applied fertilizer or the base uptake, for example by split application of fertilizer, better seedbed preparation, water control, etc.

1009 – 1012 The required nitrogen uptake for the potential production situation is obtained by multiplying total dry weight of crop residues by their minimum nitrogen concentration and adding the product of dry weight of storage organs and their minimum nitrogen concentration. The calculation method for the nitrogen requirement in the water – limited production situation is identical to that for the potential production situation, but the amounts of leaves, stems and storage organs may be smaller due to water stress. The amount of nitrogen to be applied as fertilizer for potential or water – limited production is determined from the nitrogen requirements in these production situations minus the base uptake, divided by the recovery fraction of applied fertilizer.

1014 – 1024 In the case that no fertilizer is applied, the nitrogen uptake (base uptake) in Production Situation 3 is a fraction of the nitrogen uptake in the potential production situation. The amount of plant organs produced when no fertilizers are applied is reduced in proportion to this fraction. When, in addition to nitrogen, water also limits crop growth at some stage, the amounts of plant organs produced are assumed equal to the minimum amount found at either situation. The required nitrogen fertilizer application is calculated, both for the potential and for the water — limited production situation.

1029 – 1044 If leguminous crops are considered (CROPT = 4), the nitrogen requirement is largely met by symbiotic nitrogen fixation. Therefore phosphorus is the nutrient limiting the growth of legumes. The total phosphorus requirement, the required phosphorus fertilizer application for the potential and the water – limited production situations and the amount of plant organs produced without fertilizer application are calculated in the same way as described for nitrogen (lines 1009 – 1024).

1047 – 1051 Write in the output file on unit IPS: the amount of nitrogen or phosphorus to be applied for potential and water – limited production; the base uptake of nitrogen and phosphorus for a crop with a growth period of 120 days; the recovery fraction of applied nitrogen and phosphorus fertilizer; the minimum and maximum groundwater depth; the required fertilizer nutrient application for potential and water – limited production; the total water use by transpiration of crops growing in the potential and water – limited production situations; total surface runoff; dry weight of leaves produced in the potential and nutrient – limited production situations; dry weight of stems produced in the potential and nutrient – limited production situations; dry weight of storage organs produced in the potential and nutrient – limited production situations; dry weight of leaves in the water — limited and water — plus — nutrient — limited production situations; dry weight of stems in the water — limited and water — plus — nutrient — limited production situations; dry weight of storage organs in the water and water — plus — nutrient — limited production situations. An example of this output file is shown in Table 97.

Table 97. Output file.

Texture class 6. Start day 100

Crop Maize Climate N. Tista Flood Plain

DAY	WRT	WLV	WST	WSO	LAI	ZT	SM	DVS
100	8.	8.	4.	0.	0.032	200.000	0.343	0.000
109	184.	267.	142.	0.	0.880	200.000	0.343	0.250
119	746.	1520.	996.	0.	3.458	200.000	0.343	0.503
129	1117.	2546.	2556.	0.	4.582	200.000	0.343	0.757
139	1240.	2818.	4273.	459.	4.798	200.000	0.343	1.011
149	1240.	2818.	4273.	2679.	4.798	200.000	0.343	1.288
159	1191.	2818.	4104.	4490.	4.798	200.000	0.343	1.567
169	973.	2551.	3353.	6011.	3.919	200.000	0.343	1.848
176	862.	1862.	2970.	6816.	2.279	200.000	0.343	2.018

Texture class 6. Start day 100

Crop Maize Climate N. Tista Flood Plain

DAY	WRT	WLV	WST	WSO	LAI	ZT	SM	DVS
100	8.	8.	4.	0.	0.032	200.000	0.343	0.000
109	184.	267.	142.	0.	0.880	194.639	0.216	0.250
119	647.	1227.	805.	0.	2.912	200.000	0.137	0.503
129	825.	1547.	1629.	0.	2.885	81.156	0.420	0.757
139	896.	1488.	2447.	279.	2.581	10.668	0.352	1.011
149	896.	1193.	2447.	828.	1.750	7.087	0.398	1.288
159	861.	864.	2350.	1384.	1.226	0.000	0.490	1.567
169	703.	717.	1920.	1384.	0.744	2.365	0.460	1.848
176	.623.	606.	1701.	1385.	0.646	2.579	0.457	2.018

Table 97. (continued)

nmaxo	= 311.		pmaxo	=	0.
nmaxw	= 51.		pmaxw	=	0.
nbas	= 30.		pbas	=	5.
nrec	= 0.25		prec	=	0.10
ztmi	= 0.		ztma	=	200.
npeo	= 311.		npew	=	51.
wuseo	= 25.5		wusew	=	9.2
			tsr	=	0.0
LVOPT	= 2818.		LVBAS1	=	545.
STOPT	= 4273.		STBAS1	=	826.
SOOPT	= 6816.		SOBAS1	=	1318.
TWLV	= 1926.		LVBAS2	=	545.
TWST	= 2447.		STBAS2	=	826.
TWSO	= 1385.		SOBAS2	=	1318.

9.1.3 Variable data base of the simulation model

In the preceding subsection a description was given of all the relations and data that are used in the simulation model. To apply the model it is not necessary to specify all these data and relations. Therefore a summary is given here of only those data that have to be changed to adapt the model to a particular situation. If no specific data are available, the default values can be used or intelligent guesses may be made. For each part of the model, the data and the specifications of the data files are given. A list of the variables with their definition is given in Subsection 9.1.5 and more information is supplied in Subsection 9.1.2.

WOFAUTN

Lines 36 – 50:
CLIMT, SOILTY, SURROT, SALT, ZTMIT, ZTMAT, LUSTXT, CONV, HVL, IDISR, TAB1, TAB2.
It is assumed in the present set-up that the land unit determines a unique combination of the texture classes, climatic data, surface roughness, slope and clod angle, and minimum and maximum groundwater depth from these arrays. That is of course not always so and in that case program statements have to be changed. The relative area of a land unit with a particular texture class is specified in LUSTXT. If the harvested product is required in an other form (Subsection 9.1.2), other data for CONV have to be introduced. Sometimes more specific information on harvest losses (HVL) or on the effects of pests and diseases (TAB1, TAB2) may be available, which can replace the data given here.

lines 54 – 59:
The numbers of the land units and the crop species or cultivars (Table 91) and the day number of the start of crop growth for which the calculations have to be carried out, are read from the file WOFOST.IN according to FORMAT no. 1002 (line 62).
SOILDATA.RDM, RIJTEMA.RDM and CLIMEAN.RDM are random access files containing data on soil characteristics, capillary rise and climate, respectively. To change the contents of these files, first the sequential files with the same data, called SOILDATA.DAT, RIJTEMA.DAT and CLIMEAN.DAT, must be changed. The contents of these files are described later in subroutines SOILR, CRDAT and CLIMIN. With the program RANDOM, these sequential files are transformed into the random access files SOILDATA.RDM, RIJTEMA.RDM and CLIMEAN.RDM. After that, the calculations of crop growth can be executed with the new data set. PLANTDAT.RDM is a random access file containing data on crop characteristics. To change this file it is necessary to change first the sequential file containing the same data, called PLANTDAT.DAT. The contents of this file are described in subroutine PLANTR. With the program RANDOMP this sequential file is transformed into the random access file PLANTDAT.RDM. After that, the calculations of crop growth can be executed with the new data set.
RESULT.DAT is the output file (lines 59 and 149 – 151), containing the land unit, the texture class, the relative area of the land unit with a particular texture class, the starting date of crop growth, total transpiration for potential and for water – limited production and total surface runoff during the crop – growth period and the additional yield per unit applied nitrogen or phosphorus (legumes only) fertilizer nutrient, the crop growth duration and the production of leaves, stems and storage organs in the potential and water – limited production situation and in the nutrient – limited and water – plus – nutrient – limited production situation, both without and with reduction for the effects of pests and diseases.

lines 88 – 91:
NBASE, NREC, PBASE, PREC
The base uptake and the recovery fraction for both phosphorus and nitrogen will often depend on soil type or soil texture, land unit or region; then a data array may be given, preferably based on experimental evidence.

BLOCK DATA (lines 162 – 207)

lines 194 – 200:
These data are default values that are used if no data for these variables can be derived from PLANTDAT.RDM or SOILDAT.RDM. It is not necessary to change these values, unless to do calculations for a specific combination of

data that is not yet available via data files SOILDAT.RDM and PLANTDAT.RDM.

lines 202 – 206:
The relation between air temperature and the reduction factor for gross assimilation rate and that between the crop growth period and the base uptake should be changed only if other pertinent data are available.

SUBROUTINE SOILR (lines 234 – 275)

Soil characteristics per texture class are read from SOILDATA.RDM. The sequential file SOILDATA.DAT (Table 92) contains data for 14 soil texture classes. Normally this file does not need any change except for very particular soil types. A complete description of the required soil characteristics per texture class is given in Subsection 9.1.2. When alternative soil characteristics are inserted in the sequential file, it has to be transformed first into the random access file SOILDATA.RDM with program RANDOM, before calculations can be executed for these new data.

SUBROUTINE PLANTR (lines 277 – 332)

Data on crop characteristics per species or per cultivar are read from random access file PLANTDAT.RDM. To change this file, first modifications must be introduced in the sequential file PLANTDAT.DAT. After that the sequential file is transformed into the random access file PLANTDAT.RDM, with the program RANDOMP. For any crop species or cultivar the crop characteristics, as specified in Table 96 for wheat, as an example, are required. A complete description of these crop characteristics is given in Subsection 9.1.2. In Table 95 a part of the file PLANTDAT.DAT is given that refers to the high yielding rice variety IR8 and wheat, respectively.

SUBROUTINE CRDAT (lines 334 – 375)

Data on the rate of capillary rise as a function of the distance between the bottom of the root zone and the groundwater level, on one hand, and the matric suction in the rooted zone on the other hand (RIJTEM(10,8)), are read from random access file RIJTEMA.RDM. The sequential file RIJTEMA.DAT (Table 93) contains data for the 14 soil texture classes (Tables 21 – 34). Normally this file does not need any change except for very particular soil types. With the program RANDOM this sequential file has to be transformed into the random access file RIJTEMA.RDM before calculations can be executed for new data.

SUBROUTINE CLIMIN (lines 377 – 453)

Climate data are read from random access file CLIMEAN.RDM. The sequential file CLIMEAN.DAT (Table 94) contains data for six regions of Bangladesh. A complete description of the required climatic data per region is given in Subsection 9.1.2. When other climatic data are inserted in the sequential file, it has to be transformed first into the random access file CLIMEAN.RDM with program RANDOM before calculations can be executed with these new data.

SUBROUTINE APPLE (lines 455 – 829)

In this part of the program, the dynamic calculations of the water balance and crop growth are executed. For each time step, i.e. each day in the present version, all relevant calculations are updated. This continues until the development stage of the crop equals two. All data used for these calculations are derived from subroutines SOILR, CRDAT, PLANTR and CLIMIN, which are treated above.

lines 551 – 556, 808 – 810, 817:

Every ten days, day number, dry weight of roots, leaves, stems and storage organs, leaf area index, groundwater depth, volumetric soil – moisture content in the root zone and development stage are stored on unit ISP, i.e. are printed (Table 97), because ISP is set equal to one via line 86 in program WOFAUTN. When the calculations of both the potential and water – limited crop growth calculations are completed, the remainder of Table 97 follows from subroutine NUTRI.

SUBROUTINE NUTRI (lines 968 – 1067)

The required data for the calculations in this subroutine are specified in program WOFAUTN. After completion of the crop growth simulation in subroutine APPLE, the resulting data on crop growth are printed. Below these tables, the results calculated in subroutine NUTRI are printed, i.e. the amount of fertilizer nutrients to be applied for potential and water – limited production, respectively, the base uptake and recovery fraction per nutrient, the minimum and maximum groundwater depth, the additional dry – matter production per unit applied nutrient, the total water use in transpiration, the total surface runoff, the dry weights per plant organ for the various production situations. An example of this output file is shown in Table 97. More detailed information about the printed variables can be found in Subsection 9.1.2.

9.1.4 Listing of the model (see appendix A)

9.1.5 *List of acronyms*

Acronym	Description	Units
ADMI	above – ground dry – matter increase	kg ha^{-1} d^{-1}
AIRDUC	indicates presence of air ducts in plant (logical)	unitless
ASRC	carbohydrates available for dry – matter increase	kg ha^{-1} d^{-1}
CCR(I)	values of rate of capillary rise in Tables 21 to 34 per soil texture class	cm d^{-1}
CLIM	variable indicating particular climatic data set	unitless
CLIMT	array containing variables indicating climatic data set per land unit	unitless
CLM	variable indicating climatic data set	unitless
CLMNAM	name by which climatic data set is designated (alphanumeric)	unitless
CLODAN	clod/furrow angle	degrees
CLODAR	clod/furrow angle	radians
CMH(I)	values of matric head used in Tables 21 to 34 per soil texture class	cm
CONV (ICROP)	conversion factor from calculated dry – matter production to marketable product per crop species	unitless
CRMD	rate of capillary rise or of natural drainage	cm d^{-1}
CROP	crop species or cultivar	unitless
CROPT	crop type: C_3 (tropical or temperate), C_4, leguminous	unitless
CRPNAM	name of crop or cultivar (alphanumeric)	unitless
CVF	average efficiency of conversion of assimilates into plant dry matter	kg kg^{-1}
CVL	efficiency of conversion of assimilates into leaf dry matter	kg kg^{-1}
CVO	efficiency of conversion of assimilates into storage organ dry matter	kg kg^{-1}
CVR	efficiency of conversion of assimilates into root dry matter	kg kg^{-1}
CVS	efficiency of conversion of assimilates into stem dry matter	kg kg^{-1}
D	distance between groundwater and rooting depth	cm
DAGLEN	day length, in subroutine DAYLEN	h
DALV	death rate of leaves as a result of ageing	kg ha^{-1} d^{-1}
DD	drain depth of artificial drainage system	cm

Acronym	Description	Units
DECL	declination of the sun	degrees
DELT	time interval of integration	d
DISFAC	reduction factor for crop production due to pests and diseases	unitless
DL	day length, in subroutine APPLE	h
DLC	lower threshold day length for development	h
DLO	optimum day length for development	h
DMAX	rate of artificial drainage	$cm\ d^{-1}$
DMI	rate of dry – matter increase of the crop	$kg\ ha^{-1}\ d^{-1}$
DRA	drain radius	cm
DRAINS	presence or absence of drains (logical)	unitless
DRLV	total death rate of leaves	$kg\ ha^{-1}\ d^{-1}$
DRRT	death rate of roots	$kg\ ha^{-1}\ d^{-1}$
DRSO	death rate of storage organs	$kg\ ha^{-1}\ d^{-1}$
DRST	death rate of stems	$kg\ ha^{-1}\ d^{-1}$
DSL	variable indicating development rate being a function of temperature	d^{-1}
DSLV	death rate of leaves due to water stress or due to a high LAI	$kg\ ha^{-1}\ d^{-1}$
DSLV1	potential death rate of leaves due to water stress	$kg\ ha^{-1}\ d^{-1}$
DSLV2	potential death rate of leaves due to a high LAI	$kg\ ha^{-1}\ d^{-1}$
DSP	drain spacing	cm
DVR	actual development rate of the crop	d^{-1}
DVRC	maximum development rate of the crop	d^{-1}
DVRC1	maximum pre – anthesis development rate of the crop	d^{-1}
DVRC2	maximum post – anthesis development rate of the crop	d^{-1}
DVRED	reduction factor for development rate; function of day length	unitless
DVRET	reduction factor for development rate; function of temperature	unitless
DVRETB	table of reduction factor for development rate; function of temperature	unitless
DVS	development stage of the crop	unitless
DWLV	dry weight of dead leaves	$kg\ ha^{-1}$
DWRT	dry weight of dead roots	$kg\ ha^{-1}$
DWSO	dry weight of dead storage organs	$kg\ ha^{-1}$
DWST	dry weight of dead stems	$kg\ ha^{-1}$
DZ	rate of change in groundwater depth	$cm\ d^{-1}$
E	actual soil evaporation rate	$cm\ d^{-1}$
E0	potential soil evaporation rate	$cm\ d^{-1}$

Acronym	Description	Units
E0TB(I)	table of potential soil evaporation rate versus day number	unitless
EL	evaporation rate from surface water layer	$cm\ d^{-1}$
EM	maximum soil evaporation rate	$cm\ d^{-1}$
ET0	potential evapotranspiration rate	$cm\ d^{-1}$
ET0TB(I)	table of potential evapotranspiration rate versus day number	unitless
EXC	extinction coefficient for total radiation	unitless
FINT	fraction of radiation intercepted by the crop	unitless
FL	fraction of shoot dry – matter increase partitioned to leaves	unitless
FLTB	table of FL versus development stage	unitless
FO	fraction of shoot dry – matter increase partitioned to storage organs	unitless
FOTB	table of FO versus development stage	unitless
FR	fraction of total dry – matter increase partitioned to roots	unitless
FRBASO	base uptake as fraction of total nutrient uptake in the potential production situation	unitless
FRTB	table of FR versus development stage	unitless
FS	fraction of shoot dry – matter increase partitioned to stems	unitless
FSTB	table of FS versus development stage	unitless
GASS	actual gross assimilation rate of the canopy expressed in CH_2O	$kg\ ha^{-1}\ d^{-1}$
GDUR	variable indicating crop – growth duration	d
GDUR0	counter for crop – growth duration	d
GEOP	geographical latitude of location, in subroutine CLIMIN	degrees
GP	geographical latitude of location, in subroutine DAYLEN	degrees
GRLV	rate of increase in leaf dry matter	$kg\ ha^{-1}\ d^{-1}$
GRRT	rate of increase in root dry matter	$kg\ ha^{-1}\ d^{-1}$
GRSO	rate of increase in storage – organ dry matter	$kg\ ha^{-1}\ d^{-1}$
GRST	rate of increase in stem dry matter	$kg\ ha^{-1}\ d^{-1}$
GWLV	net rate of increase in leaf dry matter	$kg\ ha^{-1}\ d^{-1}$
GWRT	net rate of increase in root dry matter	$kg\ ha^{-1}\ d^{-1}$
GWSO	net rate of increase in storage – organ dry matter	$kg\ ha^{-1}\ d^{-1}$
GWST	net rate of increase in stem dry matter	$kg\ ha^{-1}\ d^{-1}$
HVL(ICROP)	fraction harvest losses per crop species	unitless

Acronym	Description	Units
HRD	difference between matric head and gravity head at rooting depth	cm
ICROP	rank number of crop species	unitless
IDAG	number of Julian calendar day in subroutine RAINDA	unitless
IDAY	number of Julian calendar day	unitless
IDISR (ICROP)	subdivision of crops in species being more and less susceptible to pests and diseases (logical)	
IHP(ICROP)	variable indicating plant organ 1. leaves; 2. stems; 3. storage organs	unitless
IL	counter for days after emergence	d
IOPT	parameter allowing omission of crop – growth simulation without water balance	unitless
IPS	output control parameter, determining the unit on which the output of the crop – growth simulation is written	unitless
IRAND	table containing 30 randomized (day) numbers	unitless
ISPAN	life span of leaves	d
IWB	parameter allowing choice between crop growth simulation with (1) or without (0) water balance	
K0	saturated hydraulic conductivity	cm d^{-1}
KMH	hydraulic conductivity at MH	cm d^{-1}
LA(I)	array containing leaf area growth per day	ha ha^{-1}
LAI	green area index (leaf, pod and stem)	ha ha^{-1}
LASUM	total green leaf area	ha ha^{-1}
LUNIT	rank number of land unit	unitless
LUSTXT	array containing relative area of land unit with particular soil texture class	unitless
LV(I)	array containing leaf dry weight growth per day	kg ha^{-1}
LVBAS1	nutrient – limited dry – matter production of leaves	kg ha^{-1}
LVBAS2	water – plus – nutrient – limited dry – matter production of leaves	kg ha^{-1}
LVOPT	potential dry – matter production of leaves	kg ha^{-1}
MANF1	factor accounting for effect of farmers capability on recovery fraction	unitless
MANF2	factor accounting for effect of farmers capability on base uptake	unitless
MH	actual matric head in root zone	cm
MHMAX	texture – specific suction limit	cm

Acronym	Description	Units
MRES	maintenance respiration rate crop at average air temperature	kg ha^{-1} d^{-1}
MT	hydraulic head midway between drains	cm
NBAS	actual base uptake of nitrogen, function of crop – growth duration and farmers ability	kg ha^{-1}
NBASE	base uptake of nitrogen by crop species with growth duration of 120 days	kg ha^{-1}
NEO	required fertilizer – nutrient application for the potential production situation	kg ha^{-1}
NEW	required fertilizer – nutrient application for the water – limited production situation	kg ha^{-1}
NEOC	additional yield per unit applied fertilizer – nutrient for the potential production situation	kg kg^{-1}
NEWC	additional yield per unit applied fertilizer – nutrient for the water – limited production situation	kg kg^{-1}
NMAXO	required nitrogen fertilizer for the potential production situation	kg ha^{-1}
NMAXW	required nitrogen fertilizer for the water – limited production situation	kg ha^{-1}
NOPT	required nitrogen uptake for the potential production situation	kg ha^{-1}
NPERSO	minimum nitrogen concentration in storage organs	kg kg^{-1}
NPERVE	minimum nitrogen concentration in crop residues	kg kg^{-1}
NREC	recovery fraction of applied nitrogen fertilizer	kg kg^{-1}
NWAB	required nitrogen uptake for the water – limited production situation	kg ha^{-1}
OUTPUT(I)	array containing dry matter production per plant organ for the various production situations	kg ha^{-1}
PBAS	actual base uptake of phosphorus, function of crop growth duration and farmers ability	kg ha^{-1}
PBASE	base uptake of phosphorus by crop species with growth duration of 120 days	kg ha^{-1}
PERDL	maximum relative death rate of leaves due to water stress	d^{-1}
PERRT	relative death rate of roots	d^{-1}
PERSO	relative death rate of storage organs	d^{-1}
PERST	relative death rate of stems	d^{-1}

Acronym	Description	Units
PGASS	potential gross assimilation rate of closed canopy expressed in CH_2O	kg ha^{-1} d^{-1}
PGS3TB(I)	table of potential gross CO_2 assimilation rate of a closed C_3 crop versus day number	kg ha^{-1} d^{-1}
PGS4TB(I)	table of potential gross CO_2 assimilation rate of a closed C_4 crop versus day number	kg ha^{-1} d^{-1}
PMAXO	required phosphorus fertilizer for the potential production situation	kg ha^{-1}
PMAXW	required phosphorus fertilizer for the water – limited production situation	kg ha^{-1}
POPT	required phosphorus uptake for the potential production situation	kg ha^{-1}
PPERSO	minimum phosphorus concentration in storage organs	kg kg^{-1}
PPERVE	minimum phosphorus concentration in crop residues	kg kg^{-1}
PREC	recovery fraction of applied phosphorus fertilizer	kg kg^{-1}
PWAB	required phosphorus uptake for the water – limited production situation	kg ha^{-1}
Q10	increase in relative maintenance respiration rate per 10 °C increase in temperature	unitless
R(I,J)	array used in subroutine RIJTEMA, containing distance between bottom of rooting zone and groundwater level per texture class (Tables 21 to 34)	cm
RAIN	daily rainfall	cm d^{-1}
RAIND(I)	array containing number of rainy days per month	d $month^{-1}$
RAINTB(I)	array containing rainfall per month	cm $month^{-1}$
RD	rooting depth	cm
RDA(I)	array containing number of rainy days per month,in subroutine RAINDA	d $month^{-1}$
RDM	maximum rooting depth	cm
REC	recovery fraction of applied nutrient, function of type of nutrient and of farmers ability	kg kg^{-1}
RIJTEM(IK,I)	array containing distance between bottom of rooting zone and groundwater level per texture class (Tables 21 – 34)	cm
RMH	actual matric head in rooting zone	cm
RML	relative maintenance respiration rate of leaves	d^{-1}

Acronym	Description	Units
RMO	relative maintenance respiration rate of storage organs	d^{-1}
RMR	relative maintenance respiration rate of roots	d^{-1}
RMRES	maintenance respiration rate of crop at temperature of 25 °C	kg ha^{-1} d^{-1}
RMS	relative maintenance respiration rate of stems	d^{-1}
RN	daily rainfall, in subroutine RAINDA	cm d^{-1}
RR	vertical extension rate of root system	cm d^{-1}
RSM	rate of change of soil – moisture content in rooting zone	cm^3 cm^{-3} d^{-1}
RTB(I)	array containing rainfall per month, in subroutine RAINDA	cm $month^{-1}$
RTDF	counter for number of days with RD less than 10 cm	unitless
S	actual sorptivity of the soil	cm $d^{-0.5}$
S0	standard sorptivity of the soil	cm $d^{-0.5}$
SAL	slope angle of the land	degrees
SALR	slope angle of the land	radians
SALT	array containing slope angle of the land per land unit	degrees
SLA	specific leaf area	ha kg^{-1}
SLATB	table containing specific leaf area as a function of development stage of the crop	unitless
SM	actual soil – moisture content	cm^3 cm^{-3}
SM0	total pore space	cm^3 cm^{-3}
SM10	soil – moisture content at air content of 0.10 cm^3 cm^{-3}	cm^3 cm^{-3}
SMCR	critical soil – moisture content	cm^3 cm^{-3}
SMF	soil – moisture content fixed at moment groundwater enters the root zone	cm^3 cm^{-3}
SMFC	soil – moisture content at field capacity	cm^3 cm^{-3}
SMFCF	soil – moisture content at fixed field capacity, i.e. matric suction of 100 cm	cm^3 cm^{-3}
SMW	soil – moisture content at wilting point	cm^3 cm^{-3}
SOBAS1	nutrient – limited dry – matter production of storage organs	kg ha^{-1}
SOBAS2	water – plus – nutrient – limited dry – matter production of storage organs	kg ha^{-1}
SOILTY	table specifying soil texture class per land unit	unitless
SOOPT	potential dry – matter production of storage organs	kg ha^{-1}

Acronym	Description	Units
SOPE	transmission zone conductivity	cm d^{-1}
SPA	specific pod area	ha kg^{-1}
SPAN	life span of leaves	d
SR	rate of surface run-off	cm d^{-1}
SS	actual surface storage	cm
SSA	specific stem area	ha kg^{-1}
SSMAX	maximum surface storage capacity	cm
SSR	actual surface storage available in one time interval (= SS / DELT)	cm d^{-1}
STARTD	starting date of crop growth	d
STBAS1	nutrient – limited dry – matter production of stems	kg ha^{-1}
STBAS2	water – plus – nutrient – limited dry – matter production of stems	kg ha^{-1}
STOPT	potential dry – matter production of stems	kg ha^{-1}
SURRO	surface roughness	cm
SURROT	table specifying surface roughness per land unit	cm
SWDEP	soil – water depletion fraction	unitless
SWDPTB	table of soil – water depletion fraction versus potential transpiration rate	unitless
T	actual transpiration rate	cm d^{-1}
T0	potential transpiration rate	cm d^{-1}
TAB1	relation between relative crop production and reduction in production due to pests and diseases (less susceptible)	unitless
TAB2	relation between relative crop production and reduction in production due to pests and diseases (more susceptible)	unitless
TADW	total above – ground dry weight (living)	kg ha^{-1}
TBIRR(I)	array containing effective irrigation rate per month	cm $month^{-1}$
TCC3M	relation between air temperature and the reduction factor for gross assimilation rate (C_3 crop of temperate regions)	unitless
TCC3T	relation between air temperature and the reduction factor for gross assimilation rate (C_3 crop of tropical regions)	unitless
TCC4	relation between air temperature and the reduction factor for gross assimilation rate (C_4 crop)	unitless
TDW	total crop dry weight (living)	kg ha^{-1}
TEFF	temperature effect on maintenance respiration	unitless

Acronym	Description	Units
TEMP	average daily temperature	°C
TEMPTB(I)	table of average temperature versus day number	unitless
TEXTCL	soil texture class (alphanumeric)	unitless
TM	maximum transpiration rate	$cm\ d^{-1}$
TSC1	texture – specific empirical constant 1 in matric head – hydraulic conductivity relation	$cm^{2.4}\ d^{-1}$
TSC2	texture – specific empirical constant 2 in matric head – hydraulic conductivity relation	cm^{-1}
TSGF	texture – specific geometry factor in matric head – soil – moisture content relation	cm^{-1}
TSR	total surface runoff	cm
TWLV	total dry weight of leaf tissue	$kg\ ha^{-1}$
TWRT	total dry weight of root tissue	$kg\ ha^{-1}$
TWSO	total dry weight of storage organs	$kg\ ha^{-1}$
TWST	total dry weight of stem tissue	$kg\ ha^{-1}$
UPTB	table containing relation between growth period and relative base uptake	unitless
WLV	total dry weight of living leaves	$kg\ ha^{-1}$
WRT	total dry weight of living roots	$kg\ ha^{-1}$
WSO	total dry weight of living storage organs	$kg\ ha^{-1}$
WST	total dry weight of living stems	$kg\ ha^{-1}$
WUSE	total water use by transpiration over crop – growth cycle	cm
WUSEO	total water use by transpiration during crop – growth cycle for the potential production situation	cm
WUSEW	total water use by transpiration during crop – growth cycle for the water – limited production situation	cm
XIE	effective irrigation rate	$cm\ d^{-1}$
XIM	actual infiltration rate	$cm\ d^{-1}$
XIMMAX	maximum infiltration rate	$cm\ d^{-1}$
ZT	actual groundwater depth	cm
ZTF	groundwater depth fixed at moment that ground – water enters root zone	cm
ZTMA	maximum groundwater depth	cm
ZTMAT	array containing maximum groundwater depth per land unit	cm
ZTMI	minimum groundwater depth	cm
ZTMIT	array containing minimum groundwater depth per land unit	cm
ZTMRD	distance between groundwater depth and rooting depth	cm

10. ANSWERS TO EXERCISES

Exercise 1

The results are obtained from Table 1 (for C_3 species) and Table 2 (for C_4 species) through interpolation.

A general formulation to obtain a value y (gross assimilation in this case) corresponding to a value x (latitude in this case) by interpolation between two points (x_1, y_1) and (x_2, y_2) is:

$$y = y_1 + (y_2 - y_1) \cdot (x - x_1)/(x_2 - x_1)$$

Exercise 2

For LAI = 1.5, Equation 5 yields: $f_h = 1 - e^{-0.8 \times 1.5} = 0.7$

Multiplying gross assimilation obtained in Exercise 1 with this reduction factor results in the values for an LAI = 1.5.

Exercise 3

Follow the calculation scheme as given in Table 6, substituting your own latitude in the interpolation procedure.

Exercise 4

Development rate (d^{-1}) for specified temperatures

	Temperature (°C)			
	10	17.5	25	32
o	0.0091	0.0143	0.0192	0.0213
*	0.0098	0.0147	0.0192	0.0222
x	0.0118	0.0192	0.0238	0.0233
•	0.0143	0.0250	0.0303	0.0333
□	—	0.0114	0.0192	0.0250

Notice that the relation between temperature and development rate is linear over a wide range of temperatures.

Exercise 5

The post-anthesis period for both cultivars requires a TU value of 650 d °C.

At 20 °C the duration of the post-anthesis period is 650/20 = 32.5 days, hence the development rate is 1/32.5 = 0.031 d^{-1}.

At 25 °C the duration is 650/25 = 26 days, and the development rate equals 1/26 = 0.038 d^{-1}.

Exercise 6

At Phabujon, the average maximum air temperature is 27 °C; the average minimum air temperature is 20 °C. Therefore, average temperature equals (27 + 20)/2 = 23.5 °C. At Chang Khian, maximum and minimum air temperatures are 24 and 17 °C, respectively, so average temperature is 20.5 °C.

Location	Phabujon	Chang Khian
emergence	8 June	5 June
anthesis	16 September	1 October
duration of pre-anthesis period	100 d	118 d
maturity	13 October	2 November
duration of post-anthesis period	27 d	32 d
average temperature	23.5 °C	20.5 °C
$TU_{pre\text{-}anthesis}$	100 × 23.5 = 2350 d°C	118 × 20.5 = 2419 d °C
$TU_{post\text{-}anthesis}$	27 × 23.5 = 634.5 d °C	32 × 20.5 = 656 d °C
development $rate_{pre\text{-}anthesis}$	$1/100 = 0.01\ d^{-1}$	$1/118 = 0.0085\ d^{-1}$
development $rate_{post\text{-}anthesis}$	$1/27 = 0.037\ d^{-1}$	$1/32 = 0.0313\ d^{-1}$

Exercise 7, see page 387 and 388

Exercise 8

$TU_{pre\text{-}anthesis} = 760$ d °C
$TU_{post\text{-}anthesis} = 660$ d °C
$T_{threshold} = 10$ °C

Development stage in pre-silking period (ending when DVS = 1.0) follows from:

$$DVS = \sum_{i=t_0}^{t_1} (T_a - 10)/TU_{pre\text{-}anthesis}$$

Development stage in post-silking period follows from:

$$DVS = 1 + \sum_{i=t_1}^{t_2} (T_a - 10)/TU_{post\text{-}anthesis}$$

where
T_a is average air temperature on day i
t_0 is day of emergence
t_1 is day of silking
t_2 is day of maturity.

Exercise 7

Variety ''A''

harvest day	total biomass	total increment	development stage end	development stage average	fraction to leaf	fraction to stem	leaf weight	leaf increment	stem weight	stem increment	grain weight	grain increment
0	50		0				47.5		2.5		0	
		50		0.053	0.765	0.235		38.3		11.7		0
10	100		0.105				85.8		14.2		0	
		100		0.158	0.571	0.429		57.1		42.9		0
20	200		0.211				142.9		57.1		0	
		200		0.263	0.544	0.456		108.8		91.2		0
30	400		0.316				251.7		148.3		0	
		400		0.368	0.533	0.467		213.2		186.8		0
40	800		0.421				464.9		335.1		0	
		600		0.474	0.523	0.477		313.8		286.2		0
50	1400		0.526				778.7		621.3		0	
		800		0.579	0.465	0.535		372.0		428.0		0
60	2200		0.632				1150.7		1049.3		0	
		1100		0.685	0.366	0.634		402.6		697.4		0
70	3300		0.737				1553.3		1746.7		0	
		1300		0.790	0.246	0.754						
								319.8		980.2		0
80	4600		0.842				1873.1		2726.9		0	
		1700		0.895	0.123	0.877		209.1		1490.9		0
90	6300		0.947				2082.2		4217.8		0	
		2100		> 1.0	0.015*	0.485*						
								31.5		1018.5		1050*
100	8400		> 1.0				2113.7		5236.3		1050	
		2100		> 1.0	0	0		0		0		2100
110	10500										3150	
		1400										1400
120	11900										4550	
		600										600
125	12500										5150	

* Anthesis is at Day 95, so between Day 90 and Day 95 there is still growth of leaves and stems. The average development stage is that at day 92.5, i.e. 0.974. According to Figure 12 the fraction allocated to the leaves equals 0.03. In 10 days, total dry-weight increment equals 2100 kg ha^{-1}, so in 5 days it is 1050 kg ha^{-1}. The increment in leaf dry weight is therefore: $0.03 \times 1050 = 31.5$ kg ha^{-1}.

Variety ''B''

harvest	total	total	development stage		fraction to		leaf		stem		grain	
day	biomas	increment	end	average	leaf	stem	weight	increment	weight	increment	weight	increment
0	50	50	0	0.063			47.5		2.5		0	
10	100	100	0.125	0.188	0.731	0.269	84.1	36.6	15.9	13.4	0	0
20	200	200	0.250	0.313	0.556	0.444	139.7	55.6	60.3	44.4	0	0
30	400	400	0.375	0.438	0.539	0.461	247.4	107.7	152.6	92.3	0	0
40	800	600	0.500	0.563	0.526	0.474	457.9	210.5	342.1	189.5	0	0
50	1400	800	0.625	0.688	0.476	0.524	743.7	285.8	656.3	314.2	0	0
60	2200	1100	0.750	0.813	0.363	0.637	1034.1	290.4	1165.9	509.6	0	0
70	3300	1300	0.875	0.938	0.219	0.781	1275.0	240.9	2025.0	859.1	0	0
80	4600	1700	1.000	>1.0	0.073	0.927	1369.9	94.9	3230.1	1205.1	0	0
90	6300	1800	>1.0		0	0		0		0	1700	1700
100	8100	1400									3500	1800
110	9500										4900	1400

Exercise 8 (continued)

Date	T_a—10	$\Sigma(T_a$—10)	DVS	Date	T_a—10	$\Sigma(T_a$—10)	$\Sigma(T_a$—10)	DVS
June 1	6	6	0.0	July 28	16	579		
2	5	11		29	15	594		
3	4	15		30	14	608		0.8
4	6	21		31	15	623		
5	8	29		Aug. 1	14	637		
6	7	36		2	16	653		
7	6	42		3	15	668		
8	3	45		4	16	684		
9	0	45		5	17	701		
10	−1	45		6	14	715		
11	2	47		7	15	730		
12	6	53		8	14	744		
13	10	63		9	16	760		1.0
14	13	76	0.1	10	17	777	17	
15	12	88		11	18	795	35	
16	9	97		12	20	815	55	
17	8	105		13	21	836	76	
18	8	113		14	18	854	94	
19	6	119		15	21	875	115	
20	10	129		16	20	895	135	
21	12	141		17	19	914	154	
22	11	152	0.2	18	18	932	172	
23	14	166		19	21	953	193	
24	11	177		20	22	975	215	
25	10	187		21	21	996	236	
26	9	196		22	20	1016	256	
27	10	206		23	19	1035	275	
28	9	215		24	19	1054	294	
29	13	228	0.3	25	20	1074	314	
30	8	236		26	16	1090	330	1.5
July 1	10	246		27	23	1113	353	
2	12	258		28	18	1131	371	
3	11	269		29	19	1150	390	
4	9	278		30	16	1166	406	
5	8	286		31	23	1189	429	
6	8	294		Sept. 1	23	1212	452	
7	10	304	0.4	2	20	1232	472	
8	9	313		3	20	1252	492	
9	11	324		4	19	1271	511	
10	11	335		5	18	1289	529	
11	12	347		6	20	1309	549	
12	13	360		7	19	1328	568	
13	20	380	0.5	8	18	1346	586	
14	10	390		9	19	1365	605	
15	14	404		10	18	1383	623	
16	14	418		11	17	1400	640	

17	13	431		12	20	1420	660	2.0
18	12	443		13	16	1436	676	
19	13	456	0.6	14	15	1451	691	
20	10	466		15	16	1467	707	
21	12	478		16	15	1482	722	
22	13	491		17	14	1496	736	
23	14	505						
24	13	518						
25	14	532	0.7					
26	16	548						
27	15	563						

For emergence on 15 June, the calculated temperature sum has to be diminished by 76 d °C, which is the temperature sum on 14 June.

Emergence date	1 June	15 June
development stage	date	date
0.	1 June	15 June
0.2	22 June	29 June
0.4	7 July	13 July
0.6	19 July	25 July
0.8	30 July	4 Aug.
1.0	9 Aug.	13 Aug.
1.5	26 Aug.	30 Aug.
2.0	12 Sept.	17 Sept.
days between emergence and silking	69 d	59 d
average development rate	0.01449 d^{-1}	0.01695 d^{-1}
days between silking and maturity	34 d	35 d
average development rate	0.02941 d^{-1}	0.02857 d^{-1}

Exercise 9

DVS	Days after germination	Increase in leaf weight	Total leaf weight	Increase in stem weight	Total stem weight	Increase in total weight	Total dry weight
0.0	0	—	—	—	—	—	—
0.154	10		200		0		200
		1200		300		1500	
0.308	20		1400		300		1700
		1400		500		1900	
0.462	30		2800		800		3600
		1700		1200		2900	
0.615	40		4500		2000		6500
		1200		1800		3000	

0.769	50	5700	3800	9500

DVS is calculated taking into account a 65 day period between emergence and silking at a constant temperature.

The fraction of the total weight increment allocated to the leaves in relation to the average development stage of the crop is given in the following table (Figure 76).

DVS	fraction to leaf
0.077	1.0
0.231	0.8
0.385	0.737
0.539	0.586
0.692	0.400

Exercise 10

Q in kg ha^{-1}. Δt in days. Then dimension of R_q is kg ha^{-1} d^{-1}.

$Q_{t+\Delta t} = Q_t + R_q \cdot \Delta t = 200 + 15 \times 10 = 350$ kg ha^{-1}

Exercise 11

$Q_{t+\Delta t} = Q_t + R_q \cdot \Delta t = Q_t + a \cdot Q_t \cdot \Delta t = Q_t \cdot (1 + a \cdot \Delta t)$

Numerical values: $Q_0 = 5$ kg ha^{-1}; $a = 0.1$ d^{-1}.

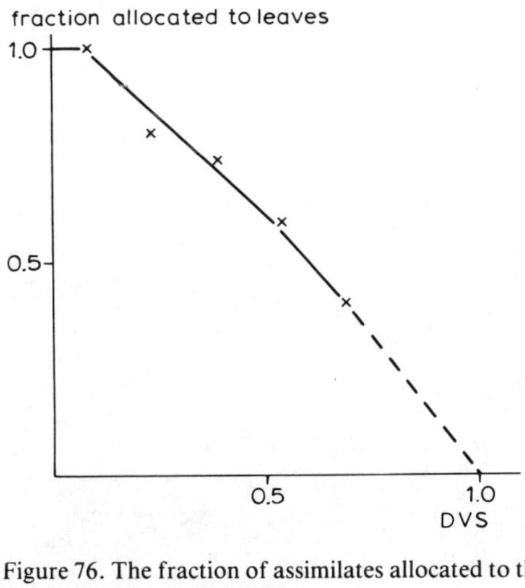

Figure 76. The fraction of assimilates allocated to the leaf blades as a function of. development stage for maize.

$\Delta t = 5$

day	Q_t	$R_q \cdot \Delta t$
0	5	2.5
5	7.5	3.75
10	11.25	5.625
15	16.875	8.438
20	25.313	12.656
25	37.969	18.984
30	56.953	

$\Delta t = 3$

day	Q_t	$R_q \cdot \Delta t$
0	5	1.5
3	6.5	1.95
6	8.45	2.535
9	10.985	3.296
12	14.281	4.285
15	18.565	5.569
18	24.134	7.240
21	31.374	9.412
24	40.787	12.236
27	53.022	15.907
30	68.929	

$\Delta t = 1$

day	Q_t	$R_q \cdot \Delta t$
0	5	0.50
1	5.5	0.55
2	6.05	0.605
3	6.655	0.6655
4	7.3205	0.7321
5	8.0526	0.8053
6	8.8579	0.8858
7	9.7437	0.9744
8	10.7181	1.0718
9	11.7899	1.1790
10	12.9689	1.2969
11	14.2658	1.4266
12	15.6924	1.5692
13	17.2616	1.7262
14	18.9878	1.8988
15	20.8866	2.0887
16	22.9753	2.2975
17	25.2728	2.5273
18	27.8001	2.7800
19	30.5801	3.0580
20	33.6381	3.3638
21	37.0019	3.7002
22	40.7021	4.0702
23	44.7723	4.4772
24	49.2495	4.9250
25	54.1745	5.4175
26	59.5920	5.9592
27	65.5512	6.5551
28	72.1063	7.2106
29	79.3169	
		7.9317

The different values for Q_{30} that result from the use of different time steps point to the fact that the smaller the time step, the more accurate the result will be. If the time step is five days, the assumption that Q_t is constant over that period is more inaccurate than if the time step is 3 days. In general, the solution improves as the time step decreases.
The phenomenon described here, where the growth rate is proportional to the quantity present, is called exponential growth. A general expression for the quantity after n time intervals can be found as follows:

$$Q_1 = Q_0 \bullet (1 + \Delta t \bullet a)$$
$$Q_2 = Q_1 \bullet (1 + \Delta t \bullet a)$$
$$Q_n = Q_{n-1} \bullet (1 + \Delta t \bullet a)$$

The expression for Q_n can be transformed into:

$$Q_n = Q_0 (1 + \Delta t \bullet a)^n$$

Q_n is the value after n steps of Δt, hence n can be substituted by TIME/Δt:

$$Q_n = Q_0 \bullet (1 + \Delta t \bullet a)^{TIME/\Delta t}$$

or

$$Q_n = Q_0 \bullet ((1 + 1/X)^X)^{a \bullet TIME}$$

with $X = 1/(\Delta t \bullet a)$

These manipulations are helpful to arrive at a so-called analytical solution for the equation describing exponential growth.
When Δt approaches zero, X approaches to infinity and the expression for Q_n approaches:

$$Q = Q_0 \bullet e^{a \bullet TIME}$$

The number e, the base of the natural logarithm stands for:

$$e = \lim_{X \to \infty} (1 + 1/X)^X = 2.7182$$

The equation with e allows calculation on a pocket-calculator of the exact solution for Q at 30 days:

$$Q_{30} = Q_0 \bullet e^{0.1 \times 30}$$

$$e^{0.1 \times 30} = e^3 = 20.09$$

$$Q_{30} = 100.43$$

This result shows that time steps of 1 day still lead to underestimation of Q at day 30.

Exercise 12

	1 T_a	2 TSUM	3 DVS	4 GASS	5 MRES	6 ASAG	7 DMI	8 FR	9 IWRT	10 WRT	11 FL	12 IWLV	13 WLV	14 FS	15 IWST	16 WST	17 FG
Initial		0.	0.							40.			100.			0.	
10 day period																	
Jan. 3	23.4	234	0.16	54.5	2.1	52.4	36.7	0.30	13.9	179	0.395	14.5	245	0.225	8.3	83	0.
Feb. 1	24.5	479	0.32	139.8	7.6	132.2	92.5	0.19	17.6	355	0.44	40.7	652	0.37	34.2	425	0.
Feb. 2	24.5	724	0.48	268.6	21.5	247.3	171.3	0.075	12.8	483	0.485	83.1	1483	0.44	75.4	1179	0.
Feb. 3	24.5	969	0.65	345.9	47.2	298.7	209.1	0.07	14.6	629	0.44	92.0	2403	0.49	102.5	2204	0.
March 1	24.8	1217	0.81	393.0	78.5	314.5	220.1	0.065	14.3	772	0.29	63.8	3041	0.645	142.0	3624	0.
March 2	24.8	1465	0.98	393.0	111.7	281.3	196.9	0.04	7.9	849	0.10	19.7	3238	0.36	70.9	4333	0.5
March 3/1	24.8	1490	1.0	393.0	14.3	25.0	17.5	0.	0.	849	0.01	18.0	3256	0.04	72.0	4405	0.95
March 3/9	24.8	1713	1.27	393.0	97.2	295.8	236.6	0.	0.	849	0.	65.0	2671	0.	0.	4405	1.
April 1	25.3	1966	1.58	447.0	112.7	334.3	267.4	0.	0.	849	0.	−53.3	2138	0.	0.	4405	1.
April 2	25.3	2219	1.90	447.0	134.1	312.9	250.3	0.	0.	849	0.	−42.6	1712	0.	0.	4405	1.
April 3/3	25.3	2295	2.0	433.7	146.3	287.4	229.9	0.	o.	849	0.	−34.0	1610	0.	0.	4405	1.

	18 IWGR	19 WGR	20 LAI	21 TADW	22 TDW	23 TDWL	24 Line no.	
Initial		0.	0.25	100.	140.	140.	a	
10 day period								
Jan. 3	0.	0.	.6125	328	507	507	b	
Feb. 1	0.	0.	1.63	1077	1432	1432	c	
Feb. 2	0.	0.	3.66	2662	3145	3145	d	
Feb. 3	0.	0.	6.0	4607	5236	5236	e	
March 1	0.	0.	7.6	6665	7437	7437	f	
March 2	112.4	1124	8.1	8695	9544	9544	g	$112.4 = 196.9 \times 0.5 \times 0.8/0.7$
March 3/1	97.0	1221	8.1	8876	9725	972	h	
March 3/9	236.6	3351	6.7	11006	11855	11270	i	
April 1	267.4	6025	5.3	13680	14529	13411	j	
April 2	250.3	8528	4.3	16183	17032	14639	k	
April 3/3	229.9	9218	4.0	16872	17721	16075	l	

Exercise 13

The solar constant expressed in $kJ\ m^{-2}s^{-1}$ equals 1.4

i.e.

$1.4 \times 60\ kJ\ m^{-2}\ min^{-1} = 1.4 \times 60 / 10\,000\ kJ\ cm^{-2}\ min^{-1} = 1.4 \times 60 \times 1000 / 10\,000\ J\ cm^{-2}\ min^{-1} = 1.4 \times 60 / 10\ J\ cm^{-2}\ min^{-1} = 1.4 \times 6 / 4.182\ cal\ cm^{-2}\ min^{-1} = 2\ cal\ cm^{-2}\ min^{-1}$.

Exercise 14

$R_I = R_A \cdot (a_A + b_A \cdot nN^{-1})$, with $a_A = 0.29$ and $b_A = 0.42$

Month	n (h)	N (h)	R_A ($MJ\ m^{-2}\ d^{-1}$)	R_I ($MJ\ m^{-2}\ d^{-1}$)
Jan.	7.1	11.4	29.9	16.5
Feb.	5.9	11.7	32.9	16.5
Mar.	5.9	12.0	35.8	17.6
Apr.	5.1	12.5	37.5	17.3
May	4.4	12.8	37.7	16.4
June	4.0	13.0	37.5	15.7
July	3.7	12.9	37.4	15.4
Aug.	2.9	12.7	37.3	14.4
Sept.	3.3	12.3	36.4	14.7
Oct.	3.9	11.8	33.8	14.5
Nov.	4.9	11.5	30.7	14.4
Dec.	5.9	11.3	28.8	14.7

Exercise 15

$R_B = \sigma \cdot T_k \cdot (0.56 - 0.079\ e_a^{0.5}) \cdot (0.1 + 0.9\ nN^{-1})$

Analysis of units:

$J\ m^{-2}\ d^{-1} = J\ m^{-2}\ d^{-1}\ K^{-4} \cdot K^4 \cdot$ (unitless) $\cdot$ (unitless)

Hence, $(0.56 - 0.079\ e_a^{0.5})$ is unitless; because e_a is in mbar the factor 0.079 is in $mbar^{-0.5}$.

With e_a in $kg\ m^{-3}$ (at 20 °C) the factor would be:

$0.079\ mbar^{-0.5} \times (kg\ m^{-3}\ mbar^{-1})^{-0.5} = 0.079 \times ((20 + 273)/0.217)^{-0.5} = 0.00215\ (kg\ m^{-3})^{-0.5}$

With e_a in mm Hg, the factor would be:

$0.079\ mbar^{-0.5} \times (mm\ Hg\ mbar^{-1})^{-0.5} = 0.079 \times (0.75)^{-0.5} = 0.091\ mm\ Hg^{-0.5}$

Exercise 16

$R_B = \sigma \cdot T_k^4 \cdot (0.56 - 0.079\ e_a^{0.5}) \cdot (0.1 + 0.9\ nN^{-1})$

Month	$(0.56 - 0.79 e_a^{0.5})$	$\sigma \cdot T_k^4$ ($MJ\ m^{-2} d^{-1}$)	$0.1 + 0.9\ nN^{-1}$	R_B ($MJ\ m^{-2}\ d^{-1}$)
Jan.	0.1853	39.16	0.6605	4.79
Feb.	0.1603	39.90	0.5539	3.54
Mar.	0.1427	40.71	0.5425	3.15
Apr.	0.1287	41.19	0.4672	2.48

Month	$(0.56 - 0.79e_a^{0.5})$	$\sigma \cdot T_k^4$ ($MJ\ m^{-2}d^{-1}$)	$0.1 + 0.9\ nN^{-1}$	R_B ($MJ\ m^{-2}\ d^{-1}$)
May	0.1259	40.92	0.4094	2.11
June	0.1331	40.60	0.3769	2.04
July	0.1368	40.33	0.3581	1.98
Aug.	0.1259	40.22	0.3055	1.55
Sept.	0.1287	40.33	0.3415	1.77
Oct.	0.1324	40.28	0.3975	2.12
Nov.	0.1495	39.90	0.4835	2.88
Dec.	0.1836	38.95	0.5699	4.08

Exercise 17

$R_N = R_I \cdot (1 - r) - R_B$

with $r = 0.05$ for water.

Month	R_I ($MJ\ m^{-2}d^{-1}$)	R_B ($MJ\ m^{-2}d^{-1}$)	R_N ($MJ\ m^{-2}d^{-1}$)
Jan.	16.5	4.79	10.9
Feb.	16.5	3.54	12.1
Mar.	17.6	3.15	13.6
Apr.	17.3	2.48	14.0
May	16.4	2.11	13.5
June	15.7	2.04	12.9
July	15.4	1.98	12.7
Aug.	14.4	1.55	12.1
Sept.	14.7	1.77	12.2
Oct.	14.5	2.12	11.7
Nov.	14.4	2.88	10.8
Dec.	14.7	4.08	9.9

Exercise 18

$h_u = a_u \cdot (1 + b_u \cdot u)$

with $a_u = 6.4 \times 10^5$ and $b_u = 0.54$

Month	u (ms^{-1})	h_u ($MJ\ m^{-2}\ d^{-1}\ {}^\circ C^{-1}$)
Jan.	2.0	1.33
Feb.	2.5	1.50
Mar.	2.8	1.61
Apr.	2.0	1.57
May	2.5	1.50
June	2.5	1.50
July	2.4	1.47

Month	u (ms^{-1})	h_u ($MJ\ m^{-2}\ d^{-1}\ °C^{-1}$)
Aug.	2.5	1.50
Sept.	2.3	1.43
Oct.	2.1	1.37
Nov.	2.1	1.37
Dec.	1.4	1.12

Exercise 19

A psychrometer is an instrument to measure the humidity of the air and belongs thus to the group of hygrometers. The psychrometer always consists of two thermometers, one normally measuring air temperature, the other is kept wet, and is thus cooled by evaporation from the wet surface; it registers the so-called wet bulb temperature T_w. If shielded from radiation, the behaviour of the wet bulb can be described by Equation 11.

$$H + LE = 0.$$

Substituting Equations 12 and 14 yields:

$$h_u(T_w - T_a) + k_u(e_s - e_a) \cdot L = 0$$

Hence

$$e_a = e_s + h_u \cdot k_u^{-1} \cdot L^{-1} \cdot (T_w - T_a)$$

This equation provides thus the possibility to determine e_a as a function of the temperature difference between the dry and the wet bulb. For this purpose nomograms or tables are available.

The units of the psychrometer constant are:

$$(\gamma) = (h_u) \cdot (k_u^{-1}) \cdot (L^{-1}) = J\ m^{-2} d^{-1} °C^{-1} \cdot (kg\ m^{-2} d^{-1} mbar^{-1})^{-1} \cdot$$

$$(J\ kg^{-1})^{-1} = mbar\ °C^{-1}$$

Exercise 20

The sensible heat loss is:

$$H = h_u(T_s - T_a)$$

The evaporative heat loss is:

$$LE = k_u(e_s - e_a)$$

January: $R_I = 1.65 \times 10^7$ J m^{-2} d^{-1}
$R_N = 10.9 \times 10^6$ J m^{-2} d^{-1}
$h_u = 13.3 \times 10^5$ J m^{-2} d^{-1} °C^{-1}
$k_u = 0.8225$ kg m^{-2} d^{-1} mbar^{-1}
$L \times k_u = 20.2 \times 10^5$ J m^{-2} d^{-1} mbar^{-1}
$T_a = 26.0$ °C
$e_a = 22.5$ mbar

$T_s = 20$ °C; e_s read from Figure 20 equals 23.4 mbar.

H = 13.3×10^5 (20 - 26)	=	-79.8×10^5 J m^{-2} d^{-1}
LE = 20.2×10^5 (23.4 - 22.5)	=	18.2×10^5 J m^{-2} d^{-1}
Balance	=	-61.4×10^5 J m^{-2} d^{-1}

$T_s = 21$ °C; e_s read from Figure 20 equals 24.9 mbar.

H = 13.3×10^5 (21 - 26)	=	-66.5×10^5 J m^{-2} d^{-1}
LE = 20.2×10^5 (24.9 - 22.5)	=	48.5×10^5 J m^{-2} d^{-1}
Balance	=	-18.0×10^5 J m^{-2} d^{-1}

$T_s = 22$ °C; e_s read from Figure 20 equals 26.5 mbar.

H = 13.3×10^5 (22 - 26)	=	-53.2×10^5 J m^{-2} d^{-1}
LE = 20.3×10^5 (26.5 - 22.5)	=	81.2×10^5 J m^{-2} d^{-1}
Balance	=	28×10^5 J m^{-2} d^{-1}

$T_s = 23$ °C; e_s read from Figure 20 equals 28.2 mbar.

H = 13.3×10^5 (23 - 26)	=	-39.9×10^5 J m^{-2} d^{-1}
LE = 20.2×10^5 (28.2 - 22.5)	=	115.1×10^5 J m^{-2} d^{-1}
Balance	=	75.2×10^5 J m^{-2} d^{-1}

$T_s = 24.0$ °C; e_s read from Figure 20 equals 29.9 mbar.

H = 13.3×10^5 (24 - 26)	=	-26.6×10^5 J m^{-2} d^{-1}
LE = 20.2×10^5 (29.9 - 22.5)	=	149.5×10^5 J m^{-2} d^{-1}
Balance	=	122.9×10^5 J m^{-2} d^{-1}

From Figure 77, the equilibrium temperature, that is the temperature where the balance of LE + H equals R_N is 23.7 °C.

$T_s = 23.7$ °C; e_s read from Figure 20 equals 29.4 mbar.

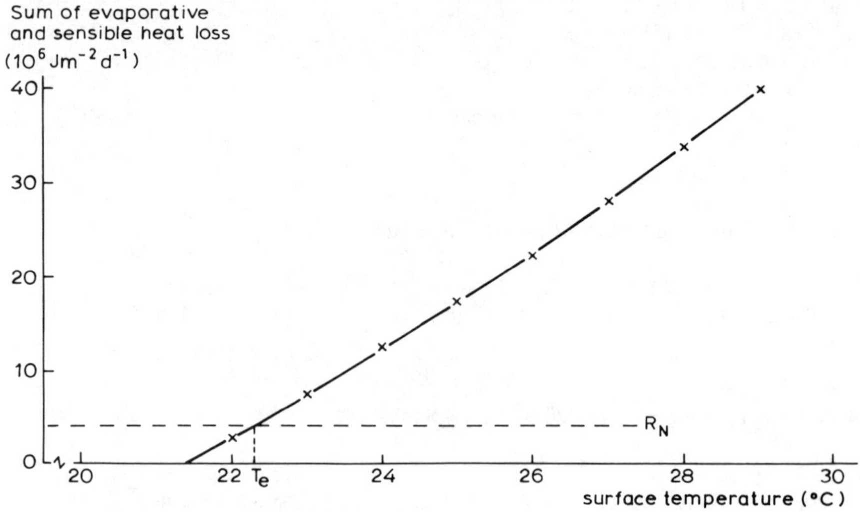

Figure 77. Illustration of the graphical analysis to determine evaporative heat loss from a water surface.

$H = 13.3 \times 10^5 (23.7 - 26.0)$	=	$-\ 30.6 \times 10^5\ J\ m^{-2}\ d^{-1}$
$LE = 20.2 \times 10^5 (29.4 - 22.5)$	=	$139.4 \times 10^5\ J\ m^{-2}\ d^{-1}$
Balance	=	$108.8 \times 10^5\ J\ m^{-2}\ d^{-1}$

At that surface temperature the evaporation rate, E, is:

$$\frac{139.4 \times 10^5}{2450 \times 10^3} = 5.69\ kg\ m^{-2}\ d^{-1} = 5.69\ mm\ d^{-1}$$

Exercise 21

The values for the dewpoint temperature can be read from Figure 22 at the appropriate vapour pressure values. Alternatively, Equation 15 can be used:

$$e_s = 6.11 \bullet e^{(17.4 \times T_s/(T_s + 239))}$$

If the dew point temperature (T_d) is substituted for T_s in this equation the resulting vapour pressure equals e_a.

Hence

$$e_a = 6.11 \times e^{(17.4 \times T_d/(T_d + 239))} \quad \text{or}$$
$$e_a/6.11 = e^{(17.4 \times T_d/(T_d + 239))} \quad \text{or}$$
$$\ln(e_a/6.11) = (17.4 \times T_d/(T_d + 239)) \quad \text{or}$$
$$(T_d + 239) \times \ln(e_a/6.11) = 17.4 \times T_d \quad \text{or}$$
$$T_d \times \ln(e_a/6.11) + 239 \times \ln(e_a/6.11) = (17.4 \times T_d) \quad \text{or}$$
$$T_d = 239 \times \ln(e_a/6.11)/(17.4 - \ln(e_a/6.11))$$

For:

e_a = 10 mbar $\ln(e_a/6.11) = 0.49$; $T_d = (239 \times 0.49)/(17.4 - 0.49) = 6.9$ °C
e_a = 15 mbar $\ln(e_a/6.11) = 0.90$; $T_d = (239 \times 0.90)/(17.4 - 0.90) = 13.0$ °C
e_a = 50 mbar $\ln(e_a/6.11) = 2.10$; $T_d = (239 \times 2.10)/(17.4 - 2.10) = 32.8$ °C

Exercise 22

The value of Δ, the slope of the saturated vapour pressure curve, can be read from Figure 22 at the appropriate temperature, i.e. read e_s at a temperature one degree below the required value and one degree above that value, and divide the difference by two:

$$\Delta(5) = (e_s(6) - e_s(4))/2$$

Alternatively, Equation 15 can again be applied:

$$e_s = 6.11 \times e^{(17.4 \times T_s/(T_s + 239))}$$

This equation describes the curve; Δ, the slope of the curve (plotted in Figure 78) equals de_s/dT, hence, from differential calculus theory:

$$\Delta = 17.4 \times 6.11 \times e^{(17.4 \times T_s/(T_s + 239)) \times (1 - T_s/T_s + 239))(T_s + 239))}$$

or

$$\Delta = 17.4 \times e_s \times (1 - T_s/(T_s + 239))/(T_s + 239)$$

T_s	e_s	Δ
2	7.06	0.51
6	7.36	0.65
10	12.29	0.82
14	16.00	1.04
18	20.67	1.30
22	26.49	1.62
26	33.69	2.00
30	42.54	2.44
34	53.35	2.98
38	66.48	3.60

Exercise 23

$$(\gamma) = \text{mbar °C}^{-1}$$

$$(\Delta) = \frac{(e_s - e_a)}{(T_s - T_d)} = \frac{\text{mbar}}{\text{°C}} = \text{mbar °C}^{-1}$$

$$\frac{(R_N)}{(h_u)} = \frac{J\,m^{-2}\,d^{-1}}{J\,m^{-2}\,d^{-1}\,°C^{-1}} = °C$$

Exercise 24

In situations were the surface has the same temperature as the air:

$$T_s = T_a = T_a + \frac{\gamma}{\gamma + \Delta} \cdot \frac{R_N}{h_u} - \frac{\Delta}{\gamma + \Delta} \cdot (T_a - T_d)$$

Hence

$$R_N = h_u\,(T_a - T_d) \cdot \Delta/\gamma$$

T_s equals T_a if, in the absence of radiation, air temperature is equal to the dewpoint temperature i.e. when the air is saturated with water vapour.

$T_s > T_a$ if the energy gain from net radiation exceeds the energy loss by evaporation, i.e.

$$R_N > h_u\,(T_a - T_d) \cdot \Delta/\gamma$$

$T_s < T_a$ if the energy loss by evaporation exceeds the gain from net radiation, i.e.

$$R_N < h_u\,(T_a - T_d) \cdot \Delta/\gamma$$

In the absence of radiation, Equation 20 becomes:

$$T_s = T_a - \frac{\Delta}{\Delta + \gamma}(T_a - T_d)$$

In this equation T_s now stands for the wet bulb temperature and T_d for the dewpoint temperature.

Hence

$$T_s = T_a - \frac{\Delta}{\Delta + \gamma} T_a + \frac{\Delta}{\Delta + \gamma} T_d$$

$$T_s = T_a\,(1 - \frac{\Delta}{\Delta + \gamma}) + \frac{\Delta}{\Delta + \gamma} T_d$$

$$T_s = \frac{1}{\Delta + \gamma}(\gamma \cdot T_a + \Delta \cdot T_d)$$

Exercise 25

The calculation is based on application of Equation 22.

Month	T_a (°C)	e_d (mbar)	Δ (mbar °C^{-1})	e_a (mbar)	R_N (MJ m^{-2}d^{-1})	h_u (MJ m^{-2}d^{-1}°C^{-1})
January	26.0	33.7	2.00	22.5	10.9	1.33
March	28.9	39.9	2.31	27.9	13.6	1.61
May	29.3	40.9	2.36	30.2	13.5	1.50
July	28.2	38.3	2.23	28.7	12.7	1.47
September	28.2	38.3	2.23	29.8	12.2	1.43
November	27.4	36.6	2.14	27.0	10.8	1.37

$\gamma = 0.66$ mbar °C^{-1}
$L = 2450 \times 10^3$ J kg^{-1}

$$E_0 = \left(\frac{1}{\Delta+\gamma}(\Delta R_N + h_u(e_d - e_a))\right)/L \quad \text{in kg m}^{-2}\text{d}^{-1} = \text{mm d}^{-1}$$

Month	E_0
January	5.63
March	6.97
May	6.48
July	5.99
September	5.56
November	5.29

Exercise 26

As explained in the text, the original Penman equation is a fair approximation of potential evapotranspiration, if the albedo is set at 0.25.

Thus R_N is different.

Month	R_N (MJ m^{-2} d^{-1})	ET_0 (mm d^{-1})
January	7.59	4.62
March	10.05	5.85
May	10.19	5.42
July	9.57	5.01
September	9.26	4.63
November	7.92	4.39

Exercise 27

A soil material consisting of 0.2 g g^{-1} clay and 0.4 g g^{-1} silt belongs to the texture class loam (according to Figure 25).

At a matric head of 1000 cm

$$SM_\psi = SM_0 \bullet e^{-\gamma \bullet (\ln\psi)^2} = SM_0 \bullet e^{-\gamma \bullet (\ln 1000)^2}$$

According to Table 16, $SM_0 = 0.503\ cm^3\ cm^{-3}$ and $\gamma = 0.0180\ cm^{-2}$ for a loam soil. This yields

$$SM_\psi = 0.503 \times e^{-0.0180 \times (\ln 1000)^2} = 0.213\ cm^3\ cm^{-3}$$

Exercise 28

From equation 27 $SM_\psi = SM_0 \bullet e^{-\gamma \bullet (\ln\psi)^2}$

According to Table 16, $SM_0 = 0.507$ and $\gamma = 0.0065$ for a silty clay.

For $SM_\psi = 0.3\ cm^3\ cm^{-3}$ this yields: $0.3 = 0.507 \times e^{-0.0065 \times (\ln\psi)^2}$.

The minus sign in the exponential function can be eliminated by interchanging the nominator and denominator on the other side of the equal sign. Hence, $e^{0.0065 \times (\ln\psi)^2} = 0.507/0.3$

$(\ln\psi)^2 = \ln (0.507/0.3)/0.0065$

$$\psi = \exp ((\ln (0.507/0.3)/0.0065)^{0.5}) = 7981.2\ cm.$$

According to Equation 29, the gravity head (g_n) is equal to the vertical distance between point n and the groundwater level.

In the present example, the groundwater table is at 1 m below soil surface and point n is 10 cm below the soil surface, so $g_n = 10 - 100 = -90$ cm.

If the soil was a loam instead of a silty clay, $SM_0 = 0.503$ and $\gamma = 0.0180$. For $SM_\psi = 0.3\ cm^3\ cm^{-3}$, the matric head would be:

$\psi = \exp ((\ln (0.503/0.3)/0.0180)^{0.5}) = 212.4$ cm.

The total hydraulic head would then be:

$H_n = \psi + g_n = 212.4 - 90 = 122.4$ cm.

Exercise 29

The hydraulic conductivity of a loam soil at a matric head of 100 cm can be calculated with

$k_\psi = k_0 \bullet e^{-\alpha x \psi}$ (Equation 30) because ψ is lower than the suction limit ψ_{max}.

Table 17 suggests for a loam soil:

$k_0 = 5.0\ cm\ d^{-1}$; $\alpha = 0.0231\ cm^{-1}$; $\psi_{max} = 300$ cm.

With $\psi = 100$ cm, $k_\psi = 5.0 \times e^{-0.0231 \times 100} = 0.50\ cm\ d^{-1}$.

With $\psi = 1000$ cm (this is well above $\psi_{max} = 300$), Equation 31 has to be used:

$k\psi = a \bullet \psi^{-1.4}$

According to Table 17, for a loam soil $a = 14.4\ cm^{2.4}\ d^{-1}$

$k_{1000} = 14.4 \times 1000^{-1.4} = 0.00091\ cm\ d^{-1}$.

Exercise 30

Tables 16 and 18 suggest that for a loam soil $SM_0 = 0.503$,
$S_0 = 11.73$ cm $d^{-0.5}$ and $A = 3.97$ cm d^{-1}.
$SM_\psi = 0.35$ cm^3 cm^{-3}, $\Delta t = 3$ h $= 0.125$ d.
Equations 33 and 34 yield:

$$S = S_0 \bullet (1 - SM_\psi/SM_0) = 11.73 \times (1 - 0.35/0.503) = 3.568 \text{ cm d}^{-0.5}.$$
$$IM_{max} = S \bullet (\Delta t)^{-0.5} + A = 3.568 \times 0.125^{-0.5} + 3.97 = 14.06 \text{ cm d}^{-1}.$$

In 0.125 d, a total of $14.06 \times 0.125 = 1.76$ cm can infiltrate. That value exceeds the rainfall of 1.5 cm, consequently the land will not flood. Substituting $\Delta t = 6$ h $=$ 0.25 d for $\Delta t = 0.125$ d yields:

$$IM_{max} = 3.568 \times 0.25^{-0.5} + 3.97 = 11.11 \text{ cm d}^{-1}$$

When the rain continues for 6 hours at the same intensity, 3 cm of rain will fall. This exceeds the $11.11 \times 0.25 = 2.78$ cm that can enter the soil in 6 hours, hence the land will flood.

Exercise 31

$E_p = E_d \cdot E_f \cdot E_i$ (Equation 36) Use Table 19.

In a small irrigation scheme (less than 1000 ha) with a predetermined schedule,
$E_i = 0.7$
When small blocks are served from unlined canals, $E_f = 0.7$
Using surface irrigation on a loam soil results in $E_d = 0.7$
E_p therefore equals $0.7 \times 0.7 \times 0.7 = 0.343$

Exercise 32

$E_m = E_0 \cdot e^{-0.4 \times LAI} = 0.5 \times e^{-0.4 \times 3} = 0.15$ cm d^{-1}

According to Tables 21-34 the capillary rise in a situation where $z_t = 100$ cm and $\psi = 1000$ cm, equals for the 14 soil classes:

1.	coarse sand	$CR < 0.02$ cm d^{-1}, so a mulch layer will be formed
2.	loamy sand	$CR < 0.02$ cm d^{-1}, so a mulch layer will be formed
3.	fine sand	$CR = 0.4$ cm d^{-1}, so a mulch layer will not be formed
4.	fine sandy loam	$CR > 0.5$ cm d^{-1}, so a mulch layer will not be formed
5.	silt loam	$CR > 0.5$ cm d^{-1}, so a mulch layer will not be formed
6.	loam	$CR = 0.5$ cm d^{-1}, so a mulch layer will not be formed
7.	loess loam	$CR = 0.2$ cm d^{-1}, so a mulch layer will not be formed
8.	sandy clay loam	$CR > 0.5$ cm d^{-1}, so a mulch layer will not be formed
9.	silty clay loam	$CR = 0.2$ cm d^{-1}, so a mulch layer will not be formed
10.	clay loam	$CR = 0.1$ cm d^{-1}, so a mulch layer will be formed
11.	light clay	$CR > 0.5$ cm d^{-1}, so a mulch layer will not be formed
12.	silty clay	$CR < 0.15$ cm d^{-1}, so a mulch layer will be formed
13.	heavy clay	$CR = 0.02$ cm d^{-1}, so a mulch layer will be formed
14.	peat	$CR < 0.06$ cm d^{-1}, so a mulch layer will be formed

Exercise 33

With slope angle $\phi = 10°$, d = 20 cm and furrow angle $\delta = 30°$

Equation 40 yields:

$$SS_{max} = 0.5 \times d \bullet \frac{\sin^2(\sigma - \phi)}{\sin \sigma} \bullet \frac{(\tan(\sigma + \phi))^{-1} + (\tan(\sigma - \phi))^{-1}}{2 \times \cos \phi \times \cos \sigma}$$

$$= 0.5 \times 20 \times \frac{\sin^2(20)}{\sin 30} \bullet \frac{(\tan(40))^{-1} + (\tan(20))^{-1}}{2 \times \cos 10 \times \cos 30} = 5.4 \text{ cm}$$

This is somewhat lower than the 5.5 cm read from Figure 26.

Exercise 34

$\Delta t = 10$ d; $P = 0.5$ cm d^{-1}; $\psi = 623$ cm; LAI $= 4.0$; $SS_t = 0$ cm; $\phi = 10°$; d $= 2$ cm; $\sigma = 30°$; $E_0 = 0.4$ cm d^{-1}; $z_t = 150$ cm

1. $IM_{max} = S_0 \bullet (1 - SM_\psi/SM_0) \bullet (\Delta t)^{-0.5} + A$
For a clay loam $SM_0 = 0.445$ cm^3 cm^{-3}; $A = 0.76$ cm d^{-1};
$S_0 = 4.70$ cm d$^{-0.5}$; $\gamma = 0.0058$

$$SM_\psi = SM_0 \bullet e^{-\gamma \bullet (\ln\psi)^2} = 0.445 \times e^{-0.0058 \times (\ln 623)^2} = 0.35 \text{ cm}^3 \text{ cm}^{-3}$$

$$IM_{max} = 4.7 \times (1 - 0.35/0.445) \times 10^{-0.5} + 0.76 = 1.08 \text{ cm d}^{-1}$$

2. SS_{max} according to Figure 28 is 0.5 to 0.6 cm; according to Equation 40, $SS_{max} = 0.54$ cm. The latter value is used in the following calculations.

3. $P = 0.5$ cm d^{-1}

4. $I_e = 0$ cm d^{-1}

5. $E_m = E_0 \times e^{-0.4 \times LAI} = 0.4 \times e^{-0.4 \times 4} = 0.08$ cm d^{-1}

6. Equation 39 is to be used to calculated E_a.

$$E_a = E_m \bullet (SM_\psi - SM_{16000}/3)/(SM_0 - SM_{16000}/3)$$

$$SM_{16000} = 0.445 \times e^{-0.0058 \times (\ln 16000)^2} = 0.258 \text{ cm}^3 \text{ cm}^{-3}$$

$$SM_0 = 0.445$$

$$E_a = 0.08 \times (0.35 - 0.258/3)/(0.445 - 0.258/3) = 0.059 \text{ cm d}^{-1}$$

7. $Q = P + I_e - E_a - IM_{max}$
$= 0.5 + 0 - 0.06 - 1.08 = -0.64$ cm d^{-1}

8. and 9. Q is less than 0 so —Q has to be compared to $SS_t/\Delta t$.
$SS_t = 0$, so $-Q$ is greater than $SS_t/\Delta t$
In that case:

$$IM = P + I_e - E_a + SS_t/\Delta t = 0.5 + 0 - 0.06 + 0 = 0.44 \text{ cm d}^{-1}$$
$$DS = SS_t/\Delta t = 0 \text{ cm d}^{-1}$$
$$SR = 0 \text{ cm d}^{-1}$$

Exercise 35

a. $\psi > g_n$, so upward water flow is possible.
b. RD = 100 cm; ψ = 300 cm; z_t = 250 cm; soil texture class: loam.

$$CR = \frac{k_0 \cdot (e^{-\alpha\cdot\psi} - e^{-\alpha\cdot(z_t - RD)})}{e^{-\alpha\cdot(z_t - RD)} - 1}$$
$$= \frac{5.0 \times (e^{-0.0231 \times 300} - e^{-0.0231 \times 150})}{e^{-0.0231 \times 150} - 1}$$
$$= 0.156 \text{ cm d}^{-1}$$

c. According to Table 26, CR will be about 0.15 cm d^{-1} (using the distance between roots and water table, 250 − 100 = 150 cm).

d. At any one moment water can flow in one direction only, i.e. D = 0 if CR is greater than 0, and vice versa.
So (CR − D) = CR = 0.15 cm d^{-1}

Exercise 36

RD = 100 cm, ψ = 100 cm, z_t = 250 cm.
In this case, g_n = 100 − 250 = − 150 cm. Water will flow downwards as
$H_n = \psi + g_n$ = 100 − 150 = − 50 cm (negative sign indicates downward flow).

$$D = k_\psi = k_0 \cdot e^{-\alpha\cdot\psi} = 5.0 \times e^{-0.0231 \times 100} = 0.5 \text{ cm d}^{-1}$$
$$CR - D = -D \text{ if } D > 0, \text{ hence } CR - D = -0.5 \text{ cm d}^{-1}$$

Exercise 37

Using the data of Exercise 34, and assuming RD = 75 cm,
$H_n = \psi - (z_t - RD)$ = 623 − (150 − 75) = 548 cm
(CR − D) = CR.
According to Table 30, CR equals 0.175 cm d^{-1} over the distance of 75 cm.

Exercise 38

Using the data of Exercise 34 and assuming ET_0 = 0.35 cm d^{-1}, E_0 = 0.4 cm d^{-1}, ψ = 623 cm and LAI = 4.0:

1. $T_0 = ET_0 - 0.1 \times E_0 = 0.35 - 0.1 \times 0.4 = 0.31 \text{ cm d}^{-1}$

2. $T_m = (1 - e^{-0.8 \times LAI}) \cdot T_0 = (1 - e^{-0.8 \times 4}) \times 0.31 = 0.297 \text{ cm d}^{-1}$

3. SM_ψ = 0.35 $cm^3\ cm^{-3}$ (Exercise 34, step 1)

4. SM_{cr} = $(1 - p) \cdot (SM_{100} - SM_{16000}) + SM_{16000}$
According to Table 20, p = 0.8
According to exercise 34, step 6, $SM_{100} = 0.393\ cm^3\ cm^{-3}$;
$SM_{16000} = 0.258\ cm^3\ cm^{-3}$
So $SM_{cr} = (1 - 0.8) \times (0.393 - 0.258) + 0.258 = 0.285\ cm^3\ cm^{-3}$

5. $SM_{10} = 0.445 \times e^{-0.0058 \times (\ln 10)^2} = 0.432\ cm^3\ cm^{-3}$

6. SM_ψ is less than SM_{10} and SM_{100}, but greater than SM_{16000} and SM_{cr}.
Hence $T = T_m$.

Exercise 39

$RSM = (IM + (CR - D) - T)/RD$ (Equation 24)
$RD = 75$ cm
$IM = 0.445\ cm\ d^{-1}$ (step 8/9, Exercise 34)
$(CR - D) = 0.175\ cm\ d^{-1}$ (Exercise 37)
$T = 0.297\ cm\ d^{-1}$ (step 6, Exercise 38)

This yields:
$RSM = (0.445 + 0.175 - 0.297)/75 = 0.00431\ cm^3\ cm^{-3}\ d^{-1}$
The moisture content at the end of the time-interval is
$SM_{\psi(t+\Delta t)} = SM_{\psi(t)} + RSM \cdot \Delta t = 0.35 + 0.00431 \times 10 = 0.393\ cm^3\ cm^{-3}$.
The matric head, ψ, at the end of the time-interval is
$\psi = e^{((\ln(SM_0/SM_\psi)/\gamma)^{0.5})} = e^{((\ln(0.445/0.393)/0.058)^{0.5})} = 102.4$ cm.

Exercise 40

RD = 50 cm, ψ = 100 cm, DD = 160 cm, L = 5000 cm, r = 8 cm, m_t = 60 cm, Δt = 10 days.
$z_t = DD - 0.5\ m_t = 160 - 0.5 \times 60 = 130$ cm

$D_{max} = k_0 \cdot m_t/(m_t + L/\pi \cdot \ln(L/(\pi \cdot r))) =$

$$5.0 \times \frac{60}{60 + 5000/3.1416 \times (\ln \frac{5000}{3.1416 \times 8})} =$$

$0.035\ cm\ d^{-1}$

CR according to Equation 45 is equal to:

$$\frac{k_0 \times (e^{-\alpha \times \psi} - e^{-\alpha \times (z_t - RD)})}{e^{-\alpha \times (z_t - RD)} - 1} = 5.0 \times$$

$$\frac{e^{-0.0231 \times 100} - e^{-0.0231 \times 80}}{e^{-0.0231 \times 80} - 1} = 0.346\ cm\ d^{-1}$$

This means $(CR - D) = 0.346\ cm\ d^{-1}$

$\Delta z = (D_{max} + (CR_t - D))/SM_o = (0.035 + 0.346)/0.503 = 0.757\ cm\ d^{-1}$

$z_{t+\Delta t} = z_t + \Delta z \bullet \Delta t = 130 + 7.6 = 137.6$ cm

$m_{(t+\Delta t)} = 2 \bullet (DD - z_{t+\Delta t}) = 2 \times (160 - 137.6) = 44.8$ cm

Exercise 41

1 kg H_2O $ha^{-1}d^{-1}$ = 1 dm^3 H_2O $ha^{-1}d^{-1}$ = $10^6 mm^3 H_2O$ $ha^{-1}d^{-1}$ = $10^6 \times 10^{-10}$ mm^3 H_2O mm^{-2} d^{-1} = 10^{-4} mm d^{-1}. Thus, 1 mm d^{-1}equals 10^4 kg ha^{-1} d^{-1} = 10 m^3 $ha^{-1}d^{-1}$.

Exercise 42

C_4 species with stomatal regulation; T_{tot} = 200 mm H_2O; RH = 75%. According to Table 36, the transpiration/assimilation ratio for a C_4 crop with regulating stomata equals 35 kg H_2O per kg CO_2 fixed. From Exercise 41, 200 mm = 2×10^6 kg ha^{-1}.

$$F_{gct} = T_{tot}/TAR = 2 \times 10^6/35 = 57\ 100 \text{ kg } CO_2 \text{ ha}^{-1}.$$

A total of 57 100 kg CO_2 ha^{-1} can thus be fixed under these conditions. In Section 2.1 it has been argued that the daily increment in dry-matter production can be estimated from gross assimilation with:

$$\Delta W = 0.7 \times (F_{gs} - 0.015\ W)$$

The average W over a growing season may be approximated by half the dry weight at harvest, i.e. 0.5 W_{tot}. If the total growing season lasts 100 days, the formulation thus becomes:

$$W_{tot} = 0.7 \times (F_{gst} - 0.75\ W_{tot})$$

F_{gct} is equal to 57 100 kg CO_2 ha^{-1}.
First conversion to CH_2O: F_{gst} = 57 100 × 30/44 = 38 932 kg CH_2O ha^{-1}

$$W_{tot} = 0.7 \times (38\ 932 - 0.75\ W_{tot}) = 27\ 252 - 0.525\ W_{tot}$$
$$1.525\ W_{tot} = 27\ 525;\ W_{tot} = 17\ 870 \text{ kg ha}^{-1}$$

For C_3 species lacking stomatal regulation, according to Table 36, TAR = 90 kg H_2O kg^{-1} CO_2

$$F_{gct} = T_{tot}/TAR = 2 \times 10^6/90 = 22\ 200 \text{ kg } CO_2 \text{ ha}^{-1}$$

Conversion to CH_2O yields: 15 136 kg ha^{-1}.
Following the same reasoning as in the previous example yields:

$$W_{tot} = 0.7 \times (15\ 136 - 0.75\ W_{tot});\quad W_{tot} = 10\ 595 - 0.525\ W_{tot}$$
$$W_{tot} = 6948 \text{ kg ha}^{-1}$$

These results show that an efficient C_4 species outyields an inefficient C_3 species by a factor of more than 2.5 under moisture limited conditions.

Exercise 43
The inverse of the slope of the line in Figure 31 is:

$$\frac{0.75 \times 10^{-6}}{2.04 \times 10^{-3}} = 0.37 \times 10^{-3} \text{ kg } CO_2 \text{ m}^{-3}.$$

(units are kg CO_2 m^{-2} s^{-1}/(m s^{-1}) = kg CO_2 m^{-3}, which indeed is a concentration).

For the conversion into ppm (v/v), the gas law is applied: at a temperature of 20 °C the volume of 1 kMol of a gas is 24 m^3.
Thus
0.37 kg CO_2 = 0.37/44 × 24 = 0.2 m^3

The drop in concentration is thus
0.2×10^{-3} m^3 m^{-3} = $0.2 \times 10^{-3} \times 10^6$ = 200 ppm (v/v).

Therefore the concentration inside the stomatal cavity is
330 — 200 = 130 ppm (v/v) CO_2.

Exercise 44
The transpiration coefficient is obtained from Figure 35 as total water use minus non-productive water use, divided by dry-matter production:
(500 − 98) mm/6.6 t ha^{-1} = 402 kg H_2O m^{-2}/0.66 kg m^{-2} = 609.1 kg H_2O kg^{-1} dry matter.

Exercise 45
Soil-moisture storage capacity for available water is (0.225—0.09) = 0.135 cm^3 cm^{-3}. Wilting point = 0.09 cm^3 cm^{-3}; 25 mm water = 2.5 cm^3 cm^{-2}. The volume of soil that can be brought to field capacity with this amount of water is equal to that amount divided by the difference between water content at field capacity and that at wilting point:

$$V_c = 2.5 /(0.225 - 0.09) = 18.5 \text{ cm}^3$$

As this is calculated in units of cm^{-2}, the soil will be at field capacity up to 18.5 cm depth.
The amount of water needed to restore the soil down to a depth of 150 cm to field capacity is

$$150 \times (0.225 - 0.09) = 20.25 \text{ cm}^3 \text{ cm}^{-2} = 202.5 \text{ mm}$$

Exercise 46

Day	WLV	DWLV
0	2000	40
1	1960	39.2
2	1921	38.4
3	1883	37.7
4	1845	36.9
5	1805	36.1
6	1769	35.4
7	1734	34.7
8	1699	34.0
9	1665	33.3
10	1632	

Using the models approach: DWLV = WLV × 0.2 = 400, so the weight of leaves after 10 days is equal to 2000 − 400 = 1600.

The difference between the approximation used in the model and the use of time steps of one day is a result of the long time interval used. This violates the assumption that the value of the integral (WLV) is constant over the entire time interval.

Exercise 47

Total seasonal transpiration equals 190.7 mm and total above-ground dry weight 11 200 kg ha^{-1}.

Calculating the transpiration coefficient on the basis of above-ground material only, yields:

transpiration coefficient = 190.7/11 200 = 190.7 kg H_2O m^{-2}/1.12 kg m^{-2} = 170.3 kg H_2O kg^{-1} dry matter.

Calculating the transpiration coefficient on the basis of total dry weight yields:

190.7 kg H_2O m^{-2}/11 900 kg ha^{-1} = 160.3 kg H_2O kg^{-1} dry matter.

Exercise 48

N_a (kg ha^{-1})	N_g (kg ha^{-1})	N_s (kg ha^{-1})	N_{tot} (kg ha^{-1})
0	49.3	19.8	69.1
50	70.3	20.8	91.1
100	76.9	32.5	109.4
150	90.8	35.2	126.0

Exercise 49

N_a	0	50	100	150
Harvest index	0.494	0.512	0.498	0.498

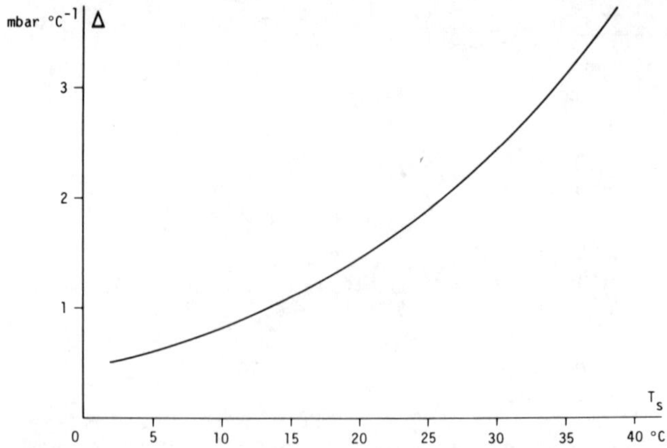

Figure 78. The relation between temperature and the slope of the saturated vapour pressure curve.

Denote grain weight by g and straw weight by s. The grain/straw ratio equals g/s, and the harvest index equals g/(g + s). Hence g/(g + s) = (g/s)/(g/s + s/s) = (g/s)/(g/s + 1)
Thus HI = (g/s)/(g/s + 1)

Exercise 50
Recovery fraction of the applied nitrogen equals (60 – 20)/100 = 0.4.

Exercise 51
See Figure 79

Exercise 52
From Exercise 49: HI = (g/s)/(1 + g/s)

For Figure 39 a g/s = 0.53, so HI = 0.53/1.53 = 0.346

39 b g/s = 1.3 , so HI = 1.3 /2.3 = 0.565

Exercise 53
Theoretical efficiencies, in kg grain per kg nitrogen, for

Figure 41a: $E_{in} = 1/(0.01 + 1.89 \times 0.004) = 57.0$

Fig 41b: $E_{in} = 1/(0.01 + 0.77 \times 0.004) = 76.5$

Exercise 54
The grain/straw ratio is calculated from the initial efficiency according to:

$$E_{in} = 410 = 1/(0.0011 + 0.0005 \times s/g)$$

$$g/s = 0.0005/(1/E_{in} - 0.0011) = 0.37$$

Exercise 55
The equation applied is

$E_{in} = 1/(n_r + n_s \times \text{shoot/root})$

with n_r and n_s being the limiting nitrogen concentrations in root and aboveground organs, respectively. n_r is approximately 0.003.
n_s varies between 0.0065 and 0.01 and root/shoot is between 0.8 and 1.75.
Initial efficiencies as function of relevant variables:

	n_s	
shoot/root	0.0065	0.01
1/0.8	89.9	64.5
1/1.75	148.9	114.8

Exercise 56

	P/N
Field 1	0.09
Field 2	0.04
Field 3	0.15

Using the rule that there is relative nitrogen shortage if P/N approaches 0.15 and that there is relative phosphorus shortage if P/N approaches 0.04, it may be expected that P fertilizer application will increase the uptake of N in Field 2 and that N fertilizer application will increase the uptake of P in Field 3. In Field 1 the P/N ratio is close to its optimum value, hence the supply of P and N is well balanced.

Exercise 57

TADW = 12 418 kg ha^{-1} (Column 21, Table 11)
WGR = 6 844 kg ha^{-1} (Column 19, Table 11)
The production of straw amounts to (TADW – WGR) = 5 574 kg ha^{-1}
In the case of well managed bunded rice (remember: this is an experiment station), the crop will always be optimally supplied with water. Therefore the production in Production Situation 2 will be equal to that calculated for Production Situation 1, i.e. 12 418 kg ha^{-1}.
If an unfertilized crop produces 2000 kg ha^{-1}, while the maximum production under Production Situation 2 is 12 418 kg ha^{-1}, there is certainly good reason to consider the use of fertilizers.
If nitrogen is at its minimum concentration, the unfertilized crop contains 1000 × 0.01 kg N in its grain and 1000 × 0.004 kg N in its straw. This correponds to (10 + 4) = 14 kg N taken up. The same quantity was found in a chemical analysis. It may therefore be concluded that nitrogen availability limits plant production.

Exercise 58

The control plots received no fertilizer at all. The corresponding rough rice yields are:

Crop year 1952-1953: 1214 kg ha^{-1}

Crop year 1953-1954: 1529 kg ha^{-1}
Crop year 1954-1955: 1756 kg ha^{-1}

The control plot in crop year 1953-1954 has probably received fertilizer(s) in the previous year, when it was included in the experiment as a fertilized plot (randomized design). The control plot in crop year 1954-1955 may have been fertilized twice, viz. in crop years 1952-1953 and 1953-1954. Carry-over of nutrients manifests itself in a higher natural soil fertility, which leads to higher control yields.

The given grain-straw ratio of 2 permits calculation of the slope, E_{in}, of the yield-uptake curve as follows:

$$E_{in} = \frac{Y}{Y \times 0.0078 + (2 \times Y) \times 0.0036} = 66.7 \text{ kg kg}^{-1}$$

in which 2 × Y stands for equal to two times the grain yield.

The nitrogen uptake from the control plots in the three consecutive years amounts to Y/E_{in}:

Uptake in crop year 1952-1953: 1214/66.7 = 18.2 kg ha^{-1}
Uptake in crop year 1953-1954: 1529/66.7 = 22.9 kg ha^{-1}
Uptake in crop year 1954-1955: 1756/66.7 = 26.3 kg ha^{-1}

Exercise 59

When addition of phosphorus no longer gives increasing yields, there is obviously P sufficiency. The entire P needs of the crop can be met by the soil; P fertilization is not needed.

In crop year 1952-1953, P fertilizer application resulted in distinctly higher yields in comparison to the control. In 1953-1954 there was less effect of P fertilizer application, presumably because some of the phosphorus applied in the first year was taken up only in the second year. Carry-over of unused phosphorus in the soil to subsequent seasons leads to P suffency after three years. Nitrogen is much more mobile in the system than phosphorus. N losses are therefore higher than P losses and carry-over of N is lower than carry-over of P. Hence the conclusions drawn from this experiment do not seem illogical.

Yield (N 37.5 – P 37.5) in 1954-1955 = 3547 kg ha^{-1}. The slope of the yield-uptake curve was 66.7 kg grain kg^{-1} N (Exercise 58), so that N uptake from plot (N 37.5 – P 37.5) amounts to 3547/66.7 = 53.2 kg ha^{-1}= $u_{m,n}$ in Equation 79. Nitrogen uptake from the control plot amounts to 26.3 kg ha^{-1} = $u_{o,n}$ in Equation 79. The quantity of nitrogen applied is 37.5 kg ha^{-1} = A_n in Equation 79.
Hence, recovery fraction R_n = (53.2 – 26.3)/37.5 = 0.72
For Field (N 75.0 – P 37.5), R_n = (62.8 – 26.3)/75 = 0.49
For Field (N 150 – P 37.5), R_n = (70.2 – 26.3)/150 = 0.29

The fact that the apparent recovery fraction decreases with increasing fertilizer application rate could indicate that at the higher application rates the harvested material no longer had the minimum N concentration, because in Section 4.1 it was shown that the application-uptake relation is a straight line in most cases, at least up to the level of 75 kg ha^{-1}. Another possibility could be that these traditional varieties respond to nitrogen application mainly by producing a more abundant vegetative growth, so that the grain/straw ratio is unfavourably affected and application of a constant grain/straw ratio in the calculation underestimates real nitrogen uptake.

The nitrogen recovery fraction of 0.29 is lower than necessary. An application method should be chosen that increases R_n to a value above, say, 0.5. Thai rice farmers are among the best in the world and will probably not apply such a high dose in one dressing. In an experiment it is, of course, perfectly in order to include such suboptimal treatments.

Exercise 60

The nitrogen fertilizers used are ammonium sulphate, urea, ammonium nitrate and sodium nitrate. Application dates are 16 July 1966 and 21 September 1966. The fertilizers were broadcast or placed either in rows at the surface, or in rows at a depth of 5 cm.

The P/N ratios at the time of primordium initiation are:

Treatment A : P/N = 0.06	Treatment E : P/N = 0.05
Treatment B : P/N = 0.07	Treatment F : P/N = 0.05
Treatment C : P/N = 0.07	Treatment G : P/N = 0.06
Treatment D : P/N = 0.09	Treatment H : P/N = 0.06

The P/N ratio in plant tissue lies normally between 0.04 and 0.15. The low ratios calculated for this experiment suggest a relative P shortage. This P shortage interferes with N uptake: P/N ratios are low in all cases.

Nitrogen fertilizer application at transplanting resulted in Lower P/N ratios, caused by higher N concentrations in the tissue and slightly lower P concentrations. Due to the fertilizer application, the N availability increased but P uptake remained essentially the same.

The grain/straw ratios at harvest time are:

Treatment A: 0.27	Treatment E: 0.34
Treatment B: 0.28	Treatment F: 0.38
Treatment C: 0.28	Treatment G: 0.39
Treatment D: 0.40	Treatment H: 0.40

Modern high yielding varieties have grain/straw ratios close to 1.0. The calculated ratios are much lower and suggest that traditional long straw varieties were used in this experiment. The relatively large proportion of straw is associated with a low initial efficiency of the yield-uptake curve. In the case of a HYV, the initial efficiency of the yield-uptake curve is typically of the order of 70. For Treatment A in this experiment it equals $3335/(3335 \times 0.0086 + 12\,410 \times 0.0032) = 48.7$.

Total nitrogen uptake under Treatment A amounts to: 3335 × 0.0086 + 12 410 × 0.0032) = 28.68 + 39.71 = 68.4 kg $ha^{-1} = u_{m,n}$.
Total nitrogen uptake from the control plot amounts to 1930 × 0.0091 + 4085 × 0.0036 = 17.56 + 14.71 = 32.3 kg $ha^{-1} = u_{o,n}$.
The nitrogen application is the same under all treatment: $A_n = 60$ kg ha^{-1}.
Equation 79 can be used to calculate the recovery fraction R_n realized under Treatment A:
$R_n = (68.4 - 32.2)/60 = 0.60$ kg kg^{-1}

The R_n values calculated for each of the treatments are as follows:

Treatment A: $R_n = 0.60$ kg kg^{-1}
Treatment B: $R_n = 0.59$ kg kg^{-1}
Treatment C: $R_n = 0.43$ kg kg^{-1}
Treatment D: $R_n = 0.16$ kg kg^{-1}
Treatment E: $R_n = 0.55$ kg kg^{-1}
Treatment F: $R_n = 0.41$ kg kg^{-1}
Treatment G: $R_n = 0.33$ kg kg^{-1}
Treatment H: $R_n = 0.27$ kg kg^{-1}

a. A recovery fraction of 0.6 is satisfactory. In both treatments, the fertilizer was placed at some depth, i.e. in the oxygen-poor sub-surface layer. In Treatment A, ammonium is to some extent adsorbed on the surface of clay minerals and protected against leaching. In Treatment B, urea N is transformed in the soil to ammonium and is equally effective as ammonium ions from an ammonium fertilizer (as in Treatment A). Subsequent transformation of ammonium N to nitrate N does not take place in the low-oxygen sub-surface soil and this prevents excessive nitrogen losses through leaching and denitrification.

b. N compounds added to the soil surface will move downward with percolating water and/or by diffusion. If nitrate ions enter the low-oxygen subsurface soil, denitrification will readily take place. Further downward transport of N compounds brings the nitrogen out of reach of the rice roots (leaching). There is, therefore, good reason to expect high losses of nitrogen under Treatments C and D and consequently low recoveries. Under Treatment C, part of the nitrogen was furnished as ammonium ions. Ammonium is not subject to denitrification and leaches less rapidly than nitrate ions because it is to some extent retained by clay minerals. It is therefore to be expected that nitrogen recovery will be disastrously low under Treatment D, and slightly higher, but not good, under Treatment C. The calculated recovery values confirm this.

c. Under Treatment D, highly mobile nitrate ions were brought into the root zone at the time of transplanting when the root system is small, and the nitrogen requirement of the plant is low. The resulting long residence time of the nitrogen in the soil must lead to high losses due to denitrification and leaching. Under Treatment H, sodium nitrate was added at a time when the crop was fully developed and in a much better position to take up the added nitrogen.

d. Under Treatment F, urea is broadcast at a time when the crop had a high demand for nitrogen and a fully developed root system, capable of taking up much nitrogen in a short time. This is not the case at the time of transplanting and consequently the observed nitrogen recovery is higher than the 0.2 suggested for urea broadcast at transplanting.

Exercise 61
TADW = 12 418 kg ha^{-1} (Column 21, Table 11)
WGR = 6 844 kg ha^{-1} (Column 19, Table 11)

Straw weights equals (12 418 – 6 844) = 5 574 kg ha^{-1}

Total minimum N uptake in Production Situation 2 equals:

$6844 \times 0.01 + 5574 \times 0.004 = 90.7$ kg N ha^{-1}($u_{m,n}$)

The initial efficiency of the yield uptake curve is calculated as WGR/total N uptake = 6844/90.7 = 75.5.
The N uptake from an unfertilized plot with a yield of 1000 kg grain is:

1000/75.5 = 13.3 kg N ha^{-1}($u_{o,n}$).

At a nitrogen recovery, R_n, of 0.55, the nitrogen fertilizer requirement, D_n, equals:

$D_n = (u_{m,n} - u_{o,n})/0.55 = (90.7 - 13.3)/0.55 = 140.7$ kg ha^{-1}

This correspondds with 140.7/0.45 = 312 kg urea.

Exercise 62
Total minimum P uptake in Production Situation 2 amounts to:

$6844 \times 0.0011 + 5574 \times 0.0005 = 10.3$ kg $ha^{-1} = u_{m,p}$

The initial efficiency of the yield uptake curve amounts to WGR/total P uptake = 6844/10.3 = 664.
P uptake from an unfertilized plot with a yield of 1000 kg grain is: 1000/664 = 1.5 kg $ha^{-1} = u_{o,p}$.
At a P recovery, R_p, of 0.02, the phosphorus fertilizer requirement, D_p, equals (10.3-1.5)/0.02 = 440 kg ha^{-1}, which is satisfied with an application of 440/0.116 = 2750 kg rock phosphate per hectare.

Exercise 64

State variables : WLVI, WI, RDI, z_t, SS_t, ψ.
Forcing variables : H_a, T_a, P, E_0, ET_0, SLA, F_{ov}, F_{cl}.
Constants : H_g, k_e, E_g, R_m, TU_{pre}, TU_{post}, T_0, p, ϕ, SM_0, γ, S_0, A, K_0, a, α, ψ_{max}, E_d, E_f, E_i, d, σ, DD, L_d, r_d.

For the calculation of production for Production Situation 3, dynamic simulation is not applied. Hence, the variables used in that calculation cannot be classified in those terms.

Exercise 65

$SM_0 = 1 - BD/SD$

		A_p	B_1	B_2
Bulk Density	(g cm^{-3})	1.1	1.3	1.3
C_m	(g g^{-1})	0.0125	0.006	0.005
SD	(g cm^{-3})	2.554	2.578	2.582
SM_0 calculated	(cm^3 cm^{-3})	0.569	0.496	0.497

The indicative SM_0 (Table 16): 0.432 cm^3 cm^{-3}

Exercise 66

The lower BD and higher SM_0 are explained by a higher organic carbon (organic matter) content of the surface horizon.

A peat which consists for 0.01 to 0.05 g g^{-1} of mineral material has an organic matter content of (1 — (0.01 to 0.05)) g g^{-1}, or O_m = 0.99 to 0.95 g g^{-1}. According to Equation 92, its specific density lies between 1.44 cm^{-3} and 1.46 g cm^{-3}. It follows from Equation 91 that the total pore space lies between 1 – 0.05/1.44 and 1 – 0.15/1.46, or between 0.965 cm^3 cm^{-3} and 0.90 cm^3 cm^{-3}. The solid fraction, $V_s/V_t = 1 - SM_0$, occupies only 0.035 to 0.10 cm^3 cm^{-3}!!

N.B.: Most mineral soils have a solid fraction occupying 0.4 to 0.6 cm^3 cm^{-3} (V_s = (1 — SM_0) × V_t). Peats have an exceptional matrix geometry. Natural peats are saturated with water. When the Roman historian Tacit (A.D. 55-118) described peatlands he wrote 'They are neither land nor water'.

Exercise 67

$SM_\psi = SM_0 \cdot e^{-\gamma\cdot(\ln\psi)^2}$ (Equation 27)

For a sandy clay loam, SM_0 – 0.432 cm^3 cm^{-3} and γ – 0.0096 cm^{-2}.

Measured: SM_{333} = 0.15 — 0.16 cm^3 cm^{-3}; SM_{16000} = 0.10 cm^3 cm^{-3}

	A_p	B_1	B_2
With γ = 0.035 cm^{-2} and SM_0 =	0.569	0.496	0.497 cm^3 cm^{-3}
Calculated SM_{333}:	0.176	0.153	0.153 cm^3 cm^{-3}

These calculated SM_{333} values are close to the measured values.

With γ = 0.017 cm^{-2} and SM_0 =	0.569	0.496	0.497 cm^3 cm^{-3}
Calculated SM_{16000}:	0.116	0.101	0.101 cm^3 cm^{-3}

These calculated SM_{16000} values are close to the measured value of 0.1 cm^3 cm^{-3}.

Exercise 68

SD = 3.72/(1.43 + 2.13 x C_m) (Equation 93)

SM_0 = 1 — BD/SD (Equation 91)

	A_1	A_3	B_{22t}	Source
C_m =	0.027	0.019	0.007 g g^{-1}	Table 52
BD =	1.2	1.4	1.5 g cm^{-3}	Table 52
SD =	2.501	2.530	2.575 g cm^{-3}	Calculated
SM_0 =	0.520	0.447	0.417 cm^3 cm^{-3}	Calculated
$\gamma = 0.03 \times \ln(SM_0/SM_{333})$				
SM_{333} =	0.19	0.17	0.22 cm^3 cm^{-3}	Table 52
γ =	0.030	0.029	0.019 cm^{-2}	Calculated
Calculated SM_{100} =	0.275	0.242	0.279 cm^3 cm^{-3}	Equation 27
$SM_{100} - SM_{16000}$ =	0.145	0.112	0.099 cm^3 cm^{-3}	Table 52
The horizon depth =	25	28	89 cm	Table 52
W_a =	3.6	3.1	8.8 cm	Equation 51
gravel (X_g) =	0	0.70	0.75 cm^3 cm^{-3}	Table 52
$(1 - X_g) \times W_a$ =	3.6	0.93	2.20 cm	Calculated

Calculated for B_{21}: $W_a = 2.8$ cm; $X_g = 0.6$ cm^3 cm^{-3}; $(1 - X_g) \times W_a = 1.1$ cm.

The quantity of ultimately available water is (3.6 + 0.96 + 2.25 + 1.1) = 7.9 cm.

If the soil were gravelless, W_a = (3.6 + 3.1 + 8.8 + 2.8) = 18.3 cm.

Exercise 69

In P9 iron mottles are observed at shallow depth, viz. in B_1 (19-57 cm), whereas in Kpuabu 1 no such iron mottling is described. Therefore, Kpuabu 1 has the deeper watertable.

Profile P9 displays iron mottles in horizon B_2 (at a depth of 57-170 cm); the observed watertable depth is not uncommon.

Exercise 70

Roots are observed in P9 in horizon B_2 (57-170 cm) with 'few coarse, common medium and many fine roots'. There are worm and ant holes throughout. Rooting is not impeded in any way. In Kpuabu 1, **very few** fine and medium roots are observed deeper than 5 cm and no mention is made of worm or ant activity. In Kpuabu 1, rooting is not exactly impeded, but it is not an ideal rooting medium either.

Many macro- and mesopores and worm holes are observed in the deepest horizon of P9. In the profile description of Kpuabu 1, such evidence of biological activity is not recorded. It is therefore concluded that biological soil homogenization is more intense in profile P9 than in Kpuabu 1.

Exercise 71

	A_1	A_3		Source
Horizon depth	25	28	cm	Table 52
Horizon volume	25×10^8	28×10^8	$cm^3 ha^{-1}$	
Bulk density	1.2	1.4	$g\ cm^{-3}$	Table 52
Weight	3×10^9	3.92×10^9	$g\ ha^{-1}$	
P content	390	370	ppm	Table 52
=	0.000390	0.000370	$kg\ kg^{-1}$	
Total P in each horizon	1.17×10^6	1.45×10^6	$g\ ha^{-1}$	
=	1170	1450	$kg\ ha^{-1}$	

Total P in upper 53 cm: $(1170 + 1450) = 2620\ kg\ ha^{-1}$

K_2O content	0.169	0.180	%	Table 52
=	0.00169	0.0018	$kg\ kg^{-1}$	
K content	0.00140	0.00149	$kg\ kg^{-1}$	
Total K in each	4.200×10^6	5.841×10^6	$g\ ha^{-1}$	
horizon =	4200	5841	$kg\ ha^{-1}$	

Total K in upper 53 cm: $(4200 + 5841) = 10041\ kg\ ha^{-1}$

Exercise 72

Total P is 335 times higher than Bray-extractable P. The high P retention capacity of soil Kpuabu 1 is not surprising: this soil is high in exchangeable aluminium. Table 47 indicates a low R_p in old soils, that are high in Fe and Al, for the same reason.

Production of 3000 kg of rice grains requires a nitrogen uptake of 3000/(50 to 70) = 43 to 60 $kg\ ha^{-1}$. With a P/N ratio of 0.1 under balanced supply, $0.1 \times (43$ to $60) = 4.3$ to 6.0 kg available P is required. For the profile horizons A_1 and A_3, available P is estimated at about 8 $kg\ ha^{-1}$ (with Bray no. 1) or 13.5 $kg\ ha^{-1}$ (with Bray no. 2). The need to apply P fertilizers seems non-existing.

Because most crops require similar or higher quantities of potassium than of N, 43 to 60 $kg\ ha^{-1}$ would have to be available.

The three horizons contain a total amount of 14.5 kg K available per ha; there seems to be a need for application of potassium fertilizers.

Exercise 73

$CEC_{total} = 11.79\ me\ (100\ g)^{-1}$ (Table 52)

$CEC_{total} = O_m \times CEC_{O_m} + (1 - O_m) \times CEC_{min}$

$org\ C = 0.027\ g\ g^{-1}$ (Table 52)

$O_m = org\ C/0.55 = 0.0489\ g\ g^{-1}$

$CEC_{O_m} = 100$ (Table 55)

$$CEC_{min} = (CEC_{total} - O_m \times CEC_{O_m})/(1 - O_m) = (11.79 - 0.0489 \times 100)/(1 - 0.0489) = 7.25\ me\ (100\ g)^{-1}$$

This is low compared to the listed CECs for pure clays in Table 55. The Kpuabu soil has a low capacity to retain and exchange added K ions.

The organic matter in the surface soil is important for the supply of cations. Although it represents only 0.0489 g g^{-1} of the surface horizon, it is responsible for 41.5% of the total CEC of the surface soil material.

The observed base saturation of less than 0.03 me me^{-1} indicates a poor overall nutrient status.

Exercise 74

The gravel content of 0.7 cm^3 cm^{-3} in layers deeper than 25 cm in Kpuabu 1 makes the overall nutrient status of this soil even poorer.

The surface soil of profile Kpuabu 1 has a CEC of 11.79 me (100 g)$^{-1}$, versus only 5.71 me (100 g)$^{-1}$ in profile P9. Note the higher organic matter content of Kpuabu 1.

Deeper than 25 cm below soil surface, only 0.3 cm^3 cm^{-3} of soil Kpuabu 1 takes part in ion exchange. The gravel fraction can be considered inert; it is removed before analysis. The conclusion is that the real capacity to retain and exchange cations is about the same in both profiles: it is very low.

Exercise 75

An amount of 10.000 kg dry matter contains 250 kg N ha^{-1}, that is equivalent to $250/0.16 = 1562.5$ kg proteins ha^{-1}. Maintenance requirement is 0.035 kg CH_2O kg^{-1} (proteins) d^{-1}. Thus $1562.5 \times 0.035 = 54.7$ kg CH_2O $ha^{-1}d^{-1}$.

An amount of 10.000 kg dry matter contains 1000 kg minerals ha^{-1}.

Maintenance requirement is 0.07 kg CH_2O kg^{-1} (minerals) d^{-1}.

Thus $1000 \times 0.07 = 70$ kg CH_2O $ha^{-1}d^{-1}$.

Total: $54.7 + 70 = 124.7$ kg CH_2O $ha^{-1}d^{-1}$.

That is 0.0125 kg CH_2O kg^{-1} (dry matter) d^{-1}, which is close to the value of 0.015 suggested in Section 2.1.

Exercise 76

For proteins, growth efficiency is 0.404, if NO_3^- is the basic component, or 0.616 if NH_4^+ is the basic component (the difference is the energy requirement for nitrate reduction).

For structural carbohydrates the growth efficiency is 0.826, and for lipids it is 0.33.

The overall growth efficiency with NO_3^- as source, is thus calculated as:

$0.404 \times f_{prot} + 0.826 \times f_{carb} + 0.33 \times f_{lip}$

and with NH_4^+ as source:

$0.616 \times f_{prot} + 0.826 \times f_{carb} + 0.33 \times f_{lip}$

a. Cassava

Growth efficiency using NO_3^-:

$0.404 \times 0.075 + 0.826 \times 0.925 + 0.33 \times 0. = 0.794$

using NH_4^+:

$0.616 \times 0.075 + 0.826 \times 0.925 + 0.33 \times 0. = 0.81$

b. Maize

Growth efficiency, using NO_3^- :

$0.404 \times 0.125 + 0.826 \times 0.875 + 0.33 \times 0. = 0.773$

using NH_4^+ :

$0.616 \times 0.125 + 0.826 \times 0.875 + 0.33 \times 0. = 0.80$

c. Soya-bean

Growth efficiency, using NO_3^- :

$0.404 \times 0.4 + 0.826 \times 0.4 + 0.33 \times 0.2 = 0.558$

using NH_4^+ :

$0.616 \times 0.4 + 0.826 \times 0.4 + 0.33 \times 0.2 = 0.643$

The corresponding growth rates are:

a. Cassava

1. $400 \times 0.794 = 317.6$ kg ha^{-1} d^{-1}
2. $400 \times 0.81 = 324.0$ kg ha^{-1} d^{-1}

b. Maize

1. $400 \times 0.773 = 309.2$ kg ha^{-1} d^{-1}
2. $400 \times 0.80 = 320.0$ kg ha^{-1} d^{-1}

c. Soya-bean

1. $400 \times 0.558 = 223.6$ kg ha^{-1} d^{-1}
2. $400 \times 0.643 = 257.2$ kg ha^{-1} d^{-1}

Exercise 77

Location	1/N	T_a
Harrow	0.0323	21.1
Ottawa	0.0345	20.6
Normandin	0.0185	15.0
Swift Current	0.0244	19.1
Lacomoe	0.0169	14.7
Beaverlodge	0.0185	13.9
Fort Vermilion	0.0208	15.3
Fort Simpson	0.0263	16.4

T_O is estimated to be 5 °C, and $b = 0.002$ d^{-1} °C^{-1} (Figure 80)
The development rate at an average air temperature of 12.5 °C is $0.002 \times (12.5 - 5) = 0.015$. It will take $1/0.015 = 67$ days from heading to maturity for this variety.

Exercise 78

Rate of development of two spring wheat cultivars from emergence to anthesis:

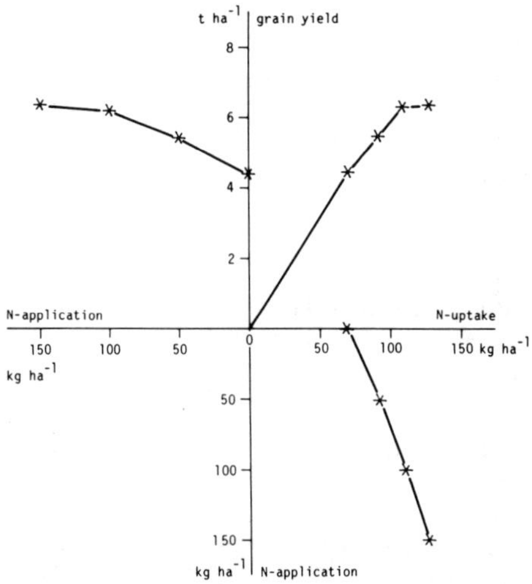

Figure 79. The relation between nitrogen uptake and grain yield, the relation between nitrogen application and nitrogen uptake, and the relation between nitrogen application and grain yield for a hypothetical grain crop.

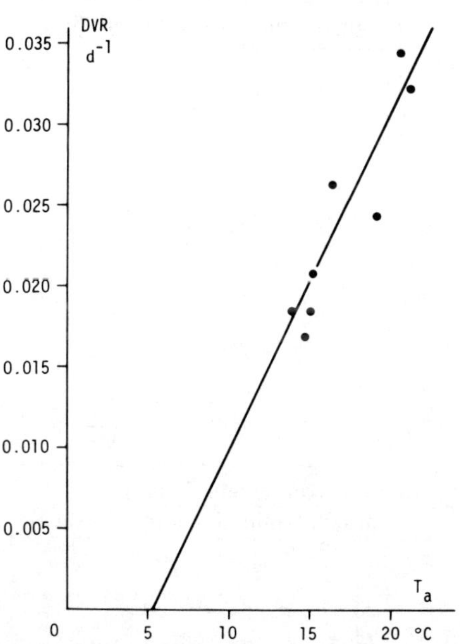

Figure 80. The relation between average temperature and the rate of development for a wheat crop.

T_a	cv. Timgalen	T_a	cv. Orca
12.5	0.01075	11.6	0.01053
14.1	0.01389	13.2	0.01282
17.0	0.01639	17.0	0.01449

Rate of development of two spring wheat cultivars from anthesis to maturity:

T_a	cv. Timgalen	T_a	cv. Orca
19.5	0.02041	17.6	0.02326
20.5	0.02083	17.6	0.025
22.1	0.02439	16.9	0.02564

These results are graphically presented in Figure 81.

Exercise 79

Maize

Time	Activity	Labour requirement ($h\ ha^{-1}$)		
		manual	draught animal	equipment (heavy)
10-20 May	1st ploughing		28	6
21-30 May	2nd ploughing		28	6
	+ harrowing		24	3
1-10 June	sowing	80		
	+ fertilization	3		
1-10 July	weed control	145		
1-10 Sept.	scaring birds	55		
11-20 Sept.	scaring birds			
21-30 Sept.	scaring birds			
1-10 Nov.	harvest	330		
	transport	126		
	processing	210		

Cassava

Time	Activity	Labour requirement ($h\ ha^{-1}$)		
		manual	draught animal	equipment (heavy)
1-10 Apr.	1st ploughing		28	6
11-20 Apr.	2nd ploughing		28	
	+ harrowing		24	
21-30 Apr.	prep. cuttings	25		

Cassava

Time	Activity	Labour requirement (h ha^{-1})		
		manual	draught animal	equipment (heavy)
	+ planting	70		
	+ fertilization	3		
11-20 May	weed control		7	2-16
1-10 June	weed control	120		
1-10 July	hilling up	85		
1-10 Nov.	harvest	180		
	transport	240		

Exercise 80
Total number of labour hours per hectare for:

rice	1343
maize	1029
cassava	810

Exercise 81
Wheat

a. 150 kg ha^{-1} sowing seed is needed for wheat. With a production target of 2500 kg per family, the relation between yield and acreage can be described by:

$$\text{Acreage} = \frac{2500}{(\text{yield} - 150)}$$ (see Figure 82)

b. Cassava
Energy content is 6.0 MJ kg^{-1}.
Family requirements are 18.5 GJ yr^{-1}.

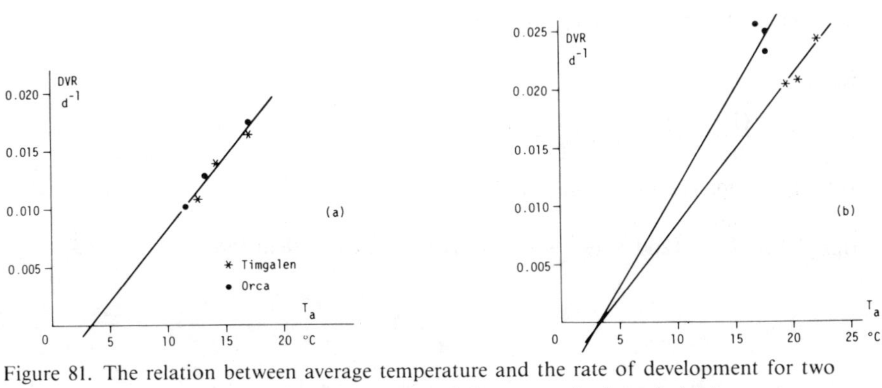

Figure 81. The relation between average temperature and the rate of development for two wheat varieties in the pre-anthesis period (a) and the post-anthesis period (b).

Edible yield has to be $\frac{18.5 \times 1000}{6} = 3100 \text{ kg yr}^{-1}$.

With a non-edible portion of 0.2, 3100/0.8 = 3875 kg yr^{-1} must be produced.

This amount of 3875 kg yr^{-1} is for consumption purposes only and does not include the amount of cassava needed for planting.

Exercise 82

Y_t is yield at time t (years elapsed since starting year).
Y_0 is yield at starting year.
Y_e is equilibrium yield i.e. yield at time equals infinity.
k is factor describing the relative annual decline in yield.

Estimates are:
$Y_0 = 2200 \text{ kg ha}^{-1}$
$Y_e = 900 \text{ kg ha}^{-1}$
At $t = 19$ year, $Y_t = 1500 \text{ kg ha}^{-1}$

Thus, $1500 = (2200 - 900) \times e^{-k \times 19} + 900$

$$600 = 1300\, e^{-k \times 19}$$

$$k = -\frac{\ln(600/1300)}{19} = 0.0407 \text{ yr}^{-1}$$

Exercise 83

At the 'Beharrungspunkt' (the equilibrium point), the yield in successive years is constant:

$$R_{n+1} = (R_n + I_r) \bullet (1 - T_c) = R_n$$

Hence

$$R_n - (1 - T_c) \times R_n = I_r \times (1 - T_c)$$

$$R_n \times T_c = I_r \times (1 - T_c)$$

$$R_n = \frac{I_r \times (1 - T_c)}{T_c} = I_r \times (\frac{1}{T_c} - 1)$$

With $I_r = 1200$ rye equivalents ha^{-1}
$T_c = 0.09$
this yields: $R_n = 1200 \times (\frac{1}{0.09} - 1) = 12133$ rye equivalents ha^{-1}

At equilibrium (that is $R_{n+1} = R_n$) the yield is equal to I_r i.e. 1200 rye equivalents ha^{-1}

After increasing 'Tätigkeit' from 0.09 to 0.16 the yields will be:

Year	yield	R_{n+1}	Year	yield	R_{n+1}
1	2133	11200	11	1363	7157
2	1984	10416	12	1337	7020
3	1859	9757	13	1315	6905
4	1753	9204	14	1297	6808
5	1665	8739	15	1281	6727
6	1590	8349	16	1268	6658
7	1528	8021	17	1257	6601
8	1475	7746	18	1248	6553
9	1431	7515	19	1240	6512
10	1394	7320	20	1234	6478

These results indicate that in the long run the yields approach again the annual **input** of 'Reichtum' into the system, however at an appreciably lower **level** of 'Reichtum'.

Exercise 84

In addition to the data given in the exercise, the 'Reichtum' at equilibrium for the cases presented has to be given. This has been calculated using the following equations:

$$R_o = \frac{Y_o}{T_c} - I_r \qquad \text{('Reichtum' at start of cycle)}$$

$$Y_0 = \frac{f \bullet I_r}{(1 + e^{-k} + e^{-2k} + e^{-3k})} + I_r$$ for four successive harvests and 'f' years of recovery

$$Y_0 = \frac{f \bullet I_r}{(1 + e^{-k} + e^{-2k})} + I_r$$ for three successive harvests and 'f' years of recovery

$$k = \ln(1 - T_c)$$

T_c = 0.08 or 0.16 yr^{-1} (the 'Tätigkeit')
I_r = 120 kg yr^{-1} (the natural input in 'Reichtum')
f = 10 or 15 years (number of fallow years)

This yields the following sequences:

$T_c/f/C_1$	'Reichtum' at t = 0	Successive yields				Total	Average	LTAV[1]	SSR[2]
		1	2	3	4				
0.08/15/4	7727	628	587	550	515	2280	570	120	0.105
0.16/15/4	4215	694	602	525	460	2281	570	120	0.105
0.08/10/4	5611	458	431	406	384	1679	420	120	0.143
0.16/10/4	3020	502	441	390	347	1680	420	120	0.143
0.08/15/3	9513	771	719	671	—	2161	720	120	0.083
0.16/15/3	5049	827	714	619	—	2160	720	120	0.083
0.08/10/3	6802	554	519	487	—	1560	520	120	0.115
0.16/10/3	3576	591	516	453	—	1563	521	120	0.115

1. Long-term average.
2. Sowing seed ratio, calculated as total seed sown ($n \times 60$) divided by the total yield.

If, as is the case here, systems are in equilibrium, there will be no change in the average 'Reichtum', meaning that in the *n* cropping years as much 'Reichtum' is extracted as is added during the whole cycle:

$$\sum_{i=1}^{n} Y_i = p \bullet I_r \qquad (p = f + n)$$

During the 15 years recovery/3 years cropping regime the average yield is highest and consequently the sowing seed ratio lowest.

Exercise 85

$$L(H_{hw}) = L_w/2 + ((H_c - H_w/2)/H_c) \cdot L_c = 0.75 + 0.375 = 1.125$$

$$L(H_{hc}) = L_c/2 + ((H_w - H_c/2)/H_w) \cdot L_w = 0.75 + 0.6667 = 1.4167$$

$$G_w/G_c = L_w/L_c \cdot e^{(-k_e \cdot (L(H_{hw}) - L(H_{hc})))}$$

$$= 1 \times e^{(-0.65 \times (1.125 - 1.4167))} = 1.21$$

$G_w + G_c = 1$
hence $G_w = 0.55$
$G_c = 0.45$

Exercise 86
See figure 83

Number of weedings	'sugar-beet-like'	'wheat-like'
0	293	6965
1	3120	—
2	9346	11800

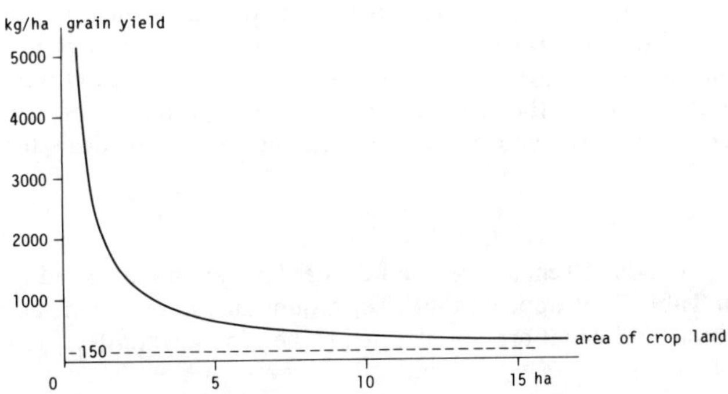

Figure 82. Grain yield required to obtain a production target of 2500 kg wheat from varying areas at crop land. The dotted line represents the amount of sowing seed (150 kg ha^{-1}) that was to be preserved.

Dry-matter production in a completely weed-free crop is higher since the weeds temporarily present will intercept a part of the radiation, moreover even in a regulary weeded crop some weeds will remain and compete with the crop.

Exercise 87

As is shown in Table 74, there is a close negative relation between the estimated relative end of the necessary weed-free period and the relative yield without weed control. Both numbers are an expression of the competitive ability of the crops mentioned in the table. This competitive ability depends on two major factors: the amount of leaf material after emergence and the ability to develop a closed canopy with sufficient height. The crops on top of the list have in common their ability to form a closed canopy in a relatively short time because of the size of their seeds and the height they will reach. The crops at the bottom of the list either have very small seeds or form a very limited canopy in time.

Exercise 88

The r values are calculated using Equation 108; with the r values the growth rate of the epidemic can be calculated with Equation 107.

Period (d)	r (d^{-2})	dP/dt (kg $ha^{-1}d^{-1}$)
40- 70	0.10	1.00×10^{-6}
70- 90	0.161	3.22×10^{-5}
90-110	0.139	6.93×10^{-4}
110-125	0.061	4.49×10^{-3}
125-135	0.092	1.47×10^{-2}

Exercise 89

See figure 84

Exercise 90

Where sophisticated pest and disease management is practiced, it may be possible to detect epidemics that occur only once in a couple of years and treat them with the appropriate chemicals, omitting treatment in years when the epidemic does not develop. Hence, in sophisticated management-systems the mean yield increase over the year is no longer the criterion, but the prospective to reclaim expected losses each growing season, based on keen observations and adapted choices of chemicals for each specific situation.

Exercise 91

Crops with a rooting depth of about 10 cm above groundwater level are not affected by the groundwater. From Table 71 it appears that rice, groundnut, mung bean, potato, onion, tobacco, chillie and kenaf are not affected if the groundwater level is 1 m below soil surface.

Crops with the deepest rooting system are best adapted to climates with irregular rainfall in the absence of irrigation, i.e. sorghum and sugar-cane.

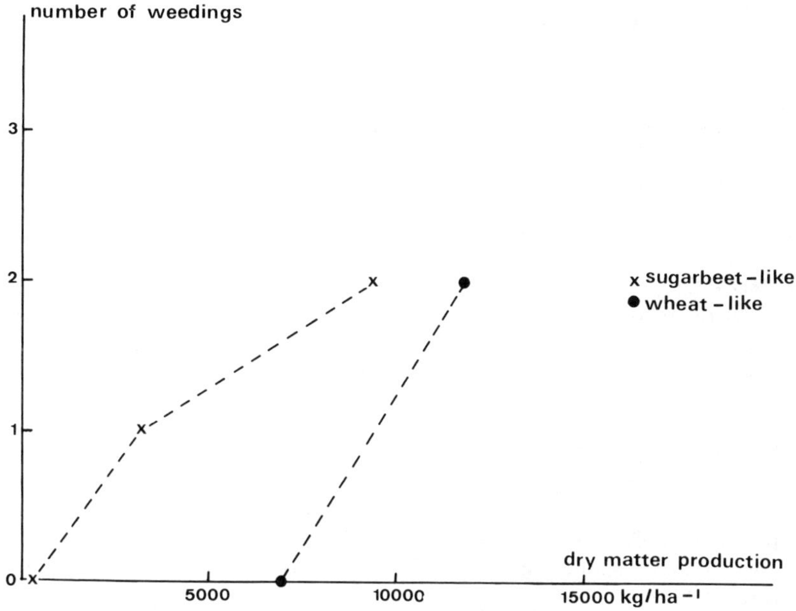

Figure 83. The relation between total dry-matter production and number of weedings for two crop types.

Exercise 92

As a result of a reduced seed/straw ratio, straw contains relatively more nitrogen, so the ratio between seed yield and nitrogen uptake will decrease. From Equation 78 it follows that the grain to nitrogen uptake ratio decreases with a reduction in the grain/straw ratio.

The average reduction in gross assimilation rate is determined by the average soil-moisture content (Table 80). The maximum yield is determined by the balance between gross assimilation rate and maintenance respiration rate (Equation 6) during seed filling. When, for example, the maximum yield under water stress is half that at optimum moisture supply, with identical vegetative growth, the straw/seed ratio is doubled. The ratio between yield and nitrogen uptake results from 1/(N concentration in seed + straw/seed • N concentration in straw). Taking wheat as example, the ratio between yield and nitrogen uptake will be $1/(0.01 + 1.25 \times 0.004) = 67$ at optimal water supply, and $1/(0.01 + 2.5 \times 0.004) = 50$ at suboptimal water supply.

Exercise 93

$T_m = 2.5\ \mathrm{mm\ d^{-1}}$; $SM_O = 0.439\ \mathrm{cm^3\ cm^{-3}}$; $\gamma = 0.0312$

$$SM_{100} = 0.439 \times e^{-0.0312 \times (\ln 100)^2} = 0.2265\ \mathrm{cm^3\ cm^{-3}}$$

$$SM_{16000} = 0.439 \times e^{-0.0312 \times (\ln 16000)^2} = 0.0236\ \mathrm{cm^3\ cm^{-3}}$$

Critical soil-moisture content:

$$SM_{cr} = (1 - p) \cdot (SM_{100} - SM_{16000}) + SM_{16000}$$

With $p = 0.80$:

$$SM_{cr} = 0.2 \times (0.2265 - 0.0236) + 0.0236 = 0.064\ cm^3\ cm^{-3}$$

In 10 day period 35, soil-moisture content was 0.12 $cm^3\ cm^{-3}$.

Amount of water, available for unrestricted crop growth:

$$W_a = RD \cdot (SM_\Psi - SM_{cr})$$
$$= 75 \times (0.12 - 0.064) = 4.2\ cm$$

The number of days with unrestricted assimilation rate at a maximum transpiration rate T_m of 2.5 mm d^{-1}, i.e. 0.25 cm d^{-1}:

$$W_a/T_m = 4.2 / 0.25 = 17\ days$$

Exercise 94

In 10 day period 36, $SM_\Psi = 0.09\ cm^3\ cm^{-3}$.

This is lower than SM_{100}, but higher than SM_{cr}. Gross assimilation is equal to that of irrigated groundnut.

In 10 day period 2, $SM_\Psi = 0.04\ cm^3\ cm^{-3}$. This is below SM_{cr} and above SM_{16000}, so the reduction in gross assimilation rate is calculated with

$$(SM_\Psi - SM_{16000})/(SM_{cr} - SM_{16000}) = (0.04 - 0.0236)/(0.064 - 0.0236) = 0.4$$

Gross assimilation in the absence of irrigation is 0.4 times gross assimilation in irrigated plots.

In 10 day period 4, $SM_\Psi = 0.035\ cm^3\ cm^{-3}$. This is below SM_{cr} and above SM_{16000}, so the reduction in gross assimilation rate is calculated as

$$(0.035 - 0.0236)/(0.064 - 0.0236) = 0.282$$

Exercise 95

Amount of water available for unrestricted crop growth:

$$W_a = RD \cdot (SM_\Psi - SM_{cr})$$

Critical soil moisture content:

$$SM_{cr} = (1 - p) \cdot (SM_{100} - SM_{16000}) + SM_{16000}$$
$$SM_{100} = 0.2265\ cm^3\ cm^{-3}$$
$$SM_{16000} = 0.0236\ cm^3\ cm^{-3}$$

Maximum transpiration rate:

$$T_m = 0.25\ cm\ d^{-1}$$

Number of days with unrestricted assimilation is W_a/T_m.

Crop	RD (cm)	p	SM_{cr} ($cm^3\ cm^{-3}$)	W_a (cm)	Number of days with unrestricted assimilation
Sorghum	150	0.885	0.047	10.95	44
Wheat	125	0.85	0.054	8.25	33
Onion	40	0.50	0.126	0.0	0

Sorghum will show the best performance.

Exercise 96

The activities required to upgrade:

- from Reclamation Level I to II: clearance A, road construction A, erosion control or bund construction (only for rice cultivation), flood control.
- from Reclamation Level II to III: clearance B, road construction B, drainage facilities, levelling/terracing.
- from Reclamation Level III to IV: irrigation facilities.

	Upgrading from Reclamation level I to II		
	earth movement ($m^3\ ha^{-1}$)	manual labour ($h\ ha^{-1}$)	mechanical labour ($h\ ha^{-1}$)
Road construction A	10	3.3	0.07
Clearance A		300	7
Erosion control[a] or	192	64	1.4
Bund construction[b]			
Number of bunds: 22	704	235	5.0
Number of bunds: 29	928	309	6.6
Flood protection	not required in the uplands		
Total labour requirement ($h\ ha^{-1}$)			
with erosion control		367	8.5
with bunds, 22 bunds ha^{-1}		538	12.1
with bunds, 29 bunds ha^{-1}		612	13.7

[a] Number of ridges $ha^{-1} = 6$
height of ridges = 0.8 m
earth movement = $0.5 \times 0.8 \times 0.8 \times 100 \times 6 = 192\ m^3$

[b] Number of bunds $= H/h_r \bullet (1.5 \text{ or } 2) = (\text{slope} \bullet a)/0.8 \times (1.5 \text{ or } 2)$
$= 11.5/0.8 \times (1.5 \text{ or } 2) = 22 \text{ or } 29.$
earth movement $= 0.5 \times 0.8 \times 0.8 \times 100 \times (22 \text{ or } 29) = 704 \text{ or } 928\ m^3\ ha^{-1}$.

	Upgrading from Reclamation level II to III		
	earth movement ($m^3\ ha^{-1}$)	manual labour ($h\ ha^{-1}$)	mechanical labour ($h\ ha^{-1}$)
Road construction B	30	10	0.2
Clearance B		600	20
Drainage facilities [c)]	800	533	14.5 (excavator) 30.8 (dragline)
Levelling/terracing [d)]	2049	1394	17.1
Total labour requirement	($h\ ha^{-1}$)	2537	51.8 or 68.1

[c] Number of terraces = 7
ditch length for drainage = (100 + 100/7) × 7 = 800 m ha^{-1}
earth movement = 800 $m^3\ ha^{-1}$

[d] Number of terraces = 7
$p = H/(2 \times N_t) = 11.5/(2 \times 7) = 0.82$ m
$q = a/(2 \times N_t) = 100/14 = 7.14$ m
earth movement = $0.5 \times p \times q \times w \times N_t = 0.5 \times 0.82 \times 7.14 \times 100 \times 7 =$ 2049 $m^3\ ha^{-1}$
average moving distance = $2/3 \times a/N_t = 2/3 \times 100/7 = 9.5$ m
According to Table 86, productivity equals approximately (sandy soil)
1.47 $m^3\ h^{-1}$ (manual total);
120 $m^3\ h^{-1}$ (bulldozer).

	Upgrading from Reclamation level III to IV		
	earth movement ($m^3\ ha^{-1}$)	manual labour ($h\ ha^{-1}$)	mechanical labour ($h\ ha^{-1}$)
Irrigation facilities	800	533	14.5 (excavator) 30.8 (dragline)
Total labour requirement	($h\ ha^{-1}$)	533	14.5 or 30.8

11. LIST OF SYMBOLS

Symbol	Description	Units	Equation No.	Acronym
a	Texture-specific empirical constant	$cm^{2.4}\ d^{-1}$	31	
a_f	Field length	m		
a_A	Empirical constant in Angström formula	unitless	8	
a_u	Empirical constant in equation for h_u	$J\ m^{-2}\ d^{-1}\ °C^{-1}$	13	
A	Transmission zone conductivity	$cm\ d^{-1}$	34	
A_x	Rate of application of nutrient x	$kg\ ha^{-1}$	79	
ASAG	Amount of assimilation products available for increase in dry weight	$kg\ CH_2O\ ha^{-1}\ d^{-1}$		
b	Width of base of ridge	m		
b_A	Empirical constant in Angström formula	unitless	8	
b_t	Constant in equation for crop development rate	$d^{-1}\ °C^{-1}$	94	
b_u	Empirical constant in equation for h_u	$s\ m^{-1}$	13	
BD	Bulk density of soil material	$g\ cm^{-3}$	89	
C_m	Carbon content of the soil	$g\ g^{-1}$	93	
C_s	Canopy conductance for water vapour	$J\ m^{-2}\ d^{-1}\ °C^{-1}$	23	
C.E.C.	Cation exchange capacity	me $(100\ g)^{-1}$		
CR	Rate of capillary rise	$cm\ d^{-1}$	24	
$(CO_2)_{ext}$	Concentration of CO_2 in the external air	$kg\ m^{-3}$	65	
$(CO_2)_{int}$	Concentration of CO_2 in the substomatal cavity	$kg\ m^{-3}$	65	
d	Surface roughness	cm	41	

Symbol	Description	Units	Equation No.	Acronym
dM_r	Amount of moisture added to soil plant-system by root growth	$mm\ d^{-1}$	72	DMR
dH_n	Difference in total hydraulic head between two points	cm	28	
d_l	Fraction of disease in lower half of the canopy	unitless	116	
d_s	Maximum relative death rate of leaves due to senescence	d^{-1}	115	DS
d_u	Fraction of disease in upper half of the canopy	unitless	117	
d_w	Maximum relative death rate of leaves due to waterstress	d^{-1}	115	
dW_r	Rate of change of soil moisture in root zone	$mm\ d^{-1}$	72	
D	Rate of natural drainage	$cm\ d^{-1}$	24	
D_{max}	Artificial drainage rate	$cm\ d^{-1}$	59	
D_N	Fertilizer nitrogen requirement	$kg\ ha^{-1}$		
D_{nr}	Thickness of non-rooted part of the potential rooting zone	mm	73	
D_r	Rooting depth	mm	75	
D_{rm}	Potential rooting depth	mm	72	
DΛ	Nutrient requirement for nutrient x	$kg\ ha^{-1}$	80	
DD	Drain depth	cm	61	
DS	Rate of decline in surface storage	$cm\ d^{-1}$	25	
DVS	Development stage	unitless		
DWLV	Death rate of leaf blades	$kg\ ha^{-1}\ d^{-1}$	77	DWLV
DWLVC	Death rate of leaves of crop	$kg\ ha^{-1}\ d^{-1}$		DWLVC
DWLVW	Death rate of leaves of weed	$kg\ ha^{-1}\ d^{-1}$		DWLVW
e_a	Actual vapour pressure of the air at standard screen height	mbar	9	
e_d	Saturated vapour pressure at air temperature	mbar	22	
e_s	Saturated vapour pressure	mbar	15	
e_w	Saturated vapour pressure at wet bulb temperature	mbar	84	
E	Evaporation rate from water surface	$kg\ H_2O\ m^{-2}\ d^{-1}$	11	

Symbol	Description	Units	Equation No.	Acronym
E_a	Actual evaporation rate from soil surface	cm d^{-1}	25	
E_d	Field application efficiency factor	unitless	36	
E_f	Field canal efficiency factor	unitless	36	
E_g	Conversion efficiency of carbohydrate into dry matter	kg kg^{-1}	6	
E_i	Conveyance efficiency factor	unitless	36	
E_{in}	Initial slope of yield-uptake curve	kg kg^{-1}	78	
E_m	Maximum soil evaporation rate	cm d^{-1}	37	
E_p	Overall efficiency factor of irrigation application	unitless	35	
E_0	Potential evaporation rate	cm d^{-1}	37	
EC_e	Electrical conductivity of saturated soil paste	mS cm^{-1}		
ET_0	Potential evapotranspiration rate	cm d^{-1}	49	
f_g	Fraction of dry-weight increment allocated to grain	unitless		FG
f_h	Fraction light intercepted by canopy	unitless	5	RA
f_l	Fraction of dry-weight increment allocated to leaf blades	unitless		FL
f_o	Fraction of day with overcast sky	unitless	4	
f_r	Fraction of dry-weight increment allocated to roots	unitless		FR
f_s	Fraction of dry-weight increment allocated to stems	unitless		FS
F	Flow rate of water	cm d^{-1}	32	
F_{cl}	Potential gross assimilation rate of a closed canopy on clear days expressed in CO_2	kg ha^{-1} d^{-1}	3	
F_d	Ratio between relative death rate of diseased and healthy leaves	unitless	112	
F_g	Maximum gross assimilation rate of a single leaf expressed in CO_2	kg ha^{-1} h^{-1}		

Symbol	Description	Units	Equation No.	Acronym
F_{gc}	Potential gross assimilation rate of closed canopy expressed in CO_2	kg ha^{-1} d^{-1}	3	
F_{gct}	Total seasonal gross assimilation of canopy expressed in CO_2	kg ha^{-1}		
F_{gs}	Potential gross assimilation rate of a closed canopy expressed in CH_2O	kg ha^{-1} d^{-1}	6	GRA
F_m	Maximum net assimilation rate of a single leaf expressed in CO_2	kg ha^{-1} h^{-1}		
F_{gst}	Total seasonal gross assimilation of canopy expressed in CH_2O	kg ha^{-1}		
F_n	Flow rate of water in soil	cm d^{-1}	28	
F_{ov}	Potential gross assimilation rate of a closed canopy on overcast days expressed in CO_2	kg ha^{-1} d^{-1}	3	
F_s	Proportionality factor for disease severity	unitless	115	FDS
g	Dry weight of grain	kg ha^{-1}	78	
g_n	Gravity head	cm	29	
G_c	Growth rate of crop	kg ha^{-1} d^{-1}	100	GC
G_t	Total growth rate of crop plus weeds	kg ha^{-1} d^{-1}	100	
G_w	Growth rate of weeds	kg ha^{-1} d^{-1}	100	GW
G_a	'Gattung', i.e. manure coefficient	unitless	99	
G_e	Grain equivalents of manure etc.	kg ha^{-1}	99	
GASS	Actual gross canopy assimilation rate expressed in CH_2O	kg ha^{-1} d^{-1}		
h_b	Height of bund	m		
h_{ic}	Rate of increase in height of crop	m d^{-1}		
h_{iw}	Rate of increase in height of weeds	m d^{-1}		
h_r	Height of ridge	m		

Symbol	Description	Units	Equation No.	Acronym
h_u	Sensible heat transfer coefficient	$J\ m^{-2}\ d^{-1}\ °C^{-1}$	12	
H	Sensible heat loss	$J\ m^{-2}\ d^{-1}$	11	
H_a	Measured total global radiation	$J\ m^{-2}\ d^{-1}$	4	
H_c	Height of crop	m	103	HC
H_g	Total global radiation on a clear day	$J\ m^{-2}\ d^{-1}$	4	
H_{hc}	Half the height of the crop	m	104	
H_{hw}	Half the height of the weeds	m	103	
H_n	Total hydraulic head	cm	29	
H_{RD}	Hydraulic head at lower root zone boundary	cm		
H_w	Height of weeds	m	103	HW
$(H_2O)_{int}$	Concentration of water vapour inside the stomatal cavity	$kg\ m^{-3}$	66	
$(H_2O)_{ext}$	Concentration of water vapour in the atmosphere	$kg\ m^{-3}$	66	
I	Radiation intensity below leaf area index L_t	$J\ m^{-2}\ d^{-1}$	102	
I_e	Effective irrigation rate	$cm\ d^{-1}$	25	
I_g	Rate of irrigation water release at project headworks	$cm\ d^{-1}$	35	
I_o	Radiation intensity above the canopy	$J\ m^{-2}\ d^{-1}$	102	
I_r	Increase in richness	$kg\ ha^{-1}$	97	
IM	Net rate of influx of moisture into root zone	$cm\ d^{-1}$	24	
IM_{max}	Maximum infiltration rate	$cm\ d^{-1}$	34	
IWGR	Rate of increase in dry-weight of grain	$kg\ ha^{-1}\ d^{-1}$		
IWLV	Rate of increase in dry-weight of leaf blades	$kg\ ha^{-1}\ d^{-1}$		
IWRT	Rate of increase in root dry-weight	$kg\ ha^{-1}\ d^{-1}$		
IWST	Rate of increase in dry-weight of stems	$kg\ ha^{-1}\ d^{-1}$		
k_e	Extinction coefficient for visible light	unitless	5	

Symbol	Description	Units	Equation No.	Acronym
k_0	Saturated hydraulic conductivity	$cm\ d^{-1}$	28	
k_r	Hydraulic conductivity in root zone	$cm\ d^{-1}$	68	
k_u	Vapour transfer coefficient	$kg\ m^{-2}\ d^{-1}\ mbar^{-1}$	14	
k_ψ	Hydraulic conductivity at matric head ψ	$cm\ d^{-1}$	30	
K	Intermediate variable used in population growth	unitless	110	
L	Latent heat of vaporization of water	$J\ kg^{-1}$	11	
L_c	Leaf area index of crop	$m^2\ m^{-2}$	100	LC
L_d	Drain spacing	cm	58	
L_n	Distance between two points	cm	28	
L_w	Leaf area index of weeds	$m^2\ m^{-2}$	100	LW
$L(H_{hc})$	Leaf area index above H_{hc}	$m^2\ m^{-2}$	104	LHHC
$L(H_{hw})$	Leaf area index above H_{hw}	$m^2\ m^{-2}$	103	LHHW
LAI	Leaf area index of vegetation	$m^2\ m^{-2}$	5	LT
LL	Distance of flow	cm	28	
m_t	Hydraulic head midway between drains at time t	cm	59	
MRES	Maintenance respiration rate expressed in CH_2O	$kg\ ha^{-1}\ d^{-1}$		
n	Actual duration of bright sunshine	h	8	
N	Duration of bright sunshine on cloudless day	h	8	
N_{ac}	Number of days from emergence to anthesis	d		
N_{am}	Number of days from anthesis to maturity	d		
N_b	Number of bunds per ha	ha^{-1}		
N_d	Length of phenological period	d	94	
N_{hm}	Number of days from heading to maturity	d		
N_r	Number of ridges per hectare	ha^{-1}		
N_s	Number of seedlings per hectare	ha^{-1}	101	
N_t	Number of terraces per hectare	ha^{-1}		

Symbol	Description	Units	Equation No.	Acronym
N_A	Actual nitrogen concentration in plant tissue	unitless		
$N_{(P-Y)}$	Minimum nitrogen concentration in crop residues	unitless		
N_Y	Minimum nitrogen concentration in marketable product	unitless		
O_m	Organic-matter content of the soil	$g\ g^{-1}$	92	
p	Soil-water depletion fraction for crop groups at maximum transpiration rate	unitless	52	
P	Actual precipitation rate	$cm\ d^{-1}$	25	
P_{gc}	Actual assimilation rate of canopy expressed in CO_2	$kg\ ha^{-1}\ d^{-1}$		
P_m	Maximum population level	unitless	107	
P_n	Net assimilation rate of leaves, expressed in CO_2	$kg\ m^{-2}\ s^{-1}$		
P_t	Population level at time t	unitless	107	
P_0	Initial population level	unitless	107	
PGASS	Potential gross canopy assimilation rate expressed in CH_2O	$kg\ ha^{-1}\ d^{-1}$		
PROPD	Disease severity	unitless		
q	Excavation width	m		
q_d	Relative death rate of leaves of disease-infected crop due to ageing	d^{-1}	112	
q_i	Relative death rate of disease infected leaves	d^{-1}	112	
q_t	Overall relative death rate of leaves	d^{-1}	112	
Q_t	Quantity at time t	$kg\ ha^{-1}$	7	
$Q_{t+\Delta t}$	Quantity at time $t + \Delta t$	$kg\ ha^{-1}$	7	
r	Relative growth rate	d^{-1}	101	R
r_a	Albedo of a surface	unitless	10	
r_d	Drain radius	cm	59	
R_A	Theoretical amount of radiation received in the absence of an atmosphere (Angot's value)	$J\ m^{-2}\ d^{-1}$	8	

Symbol	Description	Units	Equation No.	Acronym
R_B	Net outgoing longwave radiation	$J\ m^{-2}\ d^{-1}$	9	
R_{CO_2}	Resistance to CO_2 diffusion	$s\ m^{-1}$	65	
R_a	Reduction factor for dry matter accumulation	unitless	116	
R_d	Dark respiration rate of a single leaf expressed in CO_2	$kg\ ha^{-1}\ h^{-1}$		
R_{H_2O}	Resistance to water vapour diffusion	$s\ m^{-1}$	67	
R_I	Total global radiation	$J\ m^{-2}\ d^{-1}$	8	
R_m	Relative maintenance respiration rate	$kg\ kg^{-1}\ d^{-1}$	6	
R_n	'Reichtum', i.e. richness or fertility in year n	$kg\ ha^{-1}$	95	
R_N	Net incoming radiation	$J\ m^{-2}\ d^{-1}$	10	
R_q	Rate of change of quantity Q during interval Δt	$kg\ ha^{-1}\ d^{-1}$	7	
R_r	Extension rate of the roots	$mm\ d^{-1}$	73	
R_v	Rate of development	d^{-1}	94	
R_x	Recovery fraction of element X	$kg\ kg^{-1}$	79	
RD	Rooting depth	cm	24	
RDR	Relative death rate of leaves due to water shortage	d^{-1}	78	RDR
RSM	Rate of change in moisture content of the root zone	$cm^3\ cm^{-3}\ d^{-1}$	24	
s	Dry weight of straw	$kg\ ha^{-1}$		
S	Actual sorptivity	$cm\ d^{-0.5}$	33	
S_r	Matric head in root zone	cm	67	
S_0	Standard sorptivity	$cm\ d^{-0.5}$	33	
SD	Specific density of soil material	$g\ cm^{-3}$	90	
SLA	Specific leaf area	$m^2\ kg^{-1}$		
SM_a	Moisture content of air-dry soil	$cm^3\ cm^{-3}$	70	
SM_{cr}	Critical soil-moisture content	$cm^3\ cm^{-3}$	52	SMCR
SM_{fc}	Soil moisture content at field capacity	$cm^3\ cm^{-3}$	70	
SM_r	Soil moisture content in root zone	$cm^3\ cm^{-3}$	67	SMR

Symbol	Description	Units	Equation No.	Acronym
SM_w	Soil moisture content at permanent wilting point	$cm^3\ cm^{-3}$	71	
SM_ψ	Soil moisture content at matric head ψ	$cm^3\ cm^{-3}$	27	
$SM_{\psi,t}$	Soil moisture content at time t	$cm^3\ cm^{-3}$	57	
SM_0	Total pore space	$cm^3\ cm^{-3}$	27	
SM_{10}	Soil moisture content at matric head ψ = 10 cm	$cm^3\ cm^{-3}$	54	
SM_{16000}	Soil moisture content at matric head of 16 000 cm	$cm^3\ cm^{-3}$	52	
SR	Rate of surface runoff	$cm\ d^{-1}$	25	
SS_{max}	Surface storage capacity	cm	40	
SS_t	Actual surface storage at time t	cm	25	
t	Time	d		
T	Actual transpiration rate of crop	$cm\ d^{-1}$	24	
T_a	Mean daily air temperature at standard screen height	°C	12	
T_c	'Tätigkeit', i.e. activity	unitless	95	
T_d	Dewpoint of the air	°C	18	
T_k	Absolute air temperature	K	9	
T_m	Maximum transpiration rate of the crop	$cm\ d^{-1}$	48	
T_o	Threshold temperature for crop development	°C		
T_s	Temperature of evaporating surface	°C	12	
T_w	Wet bulb temperature	°C	84	
T_0	Potential transpiration rate of the crop	$cm\ d^{-1}$	49	
TADW	Total above-ground dry weight	$kg\ ha^{-1}$		
TAR	Transpiration-assimilation ratio	$kg\ kg^{-1}$ (CO_2 fixed)		
TDW	Total dry weight of crop	$kg\ ha^{-1}$		
TDWC	Total dry weight of crop	$kg\ ha^{-1}$		
TDWD	Total dry weight of dead crop material	$kg\ ha^{-1}$		
TDWL	Total dry weight of live crop material	$kg\ ha^{-1}$		
TDWW	Total dry weight of weeds	$kg\ ha^{-1}$		

Symbol	Description	Units	Equation No.	Acronym
TRC	Transpiration coefficient	kg kg^{-1} (dry matter)		
TU	Temperature sum	d °C		
$\bar{u}$	mean wind speed	m s^{-1}	13	
$u_{o,x}$	Uptake of nutrient x from non-fertilized plot	kg ha^{-1}	79	
$u_{o,n}$	Uptake of nitrogen from non-fertilized plot	kg ha^{-1}		
$u_{m,n}$	Uptake of nitrogen from fertilized plot	kg ha^{-1}		
$u_{m,x}$	Uptake of nutrient x from fertilized plot	kg ha^{-1}	79	
U_H	wind speed measured at height H	m s^{-1}	85	
V_{CO_2}	Rate of CO_2 diffusion into plant	kg m^{-2} s^{-1}	65	
V_{H_2O}	Rate of diffusion of water vapour into the atmosphere	kg m^{-2} s^{-1}	66	
V_g	Volume of soil air	cm^3	87	
V_l	Volume of soil-moisture	cm^3	87	
V_s	Volume of solid soil materials	cm^3	87	
V_t	Soil sample volume	cm^3	87	
w	Field width	m		
W	Total dry-weight of living part of crop	kg ha^{-1}	6	
W_0	Average weight of seedling	kg	101	
W_a	Quantity of ultimately available water	cm	51	
W_{nr}	Total amount of moisture in non rooted part of the potential rooting zone	mm	72	WNR
W_r	Total amount of moisture in the root zone	mm	74	
W_t	Weight of dry soil sample	g	89	
W_w	Absolute humidity of the air	kg m^{-3}	82	
WGOC	Dry weight of other organs of crop	kg ha^{-1}		
WGOW	Dry weight of other organs of weeds	kg ha^{-1}		
WGR	Dry weight of grain	kg ha^{-1}		

Symbol	Description	Units	Equation No.	Acronym
WLV	Dry weight of leaf blades	kg ha^{-1}		
WLVC	Dry weight of leaf blades of crop	kg ha^{-1}		
WLVW	Dry weight of leaf blades of weed	kg ha^{-1}		
WR	Total amount of moisture in the root zone	mm		
WRT	Dry weight of root system	kg ha^{-1}		
WST	Dry weight of stems	kg ha^{-1}		
Y_i	Weight of infected leaves	kg ha^{-1}	111	
Y_0	Yield in year 0, or at time 0	kg ha^{-1}	101	
Y_c	Yield of control plot in fertilizer experiment	kg ha^{-1}		
Y_d	Death rate of infected leaves	kg ha^{-1} d^{-1}	111	YD
Y_e	Yield at equilibrium	kg ha^{-1}		
Y_m	Weight of living leaves	kg ha^{-1}	111	
Y_n	Yield in year n	kg ha^{-1}	95	
Y_t	Yield at time t	kg ha^{-1}	101	
Y_t	Weight of diseased leaves	kg ha^{-1}	111	
Y_{Ax}	Yield of plot fertilized with A kg of nutrient x	kg ha^{-1}		
z_o	Roughness length of the surface	cm	85	
z_t	Groundwater depth at time t	cm	46	
α	Texture-specific empirical constant	cm^{-1}	30	
γ	Texture-specific geometry factor	cm^{-2}	27	
γ	Psychrometer constant	mbar $°C^{-1}$	16	
Δ	Slope of the saturation vapour pressure curve between the average air temperature and dewpoint	mbar $°C^{-1}$	18	
Δt	Duration of time interval	d	7	
ΔW	Rate of increase in dry-weight of the canopy	kg ha^{-1} d^{-1}	6	DMI
ΔW_r	Change in amount of moisture in the root zone	mm d^{-1}	74	
Δz	Rate of change in groundwater depth	cm d^{-1}	57	

Symbol	Description	Units	Equation No.	Acronym
ϵ	Initial light-use efficiency of assimilation of single leaves	kg CO_2 J^{-1}	41	
ϕ	Slope angle of the land	degree	14	
ψ	Soil suction or matric head	cm	27	
ψ_{cr}	Critical matric head	cm		
ψ_{max}	Texture-specific suction limit	cm	31	
σ	Clod or furrow angle	degree	40	
σ	Stefan Boltzman constant	J m^{-2} d^{-1} K^{-4}	9	

12. INDEX

13 REFERENCES

Agarwal, M.M., 1980. Phosphorus and potassium requirement of rice in relation to time of application. Indian Journal of agricultural Research 14: 53–56.

Andrus, J.R., 1948. Burmese economic life. Stanford University Press, Stanford, California, 362 p.

Anonymous, 1977. Agricultural statistics of Thailand; crop year 1976/77, No. 72. Ministry of Agriculture and Co-operatives, Bangkok.

Anonymous, 1980. Database of the supply module of the Thailand Agricultural Model (THAM-I). Research Report SOW-80-4, Centre for World Food Studies, Amsterdam-Wageningen.

Asseldonk, J.S.O. van, 1980. Reclamation and required technical inputs. Internal note of Centre for World Food Studies.

Bakel, P.J.T. van, 1981. Unsaturated zone and evapotranspiration. In: Evaporation in relation to hydrology. Verslagen Mededelingen Commissie Hydrologische Onderzoekingen T.N.O. 28, The Hague.

Balasubramanian, V. & L. Singh, 1982. Efficiency of nitrogen fertilizer use under rainfed maize and irrigated wheat at Kadawa, northern Nigeria. Fertilizer Research 3: 315–324.

Basak, M.N., 1962. Nutrient uptake by rice plant and its effect on yield. Agronomy Journal 54: 373–376.

Bauder, J.W. & A. Bauer, 1978. Irrigated alfalfa forage production on a well-drained sandy loam at Oakes, North Dakota 1972–1977. North Dakota Farm Research 36: 19–24.

Beek, J., 1979. Phosphate retention by soil in relation to waste disposal. Pudoc, Wageningen. 162 p.

Benneke, R.R. & R. Winterboer, 1973. Linear Programme applications to agriculture. Iowa State University Press, Ames, Iowa.

Beye, G., 1974. Etude comparative de l'action de la potasse et de la paille enfouié sur le développement et le rendement du riz, sur sol argileux de basse Casamance. Agronomie tropicale 29: 803–811.

Black, J.N., C.W. Bonython & G.A. Prescott, 1954. Solar radiation and the duration of sunshine. Quarterly Journal of the Royal Meteorological Society 80: 231–235.

Blair, G.J., C.P. Mamaril & M. Ismunadji, 1980. Sulfur deficiency in soils in the tropics as a constraint to food production. In: Soil-related constraints to food production in the tropics. IRRI, Los Baños, Philippines.

Boerboom, B.W.J., 1978. A model of dry matter distribution in cassava (Manihot esculenta Crantz). Netherlands Journal of agricultural Science 26: 267–277.

Bohn, H., B. McNeal & G. O'Connor, 1979. Soil Chemistry. John Wiley and Sons, New York.

Bouldin, D.R., R.S. Reid & P.J. Stangel, 1980. Nitrogen as a constraint to nonlegume food crop production. In: Soil-related constraints to food production in the tropics. IRRI, Los Baños, Philippines.

Bouwmeester, R.J.B. & P.G.L. Vlek, 1981. Rate control of ammonia volatilization from

rice paddies. Atmospheric Environment 15: 131–140.
Bredero, Th.J., 1965. Nitrogen, phoshorus and potassium uptake by lowland rice and its relation to yields on some Nigerian alluvial soils. Agronomy Journal 57: 421–425.
Bredero, Th.J., 1966. Fertilizer responses in swamp rice (*Oryza sativa* L.) in northern Nigeria. Experimental Agriculture 2: 33–43.
Briggs, L.J. & H.L. Shantz, 1913. The water requirements of plants. I. Investigations in the Great Plains in 1910 and 1911. USDA – Bureau of Plant Industry. Bulletin 282. 48 p.
Briggs, L.J. & H.L. Shantz, 1914. Relative water requirement of plants. Journal of agricultural Research 3: 1–65.
Brockman, J.S., C.M. Rope & M.T. Stevens, 1971. A mathematical relationship between nitrogen input and output in cut grass swards. Journal of the British Grassland Society 26: 75–77.
Brouwer, R., 1963. Some aspects of the equilibrium between overground and underground plant parts. Jaarboek IBS 1963: 31–39.
Brown, R.H., 1978. A difference in N-use efficiency in C3 and C4 plants and its implications in adaptation and evolution. Crop Science 18: 93–98.
Brunt, D., 1932. Notes on radiation in the atmosphere. I. Quarterly Journal of the Royal Meteorological Society 58: 389–420.
Buol, S.W., P.A. Sanchez, R.B. Cate & M.A. Granger, 1975. Soil fertility capability classification: a technical soil classification for fertility management. In: Bornemisza, E. & A. Alvaredo (Eds.). Soil management in tropical America. North Carolina State University, Raleigh, U.S.A.
Buringh, P. & H.D.J. van Heemst, 1979. Potential world food production. In: Linnemann, H., J. de Hoogh, M.A. Keyzer & H.D.J. van Heemst. MOIRA: Model of International relations in agriculture. Contributions to Economic Analysis 124. North-Holland Publishing Company, Amsterdam. p 19–72.
Cours, G., 1948. Le manioc à Madagascar. Mémoires de l'Institut Scientifique de Madagascar. Serie B-Tome III-Fascicule 2: 291–294, 299.
Dabin, B., 1980. Phosphorus deficiency in tropical soils as a constraint on agricultural output. In: Soil-related constraints to food production in the tropics. IRRI, Los Baños, Philippines.
Darwinkel, A., 1980a. Grain production of winterwheat in relation to nitrogen and diseases. I. Relation between nitrogen dressing and yellow rust infection. Zeitschrift für Acker- und Pflanzenbau 149: 299–308.
Darwinkel, A., 1980b. Grain production of winterwheat in relation to nitrogen and diseases. II. Relation between nitrogen dressing and mildew infection. Zeitschrift für Acker- und Pflanzenbau 149: 309–317.
Dash, R.N., S.K. Mohanty & S. Patnaik, 1982. Comparative study of reactivity and agronomic effectiveness of Indian, US, African and Middle East rock phosphates for growing rice. Fertilizer Research 3: 177–189.
Diest, A. van, 1978. Factors affecting the availability of potassium in soils. In: Potassium Research – Reviews and Trends. Proceedings of the 11th International Congress of the International Potash Institute, Bern.
Dobben, W.H. van, 1979. Aanpassingen bij het milieu. In: Alberda, Th. (Ed.), De groene aarde. Aula-boek 250. Het Spectrum, Utrecht/Antwerpen, 313–396.
Doorenbos, J. & A.H. Kassam, 1979. Yield response to water. Irrigation and Drainage

paper 33, F.A.O., Rome. 193 p.
Doorenbos, J., A.H. Kassam, C. Bentvelder & G. Uittenboogaard, 1978. Yield response to water. U.N. Economic Commission West Asia, Rome.
Doorenbos, J. & W.O. Pruitt, 1977. Guidelines for predicting crop water requirements. Irrigation and Drainage paper 24, F.A.O., Rome. 144 p.
Downton, W.J.S., 1975. The occurrence of C4 photosynthesis among plants. Photosynthetica 9: 96–105.
Driessen, P.M., 1978. Peat Soils. In: Soils and Rice. IRRI, Los Baños, Philippines.
Dufournet, R. & P. Goarin, 1957. Note sur la culture du manioc à Madagascar. Riz et Riziculture 3: 15–16.
Duncan, W.G., R.S. Loomis, W.A. Williams & R. Hanau, 1967. A model for simulating photosynthesis in plant communities. Hilgardia 38: 572–582.
Erdman, J.W., 1972. Een vergelijkend groei-analytisch onderzoek van de rijstcultivars IR8 en Acorni. CELOS-rapporten No. 78, CELOS, Paramaribo, Suriname. 30 p.
Esakkimuthu, N., K.K. Krishnamoorthy & S. Loganathan, 1975. Influence of nitrogen and potassium and method of application of potassium on yield and nutrients uptake in rice. Journal of the Indian Society of Soil Science 23: 452–457.
Farrington, J., 1975. Farm surveys in Malawi: the collection and analysis of labour data. Development Study no. 16. University of Reading. Department of Agricultural Economics and Management. 109 p.
F.A.O., 1979. Watershed development, with special reference to soil and water conservation. F.A.O. Soils Bulletin 44.
Flynn, L.M., C.W. Gehrke, M.E. Mührer, G.E. Smith & M.S. Zuber, 1957. Effects of temperature, rainfall and fertilizer on yields and composition of corn plants. Agricultural Experiment Station of Missouri. Research Bulletin 620. 24 p.
Fox, R.H., H. Talleyrand & D.R. Bouldin, 1974. Nitrogen fertilization of corn and sorghum grown in Oxisols and Ultisols in Puerto Rico. Agronomy Journal 66: 534–540.
Fox, R.H., H. Talleyrand & T.W. Scott, 1975. Effect of nitrogen fertilization on yields and nitrogen content of cassava, Llanera cultivar. Journal of Agriculture of the University of Puerto Rico 59: 115–124.
Frère, M. & G.F. Popov, 1979. Agrometeorological crop monitoring and forecasting. F.A.O. Plant Production and Protection Paper 17, F.A.O., Rome, 64 p.
Geus, J.G. de, 1973. Fertilizer guide for the tropics and subtropics. Centre d'Etude de l'Azote, Zurich.
Gifford, R.M., 1974. A comparison of potential photosynthesis, productivity and yield of plant species with differing photosynthetic metabolism. Australian Journal of Plant Physiology 1: 107–117.
Glover, J. & J.S.G. McCulloch, 1958. The empirical relation between solar radiation and hours of sunshine. Quarterly Journal of the Royal Meteorological Society 84: 172–175.
Goudriaan, J., 1977. Crop micrometeorology: a simulation study. Simulation Monographs. Pudoc, Wageningen. 249 p.
Goudriaan, J. & H. H. van Laar, 1978a. Calculation of daily totals of the gross CO2 assimilation of leaf canopies. Netherlands Journal of agricultural Science 26: 416–425.
Goudriaan, J. & H.H. van Laar, 1978b. Relations between leaf resistance, CO2-concentration and CO2-assimilation in maize, beans, Lalang grass and sunflower. Photo-

synthetica 12: 241–249.
Grace, J., 1983. Plant-atmosphere relationships. Outline studies in ecology. Chapman & Hall, London, New York. 92 p.
Greenland, D.J., 1974. Evolution and development of different types of shifting cultivation. In: Shifting cultivation and soil conservation in Africa. FAO, Rome, Soils Bulletin 24: 5–13.
Grove, T.L., K.D. Ritchey & G.C. Naderman, Jr., 1980. Nitrogen fertilization of maize on an Oxisol of the Cerrado of Brazil. Agronomy Journal 72:261–265.
Hanway, J.J., S.A. Barber, R.H. Bray, A.C. Caldwell, M. Fried, L.T. Kurtz, K. Lawton, J.T. Pesek, K. Pretty, M. Reed & F.W. Smith, 1962. North Central regional potassium studies. III. Field studies with corn. North Central regional publication no. 135. Agricultural and home economics experiment station. Iowa State University. Research Bulletin 503: 407–438.
Heemst, H.D.J. van, J.J. Merkelijn & H. van Keulen, 1981. Labour requirements in various agricultural systems. Quarterly Journal of international Agriculture, Frankfurt (Main) 20: 178–201.
Hochman, Z., 1978. Wheat in a semi-arid environment: A field and simulation study of the effects of water stress on yield. M.Sc. Thesis. University of Sydney, Australia. 172 p.
Hochman, Z., 1982. Effect of water stress with phasic development on yield of wheat grown in a semi-arid environment. Field Crops Research 5: 55–67.
Howeler, R.H., 1981. Mineral nutrition and fertilization of cassava. Centro Internacional de Agricultura Tropical, CIAT, Apartado 6713, Cali, Colombia. 52 p.
ILRI, 1972. Veldboek voor land- en waterdeskundigen. International Institute for Land Reclamation and Improvement, Wageningen.
Ismunadji, M. & Sismiyati, 1976. Studies on fertilizer nitrogen efficiency in lowland rice on an acid latosol of Indonesia. Paper presented at IRRI Annual Conference, April 12–15.
Jacob, A. & H. von Uexküll, 1958. Fertilizer use. Nutrition and manuring of tropical crops. Verlaggesellschaft für Ackerbau m.b.H., Hannover.
Jung, Jr., P.E., L.A. Peterson & L.E. Schrader, 1972. Response of irrigated corn to time, rate, and source of applied N on sandy soils. Agronomy Journal 64: 668–670.
Kamolpatana, H. & L.D. Walker, 1969. Crop labour time and motion study of selected crops in North-East Thailand. Bangkok, U.S.O.M. 65 p.
Kanapathy, K., 1974. Fertilizer experiments on shallow peat under continuous cropping with Tapioca. Malaysian agricultural Journal 49: 403–412.
Kang, B.T., F. Donkoh & K. Moody, 1977. Soil fertility management investigations on Benchmark soils in the humid low altitude tropics of West Africa: Investigations on Egbeda soil series. Agronomy Journal 69: 651–656.
Kang, B.T. & M. Yunusa, 1977. Effect of tillage methods and phosphorus fertilization on maize in the humid tropics. Agronomy Journal 69: 291–294.
Kemmler, G., 1980. Potassium deficiency in soils of the tropics as a constraint to food production. In: Soil-related constraints to food production in the tropics. IRRI, Los Baños, Philippines.
Kennes, W., 1981. Analysis of food expenditure patterns in Thailand. Staff working paper 81-5 (draft), Centre for World Food Studies, Amsterdam-Wageningen.
Keulen, H. van, 1975. Simulation of water use and herbage growth in arid regions. Simulation Monographs, Pudoc, Wageningen. 176 p.

Keulen, H. van, 1976. A calculation method for potential rice production. Contributions of the Central Research Institute for Agriculture, Bogor (Indonesia), no. 21, 26 p.
Keulen, H. van, 1977. Nitrogen requirements of rice with special reference to Java. Contributions of the Central Research Institute for Agriculture, Bogor (Indonesia), no. 30, 67 p.
Keulen, H. van, 1982. Graphical analysis of annual crop response to fertiliser application. Agricultural Systems 9: 113–126.
Keulen, H. van & H.D.J. van Heemst, 1982. Crop response to the supply of macronutrients. Verslagen Landbouwkundige Onderzoekingen (Agricultural Research Reports) 916. Pudoc, Wageningen, 46 p.
Keulen, H. van & N.G. Seligman, 1986. Simulation of water use, nitrogen nutrition and growth of a spring wheat crop. Simulation Monographs, Pudoc, Wageningen. (in prep.)
Khind, C.S. & F.N. Ponnamperuma, 1981. Effects of water regime on growth, yield, and nitrogen uptake of rice. Plant and Soil 59: 287–298.
Kirkham, D., 1958. Seepage of steady rainfall through soil into drains. Transactions of the American Geophysical Union 39: 892–908.
Kirkham, D., 1961. The upper limit for the height of water table in drainage design formulas. ISSS Proceedings of the Madison Conference 486–492.
Köhnlein, J., 1972. Einfluss der winterlichen Stickstoffauswaschung auf die Stickstoffversorgung des Getreides. Untersuchungen auf lehmiger Parabraunerde in Schleswig-Holstein. Landwirtschaftliche Forschung 25: 1–15.
Koyama, T., C. Chammek & N. Niamsrichand, 1973. Nitrogen application technology for tropical rice as determined by field experiments using 15N tracer technique. Tropical Agricultural Research Center, Tokyo. Technical Bulletin no. 3. 79 p.
Krantz, B.A., W.L. Nelson, C.D. Welch & N.S. Hall, 1949. A comparison of phosphorus utilization by crops. Soil Science 68: 171–177.
Lathwell, D.J., H.D. Dubey & R.H. Fox, 1972. Nitrogen-supplying power of some tropical soils of Puerto Rico and methods for its evaluation. Agronomy Journal 64: 763–766.
Lehr, J.J., 1950. Differences in the recovery of nitrogen from ammoniacal and nitrate sources in the reclaimed polders of the Zuyder Zee. Plant and Soil 2: 345–357.
Linnemann, H., J. de Hoogh, M.A. Keyzer & H.D.J. van Heemst, 1979. MOIRA: Model of International relations in agriculture. Contributions to Economic Analysis 124. North-Holland Publishing Company, Amsterdam. 379 p.
Loué, A., 1963. A contribution to the study of the inorganic nutrition of maize, with special attention to potassium. Fertilité 20: 5–32.
Lusanandana, B., S. Suwanwaong & P. Sanitwongse, 1966. A study on comparative efficiency of different forms of fertilizer nitrogen for rice. Technical Division, Rice Department, Ministry of Agriculture, Bangkok.
Mahapatra, I.C. & S.C. Panda, 1972. Uptake and utilization of nitrogen, phosphorus and potassium by dwarf *Indica* rices. IRC-Newsletter 21: 1–19.
Majumdar, D.K., 1973. Uptake of nitrogen and phosphate as influenced by sowing time and fertilization and recoveries of nitrogen and phosphate from the added fertilizers by rice (*Oryza sativa* L.). Indian Agriculturist 17: 69–74.
Meadows, D.L., D. Meadows & J. Randers, 1972. The limits to growth. Universe Books, New York.

Mederski, H.J., M.E. Miller & C.R. Weaver, 1973. Accumulated heat units for classifying corn hybrid maturity. Agronomy Journal 65: 743–747.
Mitra, G.N., V.A.K. Sharma & B. Ramamoorthy, 1958. Comparative studies on the potassium fixation capacities of Indian soils. Journal of the Indian Society of Soil Science 6: 1–6.
Monteith, J.L., 1965. Evaporation and environment. In: Fogg, G.E. (Ed.), The state and movement of water in living organisms. Cambridge University Press, London.
Monteith, J. L., 1973. Principles of environmental physics. William Clowes & Sons, London. 241 p.
Moormann, F.R. & N. van Breemen, 1978. Rice: Soil, Water, Land. IRRI, Los Baños, Philippines.
Moschler, W.W. & D.C. Martens, 1975. Nitrogen, phosphorus, and potassium requirements in no-tillage and conventionally tilled corn. Soil Science Society of America Proceedings 39: 886–891.
Motsara, M.R. & N.P. Datta, 1971. Rock phosphate as a fertilizer for direct application in acid soils. Journal of the Indian Society of Soil Science 19: 107–113.
Müller, P. & W. Reiher, 1966. Auswertung langjahriger Ertragsfeststellungen (dargestellt am Versuch "ewiger Roggenbau" in Halle). Albrecht Thaer Archiv 10. 47 p.
Nijholt, J.A., 1936. Opname van voedingsstoffen uit den bodem bij cassave. Buitenzorg Algemeen Proefstation voor den Landbouw. Mededeelingen no. 15, 25 p.
Nuttonson, M.Y., 1953. Phenology and thermal development as a means for a physiological classification of wheat varieties and for predicting maturity dates of wheat. American Institute of Crop Ecology, Washington DC, 188 p.
Nuttonson, M.Y., 1955. Wheat climate relationships and the use of phenology in ascertaining the thermal and photoperiodic requirements of wheat. American Institute of Crop Ecology, Washington DC, 388 p.
Odell, R.T., J.C. Dijkerman, W. van Vuure, S.W. Melsted, A.H. Beavers, P.M. Sutton, L.T. Kurtz & R. Miedema, 1974. Characteristics, classification and adaptation of soils in selected areas in Sierra Leone, West Africa. Bulletin 4, Njala University College, University of Sierra Leone.
Olson, R.V., 1980. Fate of tagged nitrogen fertilizer applied to irrigated corn. Soil Science Society of America Journal 44: 514–517.
Olson, R.A., K.D. Frank & A.F. Dreier, 1964. Controlling losses of fertilizer nitrogen from soils. Proceedings of the 8th International Congress of Soil Science, Bucharest, Romania, 1023–1032.
Oostendorp, D., 1964. Stikstofbemesting en grasgroei in het voorjaar. Landbouwkundig Tijdschrift 76: 101–110.
Orioli, G.A., I. Mogilner, W.L. Bartra & P.A. Semienchuk, 1967. Acumulacion de materia seca, N, P, K y Ca en Manihot esculenta. Bonplandia Tomo II 13: 175–180.
Patnaik, S. & S.T. Gaikawad, 1969. Efficiency of phosphorus and potassium absorbed during various growth stages of 'IR 8' for grain production. Indian Journal of agricultural Science 39: 117–121.
Patrick, Jr., W.H., W.A. Quirck, III, F.J. Peterson & M.D. Faulkner, 1967. Effect of continuous submergence versus alternate flooding and drying on growth, yield, and nitrogen uptake of rice. Agronomy Journal 59: 418–419.
Penman, H.L., 1948. Natural evaporation from open water, bare soil and grass. Proceedings of the Royal Society A193: 120–145.
Penman, H.L., 1956. Evaporation: an introductory survey. Netherlands Journal of

agricultural Science 4: 9–29.
Penman, H.L., 1963. Vegetation and hydrology. Commonwealth Bureau of Soils, Harpenden, Technical Communications no. 53. 124 p.
Penning de Vries, F.W.T., 1975a. The use of assimilates in higher plants. In: Cooper, J.P. (Ed.), Photosynthesis and productivity in different environments. Cambridge University Press, p. 459–480.
Penning de Vries, F.W.T., 1975b. The cost of maintenance processes in plant cells. Annals of Botany 39: 77–92.
Penning de Vries, F.W.T. & H. van Keulen, 1982. La production actuelle et l'action de l'azote et du phosphore. In: Penning de Vries, F.W.T. & M.A. Djitèye (Eds.). La productivité des pâturages sahéliens. Une étude des sols, des végétations et de l'exploitation de cette ressource naturelle. Pudoc, Wageningen. p. 196–226.
Penning de Vries, F.W.T., J.M. Krul & H. van Keulen, 1980. Productivity of Sahelian rangelands in relation to the availability of nitrogen and phosphorus from the soil. In: Rosswall, T. (Ed.). Nitrogen cycling in West African Ecosystems.
Penning de Vries, F.W.T. & H.H. van Laar (Eds.), 1982. Simulation of plant growth and crop production. Simulation Monographs, Pudoc, Wageningen. 308 p.
Penning de Vries, F.W.T., H.H. van Laar & M.C.M. Chardon, 1983. Bio-energetics of growth of seeds, fruits and storage organs. Symposium 'Potential Productivity of field crops under different environments'. IRRI, Manila, Philippines. p. 37–59.
Penning de Vries, F.W.T., J.M. Witlage & D. Kremer, 1979. Rates of respiration and of increase in structural dry matter in young wheat, ryegrass and maize plants in relation to temperature, to water stress and their sugar content. Annals of Botany 44: 595–609.
Puh, Y.S. & H. Khunathai, 1971. A comparative study on testing methods for determining available phosphorus of paddy soils in Thailand. FAO/UNDP. Soil Fertility Research Project, Bangkok. (Mimeograph).
Rabbinge, R. & F.H. Rijsdijk, 1982. Disease and crop physiology: a modeller's point of view. In: Ayres, P.G. (Ed.), Effects of diseases on the physiology of the growing plant. Society for Experimental Biology – Seminar Series II. Cambridge, p. 201–220.
Rabbinge, R., E.M. Drees, M. van der Graaf, F.C.M. Verberne & A. We sselo, 1981. Damage effects of cereal aphids in wheat. Netherlands Journal of Plant Pathology 87: 217–232.
Rabuffetti, A. & E. J. Kamprath, 1977. Yield, N, and S content of corn as affected by N and S fertilization on coastal plain soils. Agronomy Journal 69: 785–788.
Raghavendra, A.S. & V.S.R. Das, 1978. The occurrence of C4 photosynthesis: A supplementary list of C4 plants reported during late 1974 – mid 1977. Photosynthetica 12: 200–208.
Raschke, K., 1975. Stomatal action. Annual Review of Plant Physiology 26: 309–340.
Reddy, K.R. & W.H. Patrick, Jr., 1978. Utilization of labelled urea and ammonium sulfate by lowland rice. Agronomy Journal 70: 465–467.
Rietveld, M.R., 1978. A new method for estimating the regression coefficients in the formula relating solar radiation to sunshine. Agricultural Meteorology 19: 243–252.
Rijsdijk, F.H., 1979. Verliezen in de Nederlandse tarweteelt. Gewasbescherming 10: 196–199.
Rijsdijk, F.H. 1980. System analysis on the cross-road of plant pathology and crop

physiology. Zeitschrift für Pflantzenkrankheiten und Pflantzenschütz 8: 404–408.

Rijsdijk, F.H. & J.C. Zadoks, 1979. A data bank on crop losses: First experiences. EPPO Bulletin 9: 297–303.

Rijtema, P.E., 1965. An analysis of actual evapotranspiration. Verslagen landbouwkundige Onderzoekingen (Agricultural Research Reports) 659, Pudoc, Wageningen. 107 p.

Rijtema, P.E., 1969. Soil moisture forecasting. Nota 513, I.C.W., Wageningen. 18 p.

Ritchie, J.T., 1972. Model for predicting evapotranspiration from a row crop with incomplete cover. Water Resources Research 8: 1208–1213.

Roche, P., L. Grière, D. Babre, H. Calba & P. Fallavier, 1978. La carence en phosphore des sols intertropicaux et ses méthodes d'appreciation. Abstracts 11th International Congress of Soil Science, Volume I, 24. Edmonton, Canada.

Rose, C.W., 1966. Agricultural Physics. Pergamon Press, London. 230 p.

Roy, R.N. & B.C. Wright, 1973. Sorghum growth and nutrient uptake in relation to soil fertility. I. Dry matter accumulation patterns, yield and N-content of grain. Agronomy Journal 65: 709–711.

Roy, R.N. & B.C. Wright, 1974. Sorghum growth and nutrient uptake in relation to soil fertility. II. N, P and K uptake pattern by various plant parts. Agronomy Journal 66: 5–10.

Sadanandan, A.K., S.K. Mohanty, S. Patnaik & K.B. Mistry, 1980. 32P tracer studies on the efficiency of ammonium nitrate phosphates and polyphosphates for growing rice on different soil types. Journal of nuclear Agriculture Biology 9: 135–13.

Sanchez, P.A. & T.T. Cochrane, 1980. Soil constraints in relation to major farming systems in tropical America. In: Soil-related constraints to food production in the tropics. IRRI, Los Baños, Philippines.

Sanchez, P.A., G.E. Ramirez & M.V. de Calderon, 1973. Rice responses to nitrogen under high solar radiation and intermittent flooding in Peru. Agronomy Journal 65: 523–529.

Shantz, H.L. & L.N. Piemeisel, 1927. The water requirements of plants at Akron, Colorado. Journal of agricultural Research 34: 1093–1190.

Sibma, L., 1968. Growth of closed green crop surfaces in the Netherlands. Netherlands Journal of agricultural Science 16: 211–216.

Sieben, W.H., 1974. Effect of drainage conditions on nitrogen supply and yield of young loamy soils in the IJsselmeerpolders. Van Zee tot Land 51. 180 p. (Dutch, with English summary).

Sinclair, T.R. & C.T. de Wit, 1975. Photosynthesis and nitrogen requirements for seed production by various crops. Science 189: 565–567.

Slobbe, W.G. van, 1973. Een vergelijkend groei-analytisch onderzoek van de rijstcultivar Acorni bij verschillende plantdichtheden. CELOS- rapporten No. 86, CELOS, Paramaribo, Surinam. 27 p.

Somsak, C., 1974. Cassava production and its future in Thailand. In: FAO/UNDP International Expert Consultation on the use of improved technology for food production in rainfed areas of tropical Asia. Annex II. FAO, Rome.

Spitters, C.J.T. & R. Aerts, 1983. Simulation of competition for light and water in crop-weed associations. Aspects of applied Biology 4: 467–483.

Stolwijk, H.J., 1983. The supply component of BAM. Working paper SOW-83-8, Centre for World Food Studies, Amsterdam, Wageningen.

Strahler, A.N., 1963. The earth sciences. Harper & Row Publishers, New York. 681 p.

Stroosnijder, L. 1976. Infiltratie en herverdeling van water in grond. Verslagen Landbouwkundige Onderzoekingen (Agricultural Research Reports) 847, Pudoc, Wageningen. 213 p.

Stroosnijder, L. & D. Koné, 1982. Le bilan d'eau du sol. In: Penning de Vries, F.W.T. & M.A. Djitèye (Eds.). La productivité des pâturages sahéliens. Une étude des sols, des végétations et de l'exploitation de cette ressource naturelle. Pudoc, Wageningen, p. 133–165.

Suwanarit, A., 1975. Integrated research program on soil and fertilizer requirement for increasing yields of corn and sorghum. In: Thailand National Corn and Sorghum program. Annual Report Kasetsart University, Department of Agriculture, p. 228–266.

Tanaka, A., S.A. Navasero, C.V. Garcia, F.T. Parao & E. Ramirez, 1964. Growth habit of the rice plant in the tropics and its effect on nitrogen response. IRRI, Technical Bulletin no. 3, 80 p.

Thai Rice Department, 1956. A summary report on yield response of some Thai rice verieties to varying N and P_2O_5 combinations. Soil Fertility Branch, Ministry of Agriculture, Bangkok.

Traoré, M.-F., 1974. Etude de la fumure minerale azotée intensive des céréales et du rôle specifique de la matière organique dans la fertilité des sols au Mali. Agronomie tropicale 29: 567–586.

U.S.D.A. Soil Conservation Service, 1975. Soil Taxonomy. Agriculture Handbook no. 436, Washington D.C.

Veen, J.A. van, 1977. The behaviour of nitrogen in the soil. A computer simulation model. Ph.D. thesis, Univ. of Amsterdam. 164 p.

Veihmeyer, F.J. & A.H. Hendrickson, 1949. Methods of measuring field capacity and wilting percentages of soils. Soil Science 68: 75–94.

Viets, F.G. Jr., 1962. Fertilizers and the efficient use of water. Advances in Agronomy 14:223–264.

Vuure, W. van & R. Miedema, 1973. Soil survey of the Makeni Area, Northern Province, Sierra Leone. Njala University College, University of Sierra Leone.

Walcott, J.J., M. Chauviroj, A. Chinchest, P. Choticheuy, R. Ferraris & B.W. Norman, 1977. Long-term productivity of intensive rice cropping systems on the central plain of Thailand. Experimental Agriculture 13: 305–315.

Williams, N.H., M.D. Morse & J.R. Ruckman, 1972. Burning vs. incorporation of rice crop residues. Agronomy Journal 64: 467–468.

Wit, C.T. de, 1953. A physical theory on placement of fertilizers. Verslagen landbouwkundige Onderzoekingen (Agricultural Research Reports) 59.4. Staatsdrukkerij, 's-Gravenhage. 70 p.

Wit, C.T. de, 1957. An analysis of some pilot fertilizer experiments on paddy. Netherlands Journal of agricultural Science 5: 284–289.

Wit, C.T. de, 1958. Transpiration and crop yields. Verslagen landbouwkundige Onderzoekingen (Agricultural Research Reports) 64.6, Pudoc, Wageningen. 88 p.

Wit, C.T. de, 1965. Photosynthesis of leaf canopies. Verslagen landbouwkundige Onderzoekingen (Agricultural Research Reports 663, Pudoc, Wageningen, 57 p.

Wit, C.T. de, 1969. Een bodemvruchtbaarheidstheorie uit de eerste helft van de 19e eeuw. Landbouwkundig Tijdschrift 81: 245–251.

Wit, C.T. de, 1974. Early concepts in soil fertility. Netherlands Journal of agricultural Science 22: 319–324.

Wit, C.T. de et al., 1978. Simulation of assimilation, respiration and transpiration of crops. Simulation Monographs, Pudoc, Wageningen. 141 p.

Wit, C.T. de & J. Groudriaan, 1978. Simulation of ecological processes. Simulation Monographs, Pudoc, Wageningen. 175 p.

Wulfen, C. von, 1823. Ideeen zur Grundlage einer Statik des Landbaues. Mögliner Annalen, Band XI.

Yoshida, S., 1980. Fundamentals of rice crop science. IRRI, Los Banos, Philippines. 269 p.

Yoshida, S., F.T. Parao & H.M. Beachell, 1972. A maximum annual rice production trial in the tropics. IRC-Newsletter 21: 27–32.

Appendix A. Listing of the model

```
Pro Fortran Compiler - Version zz 1.2

Compilation of: B:RANDOM.FOR

Options:    L

      DIMENSION A(160),C(48),D(48)
     OPEN (3,'CLIMEAN.DAT')
     OPEN (4,'SOILDATA.DAT')
     OPEN (6,'RIJTEMA.DAT')
     OPEN (7,'SOILDATA.RDM', RECL=160)
     OPEN (9,'RIJTEMA.RDM' , RECL=48)
     OPEN (10,'CLIMEAN.RDM', RECL=48)
      DO 10 I=1,150
      READ(4,1000,END=21)A
10    WRITE(7,1000,REC=I)A
21    DO 30 I=1,150
      READ(6,1000,END=31)C
30    WRITE(9,1000,REC=I)C
31    DO 40 I=1,150
      READ(3,1000,END=41)D
40    WRITE(10,1000,REC=I)D
41    CONTINUE
1000  FORMAT(160A1)
      STOP
      END
```

```
Pro Fortran Compiler - Version zz 1.2

Compilation of: B:RANDOMP.FOR

Options:    L

      DIMENSION B(100)
     OPEN (5,'PLANTDAT.DAT')
     OPEN (8,'PLANTDAT.RDM',, RECL=100)
      DO 20 I=1,900
      READ(5,1000,END=21)B
20    WRITE(8,1000,REC=I)B
1000  FORMAT(120A1)
21    STOP
      END
```

```
Pro Fortran Compiler - Version zz 1.2

Compilation of: B:WOFAUTN.FOR

Options:    L

C          MAIN PROGRAM CENTRE FOR WORLD FOOD STUDIES
C          MICRO VERSION CP/M PROSPERO FORTRAN 66
C
           INCLUDE 'WOFOST.CMN'
C        DATA BLOCK OF DATABASE
         IMPLICIT REAL(A-H,J-Z),INTEGER(I)
         COMMON /PLANT/CVL,CVO,CVR,CVS,EXC
         COMMON /PLANT/PERDL,PERRT,PERST,Q10,RD
         COMMON /PLANT/RML,RMO,RMR,RMS,RR
```

```
      COMMON /PLANT/SLATB(30),SPA,SSA
      COMMON /PLANT/TBASE,TDW,CROPT,DVRC1,DLO,DLC,DVRC2,DSL
      COMMON /PLANT/FLTB(30),FRTB(30),FOTB(30),FSTB(30),SWDPTB(30)
      COMMON /PLANT/DVRETB(30),ISPAN,nperso,nperve,pperso,pperve
      COMMON /PLANT/AIRDUC,uptb(30)
C
      COMMON /SOIL/CEF,CLODAN,DD,DRA,DSP
      COMMON /SOIL/FAEF,FCEF,K0,MAMAX,RDM,S0
      COMMON /SOIL/SAL,SM0,SOPE,SURRO,ZTMI,ZTMA
      COMMON /SOIL/TSC1,TSC2,TSGF,ZT,TEXTCL
      COMMON /SOIL/SM,SMFC,SMFCF,SMW,SM10,SSMAX,RIJTEM(8,10)
      COMMON /SOIL/DRAINS,nbase,pbase,nrec,prec
C
      COMMON /CLMT/TEMPTB(30),RAINTB(12),ET0TB(30),E0TB(30),PGS4TB(30)
      COMMON /CLMT/PGS3TB(30),TBIRR(30),TCC4(30),TCC3T(30),TCC3M(30)
      COMMON /CLMT/GEOP,raind(12)
C
        common /div/manf1,manf2,twlv,twst,twso,lvbas1,stbas1,sobas1,
     1  lvbas2,stbas2,sobas2,nmax,pmax,gdur,lvopt,stopt,soopt,neo,new,
     2  wusew,wuseo,tsr
C
      LOGICAL*1 CRPNAM,CLMNAM
      COMMON /NAMES/CRPNAM(30),CLMNAM(30)
        dimension conv(40),hvl(40),idisr(40),tab1(30),tab2(30),ihp(40)
        dimension output(18),climt(5),soilty(5),surrot(5),salt(5)
        dimension ztmit(5),ztmat(5),lustxt(5,5)
        data climt/1.,2.,3.,5.,6/
        data soilty/5.,6.,9.,10.,11./
        data surrot/2.,2.,2.,2.,2./
        data salt/1.5,1.5,1.5,1.5,1.5/
        data ztmit/5*0./
        data ztmat/500.,400.,300.,200.,100./
        data lustxt/.35,.52,0.,0.,0.,.5,.05,.7,0.,0.,.25,.43,.2,.65,0.,
     1  0.,0.,0.,.1,.25,.0,.0,.1,.25,.75/
C
        data conv/1.15,6*1.4,2*.28,.75,1.,1.15,.4,.79,1.12,.78,4.55,
     1  1.14,.44,1.14,3.13,7.14,.53,4.35,16*1./
        data hvl/7*.07,2*0.,.2,.0,.07,.1,.1,3*.05,.1,.2,.07,4*.1,16*0./
        data idisr/14*1,2,2,4*1,2,2,1,2,16*1/
        data tab1/.3,.15,.7,.4,26*0./
        data tab2/.3,.3,.7,.45,26*0./
        data ihp/7*3,2*2,3,2,3*3,1,3,2,7*3,16*3/
C
C
        OPEN (3,'WOFOST.IN')
        OPEN (7,'SOILDATA.RDM', RECL=160)
        OPEN (8,'PLANTDAT.RDM', RECL=100)
        OPEN (9,'RIJTEMA.RDM' , RECL=48)
        OPEN (10,'CLIMEAN.RDM', RECL=48)
        open (15,'result.dat')
C
1       READ(3,1002,END=999)lunit,CROP,STARTD
1002    FORMAT(2F2.0,F4.0)
        icrop=crop
        ilunit=lunit
        clim=climt(ilunit)
        surro=surrot(ilunit)
        sal=salt(ilunit)
        ztmi=ztmit(ilunit)
        ztma=ztmat(ilunit)
        iopt=0
        do 10 i10=1,5
        if(lustxt(ilunit,i10).eq.0.) goto 10
        textcl=soilty(i10)
        if(iopt.gt.0) goto 5
C
        CALL SOILR
C
        CALL CRDAT
C
C
```

```
      CALL PLANTR(CROP)
C
C
      CALL CLIMIN(CLIM)
C
      IPS=1
      IDAY=STARTD
      NBASE=33.
      NREC=.2
      PBASE=5.
      PREC=.1
C
C
      CALL APPLE (IDAY,0,IPS)
      iopt=1
5     CALL SOILR
      CALL PLANTR(CROP)
      IDAY=STARTD
      CALL APPLE (IDAY,1,IPS)
      call nutri(ips)
C
      output(1)=lvopt
      output(2)=stopt
      output(3)=soopt
      output(4)=twlv
      output(5)=twst
      output(6)=twso
      output(7)=lvbas1
      output(8)=stbas1
      output(9)=sobas1
      output(10)=lvbas2
      output(11)=stbas2
      output(12)=sobas2
      do 222 i1=1,12,3
      i11=i1-1
      do 223 i2=1,3
      i3=i2-1
223   output(i1+i3)=output(i1+i3)*(1.-hvl(icrop))*1.15
      i111=ihp(icrop)+i11
222   output(i111)=output(i111)/1.15*conv(icrop)
C
      ichoic=ihp(icrop)
      goto (20,30,40),ichoic
20    neoc=(output(1)-output(7))/neo
      newc=(output(4)-output(10))/neo
      goto 50
30    neoc=(output(2)-output(8))/neo
      newc=(output(5)-output(11))/neo
      goto 50
40    neoc=(output(3)-output(9))/neo
      newc=(output(6)-output(12))/neo
50    do 224 i4=13,15
      i5=i4-6
      i6=i4-12
      if(idisr(icrop).eq.1)
     1 disfac=afgen(tab1,output(i5)/(output(i6)+1.))
      if(idisr(icrop).eq.2)
     1 disfac=afgen(tab2,output(i5)/(output(i6)+1.))
224   output(i4)=output(i5)*(1.-disfac)
      do 225 i4=16,18
      i5=i4-6
      i6=i4-15
      if(idisr(icrop).eq.1)
     1 disfac=afgen(tab1,output(i5)/(output(i6)+1.))
      if(idisr(icrop).eq.2)
     1 disfac=afgen(tab2,output(i5)/(output(i6)+1.))
225   output(i4)=output(i5)*(1.-disfac)
C
      write(15,2222)lunit,TEXTCL,lustxt(ilunit,i10),CROP,STARTD,
     1 wuseo,wusew,tsr,neoc,newc,gdur,output
2222  format(2f4.0,f5.2,2f5.0,6f7.2,/22x,6f8.0,/22x,6f8.0,/22x,6f8.0)
```

```
C
C
10        CONTINUE
          GOTO 1
999       CLOSE (3)
          close (15)
          STOP
C
          END
CC*******************************************************************
          BLOCK DATA
CC*******************************************************************
          INCLUDE 'WOFOST.CMN'
C       DATA BLOCK OF DATABASE
        IMPLICIT REAL(A-H,J-Z),INTEGER(I)
        COMMON /PLANT/CVL,CVO,CVR,CVS,EXC
        COMMON /PLANT/PERDL,PERRT,PERST,Q10,RD
        COMMON /PLANT/RML,RMO,RMR,RMS,RR
        COMMON /PLANT/SLATB(30),SPA,SSA
        COMMON /PLANT/TBASE,TDW,CROPT,DVRC1,DLO,DLC,DVRC2,DSL
        COMMON /PLANT/FLTB(30),FRTB(30),FOTB(30),FSTB(30),SWDPTB(30)
        COMMON /PLANT/DVRETB(30),ISPAN,nperso,nperve,pperso,pperve
        COMMON /PLANT/AIRDUC,uptb(30)
C
        COMMON /SOIL/CEF,CLODAN,DD,DRA,DSP
        COMMON /SOIL/FAEF,FCEF,K0,MHMAX,RDM,S0
        COMMON /SOIL/SAL,SM0,SOPE,SURRO,ZTM1,ZTMA
        COMMON /SOIL/TSC1,TSC2,TSGF,ZT,TEXTCL
        COMMON /SOIL/SM,SMFC,SMFCF,SMW,SM10,SSMAX,RIJTEM(8,10)
        COMMON /SOIL/DRAINS,nbase,pbase,nrec,prec
C
        COMMON /CLMT/TEMPTB(30),RAINTB(12),ET0TB(30),E0TB(30),PGS4TB(30)
        COMMON /CLMT/PGS3TB(30),TBIRR(30),TCC4(30),TCC3T(30),TCC3M(30)
        COMMON /CLMT/GEOP,raind(12)
C
          common /div/manf1,manf2,twlv,twst,twso,lvbas1,stbas1,sobas1,
     1    lvbas2,stbas2,sobas2,nmax,pmax,gdur,lvopt,stopt,soopt,neo,new,
     2    wusew,wuseo,tsr
C
        LOGICAL*1 CRPNAM,CLMNAM
        COMMON /NAMES/CRPNAM(30),CLMNAM(30)
C
          DATA CEF,CLODAN,DD,DRA,DSP/.65,30.,100.,10.,20./
          DATA FAEF,FCEF,K0,MHMAX,RDM,S0/.7,.7,14.5,130.,12.,13.05/
          DATA SAL,SM0,SOPE,SURRO,ZTM1,ZTMA/1.5,.455,8.88,2.,0.,900./
          DATA TSC1,TSC2,TSGF,ZT,TEXTCL/22.6,.049,.0169,150.,7./
          DATA ISPAN/60/
          DATA AIRDUC /0./,DRAINS/0./
          DATA MANF1/1./,MANF2/1./
C
          DATA TCC4/0.,0.,8.,0.,13.,1.,35.,1.,45.,0.,20*0./
          DATA TCC3M/0.,0.,3.,0.,8.,1.,25.,1.,35.,0.,20*0./
          DATA TCC3T/0.,0.,8.,0.,13.,1.,30.,1.,40.,0.,20*0./
          data uptb/.0,0.,40.,0.,50.,.25,60.,.40,70.,.55,80.,.67,90.,
     1 .77,100.,.85,120.,1.,180.,1.3,240.,1.56,300.,1.8,360.,2.,4*0./
          END
CC*******************************************************************
          FUNCTION AFGEN(TABLE,X)
CC*******************************************************************
          DIMENSION TABLE(30)
          IF(X.LE.TABLE(1))  GOTO 60
          DO 10 I=1,30,2
          II=I
          IF(TABLE(I).GE.X) GOTO 50
          IF(I.EQ.1) GOTO 10
          IF(TABLE(I).LT.TABLE(I-2)) GOTO 40
10        CONTINUE
          GOTO 45
40        I=II-2
45        IF(I.GT.30) I=29
          AFGEN=TABLE(I+1)
```

```
      GOTO 100
50    SLOPE=0.
      IF(TABLE(I-2)-TABLE(I).EQ.0.) GOTO 55
      SLOPE=(TABLE(I-1)-TABLE(I+1))/(TABLE(I-2)-TABLE(I))
55    BETA=TABLE(I+1)-SLOPE*TABLE(I)
      AFGEN=SLOPE*X+BETA
      GOTO 100
60    AFGEN=TABLE(2)
100   RETURN
      END
CC*****************************************************************
      SUBROUTINE SOILR
CC*****************************************************************
      INCLUDE 'WOFOST.CMN'
C     DATA BLOCK OF DATABASE
      IMPLICIT REAL(A-H,J-Z),INTEGER(I)
      COMMON /PLANT/CVL,CVO,CVR,CVS,EXC
      COMMON /PLANT/PERDL,PERRT,PERST,Q10,RD
      COMMON /PLANT/RML,RMO,RMR,RMS,RR
      COMMON /PLANT/SLATB(30),SPA,SSA
      COMMON /PLANT/TBASE,TDW,CROPT,DVRC1,DLO,DLC,DVRC2,DSL
      COMMON /PLANT/FLTB(30),FRTB(30),FOTB(30),FSTB(30),SWDPTB(30)
      COMMON /PLANT/DVRETB(30),ISPAN,nperso,nperve,pperso,pperve
      COMMON /PLANT/AIRDUC,uptb(30)
C
      COMMON /SOIL/CEF,CLODAN,DD,DRA,DSP
      COMMON /SOIL/FAEF,FCEF,K0,MHMAX,RDM,S0
      COMMON /SOIL/SAL,SM0,SOPE,SURRO,ZTMI,ZTMA
      COMMON /SOIL/TSC1,TSC2,TSGF,ZT,TEXTCL
      COMMON /SOIL/SM,SMFC,SMFCF,SMW,SM10,SSMAX,RIJTEM(8,10)
      COMMON /SOIL/DRAINS,nbase,pbase,nrec,prec
C
      COMMON /CLMT/TEMPTB(30),RAINTB(12),E10TB(30),E0TB(30),PGS4TB(30)
      COMMON /CLMT/PGS3TB(30),TBIRR(30),TCC4(30),TCC3T(30),TCC3M(30)
      COMMON /CLMT/GEOP,raind(12)
C
       common /div/manf1,manf2,twlv,twst,twso,lvbas1,stbas1,sobas1,
     1  lvbas2,stbas2,sobas2,nmax,pmax,gdur,lvopt,stopt,soopt,neo,new,
     2  wusew,wuseo,tsr
C
      LOGICAL*1 CRPNAM,CLMNAM
      COMMON /NAMES/CRPNAM(30),CLMNAM(30)
C
       IREC=TEXTCL
       READ(7,1000,ERR=999,REC=IREC)K0,MHMAX,SOPE,TSC1,TSC2,TSGF,SAL,
     1SURRO,CLODAN,ZT,S0,SM0
1000   FORMAT(16F10.4)
       RDM=150
       RETURN
999    WRITE(1,2000) IREC
2000   FORMAT(' CANNOT READ SOILDATA FOR TEXTURE CLASS ',I4)
       STOP
       END
CC*****************************************************************
       SUBROUTINE PLANTR(CROP)
CC*****************************************************************
       INCLUDE 'WOFOST.CMN'
C     DATA BLOCK OF DATABASE
      IMPLICIT REAL(A-H,J-Z),INTEGER(I)
      COMMON /PLANT/CVL,CVO,CVR,CVS,EXC
      COMMON /PLANT/PERDL,PERRT,PERST,Q10,RD
      COMMON /PLANT/RML,RMO,RMR,RMS,RR
      COMMON /PLANT/SLATB(30),SPA,SSA
      COMMON /PLANT/TBASE,TDW,CROPT,DVRC1,DLO,DLC,DVRC2,DSL
      COMMON /PLANT/FLTB(30),FRTB(30),FOTB(30),FSTB(30),SWDPTB(30)
      COMMON /PLANT/DVRETB(30),ISPAN,nperso,nperve,pperso,pperve
      COMMON /PLANT/AIRDUC,uptb(30)
C
      COMMON /SOIL/CEF,CLODAN,DD,DRA,DSP
      COMMON /SOIL/FAEF,FCEF,K0,MHMAX,RDM,S0
      COMMON /SOIL/SAL,SM0,SOPE,SURRO,ZTMI,ZTMA
```

```
      COMMON /SOIL/TSC1,TSC2,TSGF,Z1,TEXTCL
      COMMON /SOIL/SM,SMFC,SMFCF,SMW,SM10,SSMAX,RIJTEM(8,10)
      COMMON /SOIL/DRAINS,nbase,pbase,nrec,prec
C
      COMMON /CLMT/TEMPTB(30),RAINTB(12),ET0TB(30),E0TB(30),PGS4TB(30)
      COMMON /CLMT/PGS3TB(30),TBIRR(30),TCC4(30),TCC3T(30),TCC3M(30)
      COMMON /CLMT/GEOP,raind(12)
C
        common /div/manf1,manf2,twlv,twst,twso,lvbas1,stbas1,sobas1,
     1  lvbas2,stbas2,sobas2,nmax,pmax,gdur,lvopt,stopt,soopt,neo,new,
     2  wusew,wuseo,tsr
C
       LOGICAL*1 CRPNAM,CLMNAM
       COMMON /NAMES/CRPNAM(30),CLMNAM(30)
C
C
        IREC=1+24*(IFIX(CROP)-1)
        READ(8,1000,ERR=999,REC=IREC)CRPNAM
1000    FORMAT(72A1)
        READ(8,2000,ERR=999)CROPT,AIRDUC,Q10,EXC,CVL,CVO,CVR,CVS,SPAN,
     1RML
        ISPAN=SPAN
        READ(8,2000,ERR=999)RMO,RMR,RMS,RR,PERDL,PERRT,PERST,SPA,SSA,
     1  nperso
        READ(8,2000,ERR=999)DSL,RD,TDW,DVRC1,DLO,DLC,DVRC2,nperve,
     1  pperso,pperve
        READ(8,2000,ERR=999)FRTB
        READ(8,2000,ERR=999)FLTB
        READ(8,2000,ERR=999)FSTB
        READ(8,2000,ERR=999)FOTB
        READ(8,2000,ERR=999)SLATB
        READ(8,2000,ERR=999)DVRETB
        READ(8,2000,ERR=999)(SWDPTB(I),I=1,20)
2000    FORMAT(10F10.4)
        RETURN
999     WRITE(1,3000)CROP
3000    FORMAT(' CANNOT READ`PLANT DATA FOR CROP ',F3.0)
        STOP
        END
CL*************************************************************************
        SUBROUTINE CRDAT
CL*************************************************************************
        INCLUDE 'WOFOST.CMN'
C      DATA BLOCK OF DATABASE
       IMPLICIT REAL(A-H,J-Z),INTEGER(I)
       COMMON /PLANT/CVL,CVO,CVR,CVS,EXC
       COMMON /PLANT/PERDL,PERRT,PERST,Q10,RD
       COMMON /PLANT/RML,RMO,RMR,RMS,RR
       COMMON /PLANT/SLATB(30),SPA,SSA
       COMMON /PLANT/TBASE,TDW,CROPT,DVRC1,DLO,DLC,DVRC2,DSL
       COMMON /PLANT/FLTB(30),FRTB(30),FOTB(30),FSTB(30),SWDPTB(30)
       COMMON /PLANT/DVRETB(30),ISPAN,nperso,nperve,pperso,pperve
       COMMON /PLANT/AIRDUC,uptb(30)
C
       COMMON /SOIL/CEF,CLODAN,DD,DRA,DSP
       COMMON /SOIL/FAEF,FCEF,K0,MHMAX,RDM,S0
       COMMON /SOIL/SAL,SM0,SOPE,SURRO,ZTMI,ZTMA
       COMMON /SOIL/TSC1,TSC2,TSGF,Z1,TEXTCL
       COMMON /SOIL/SM,SMFC,SMFCF,SMW,SM10,SSMAX,RIJTEM(8,10)
       COMMON /SOIL/DRAINS,nbase,pbase,nrec,prec
C
       COMMON /CLMT/TEMPTB(30),RAINTB(12),ET0TB(30),E0TB(30),PGS4TB(30)
       COMMON /CLMT/PGS3TB(30),TBIRR(30),TCC4(30),TCC3T(30),TCC3M(30)
       COMMON /CLMT/GEOP,raind(12)
C
        common /div/manf1,manf2,twlv,twst,twso,lvbas1,stbas1,sobas1,
     1  lvbas2,stbas2,sobas2,nmax,pmax,gdur,lvopt,stopt,soopt,neo,new,
     2  wusew,wuseo,tsr
C
       LOGICAL*1 CRPNAM,CLMNAM
       COMMON /NAMES/CRPNAM(30),CLMNAM(30)
```

```
C
      IREC=1+10*(IFIX(TEXTCL)-1)
      DO 1 I=1,10
      READ(9,1000,ERR=999,REC=IREC)(RIJTEM(IK,I),IK=1,8)
1     IREC=IREC+1
1000  FORMAT(8F6.1)
      RETURN
999   WRITE(1,2000)TEXTCL
2000  FORMAT(' CANNOT READ RIJTEMA"S DATA FOR TEXTURE CLASS ',F3.0)
      STOP
      END
CC***********************************************************************
      SUBROUTINE CLIMIN(CLM)
CC***********************************************************************
      INCLUDE 'WOFOST.CMN'
C     DATA BLOCK OF DATABASE
      IMPLICIT REAL(A-H,J-Z),INTEGER(I)
      COMMON /PLANT/CVL,CVO,CVR,CVS,EXC
      COMMON /PLANT/PERDL,PERRT,PERST,Q10,RD
      COMMON /PLANT/RML,RMO,RMR,RMS,RR
      COMMON /PLANT/SLATB(30),SPA,SSA
      COMMON /PLANT/TBASE,TDW,CROPT,DVRC1,DLO,DLC,DVRC2,DSL
      COMMON /PLANT/FLTB(30),FRTB(30),FOTB(30),FSTB(30),SWDPTB(30)
      COMMON /PLANT/DVRETB(30),ISPAN,nperso,nperve,pperso,pperve
      COMMON /PLANT/AIRDUC,uptb(30)
C
      COMMON /SOIL/CEF,CLODAN,DD,DRA,DSP
      COMMON /SOIL/FAEF,FCEF,K0,MHMAX,RDM,S0
      COMMON /SOIL/SAL,SM0,SOPE,SURRO,ZTMI,ZTMA
      COMMON /SOIL/TSC1,TSC2,TSGF,ZT,TEXTCL
      COMMON /SOIL/SM,SMFC,SMFCF,SMW,SM10,SSMAX,RIJTEM(8,10)
      COMMON /SOIL/DRAINS,nbase,pbase,nrec,prec
C
      COMMON /CLMT/TEMPTB(30),RAINTB(12),ET0TB(30),E0TB(30),PGS4TB(30)
      COMMON /CLMT/PGS3TB(30),TBIRR(30),TCC4(30),TCC3T(30),TCC3M(30)
      COMMON /CLMT/GEOP,raind(12)
C
        common /div/manf1,manf2,twlv,twst,twso,lvbas1,stbas1,sobas1,
     1  lvbas2,stbas2,sobas2,nmax,pmax,gdur,lvopt,stopt,soopt,neo,new,
     2  wusew,wuseo,tsr
C
      LOGICAL*1 CRPNAM,CLMNAM
      COMMON /NAMES/CRPNAM(30),CLMNAM(30)
        DIMENSION DAYNUM(30)
C
        DATA DAYNUM/1.,0.,15.,0.,45.,0.,74.,0.,105.,0.,135.,0.,166.,0.,
     1 196.,0.,227.,0.,258.,0.,288.,0.,319.,0.,349.,0.,365.,0.,2*0./
C
        ihelp=clm-1.
        IREC=1+13*ihelp
        READ(10,1000,ERR=999,REC=IREC)CLMNAM,GEOP
1000    FORMAT(30A1,F5.1)
C
C
        DO 10 I10=1,30
        TEMPTB(I10)=DAYNUM(I10)
        ET0TB(I10)=DAYNUM(I10)
        E0TB(I10)=DAYNUM(I10)
        PGS4TB(I10)=DAYNUM(I10)
        PGS3TB(I10)=DAYNUM(I10)
        TBIRR(I10)=DAYNUM(I10)
10      CONTINUE
C
15      DO 20 I20=4,26,2
        ii20=(i20-2)/2
        READ(10,4000,END=25)TEMPTB(I20),RAINTB(II20),ET0TB(I20),
     1E0TB(I20),PGS4TB(I20),PGS3TB(I20),TBIRR(I20),raind(ii20)
4000    FORMAT(8F6.2)
        E0TB(I20)=E0TB(I20)*30
        ET0TB(I20)=ET0TB(I20)*30
20      CONTINUE
```

```
 25      TEMPTB(2)=(TEMPTB(4)+TEMPTB(26))/2.
         TEMPTB(28)=TEMPTB(2)
         ET0TB(2)=(ET0TB(4)+ET0TB(26))/2.
         ET0TB(28)=ET0TB(2)
         E0TB(2)=(E0TB(4)+E0TB(26))/2.
         E0TB(28)=E0TB(2)
         PGS4TB(2)=(PGS4TB(4)+PGS4TB(26))/2.
         PGS4TB(28)=PGS4TB(2)
         PGS3TB(2)=(PGS3TB(4)+PGS3TB(26))/2.
         PGS3TB(28)=PGS3TB(2)
         TBIRR(2)=(TBIRR(4)+TBIRR(26))/2.
         TBIRR(28)=TBIRR(2)
C
         RETURN
 999     WRITE(1,3000)CLM
 3000    FORMAT(' CANNOT READ CLIMATE DATA FOR CLIMATE ',F3.0)
         STOP
         END
CC***********************************************************************
         SUBROUTINE APPLE(iday,IWB,ISP)
CC***********************************************************************
C         **************  DATABASE INTERFACE **********
C
          INCLUDE 'WOFOST.CMN'
C       DATA BLOCK OF DATABASE
        IMPLICIT REAL(A-H,J-Z),INTEGER(I)
        COMMON /PLANT/CVL,CVO,CVR,CVS,EXL
        COMMON /PLANT/PERDL,PERRT,PERST,Q10,RD
        COMMON /PLANT/RML,RMO,RMR,RMS,RR
        COMMON /PLANT/SLATB(30),SPA,SSA
        COMMON /PLANT/TBASE,TDW,CROPT,DVRC1,DLO,DLC,DVRC2,DSL
        COMMON /PLANT/FLTB(30),FRTB(30),FOTB(30),FSTB(30),SWDPTB(30)
        COMMON /PLANT/DVRETB(30),ISPAN,nperso,nperve,pperso,pperve
        COMMON /PLANT/AIRDUC,uptb(30)
C
        COMMON /SOIL/CEF,CLODAN,DD,DRA,DSP
        COMMON /SOIL/FAEF,FCEF,K0,MHMAX,RDM,S0
        COMMON /SOIL/SAL,SM0,SOPE,SURRO,Z1MI,Z1MA
        COMMON /SOIL/TSC1,TSC2,TSGF,ZT,TEXTCL
        COMMON /SOIL/SM,SMFC,SMFCF,SMW,SM10,SSMAX,RIJTEM(8,10)
        COMMON /SOIL/DRAINS,nbase,pbase,nrec,prec
C
        COMMON /CLMT/TEMPTB(30),RAINTB(12),ET0TB(30),E0TB(30),PGS4TB(30)
        COMMON /CLMT/PGS3TB(30),TBIRR(30),TCC4(30),TCC3T(30),TCC3M(30)
        COMMON /CLMT/GEOP,raind(12)
C
          common /div/manf1,manf2,twlv,twst,twso,lvbas1,stbas1,sobas1,
       1  lvbas2,stbas2,sobas2,nmax,pmax,gdur,lvopt,stopt,soopt,neo,new,
       2  wusew,wuseo,tsr
C
        LOGICAL*1 CRPNAM,CLMNAM
        COMMON /NAMES/CRPNAM(30),CLMNAM(30)
C
C
          DIMENSION LV(170),LA(170)
C
C
C         ******  INITIALISATION  ******
C
          dz=0.
          tsr=0.
          wuse=0.
          gdur0=0.
          SM10=SM0-0.1
          SMFCF=SM0*EXP(-1.*TSGF*ALOG(100.)**2)
          SMW=SM0*EXP(-1.*TSGF*ALOG(16000.)**2)
          SM=SMFCF
          CLODAR=CLODAN*0.017444
          SALR=SAL*0.0174444
          SSMAX=0.5*SURRO*(SIN(CLODAR-SALR)**2/SIN(CLODAR)
       1*(COS(CLODAR+SALR)/SIN(CLODAR+SALR)+COS(CLODAR-SALR)/
```

```
     2SIN(CLODAR-SALR))/2/COS(CLODAR)/COS(SALR))
        DELT=1
        IL=1
        ID=0
        DO 5 I5=1,170
        LA(I5)=0
 5      LV(I5)=0.
C
        GWRT=0.
        GWLV=0.
        GWST=0.
        GWSO=0.
C
        DRRT=0.
        DRLV=0.
        DRST=0.
        DRSO=0.
C
        DWRT=0.
        DWLV=0.
        DWST=0.
        DWSO=0.
C
        TSUMA=0.
        TSUMR=0.
C
        DVS=0.
        SS=0.
        RTDF=0.
C
        FR=AFGEN(FRTB,DVS)
        FL=AFGEN(FLTB,DVS)
        FS=AFGEN(FSTB,DVS)
        FO=AFGEN(FOTB,DVS)
C
        WRT=FR*TDW
        TADW=(1.-FR)*TDW
        WLV=FL*TADW
        WST=FS*TADW
        WSO=FO*TADW
        SLA=AFGEN(SLATB,DVS)
C
        LAI=SLA*WLV+SSA*WST+SPA*WSO
C
        WRITE(ISP,1001)TEXTCL,IDAY,CRPNAM,CLMNAM
        WRITE(ISP,1000)IDAY,WRT,WLV,WST,WSO,LAI,ZT,SM,DVS
 1001   FORMAT(///5X,'Texture class ',F3.0,24X,'Start day ',I3,
     1/'      Crop     ',30A1,'   Climate    ',30A1,
     1///' DAY   WRT       WLV       WST       WSO       LAI       ',
     1' ZT        SM       DVS'/)
C
C       ******  DYNAMIC CALCULATIONS  ******
C
 1      IF(DVS.GE.2.) GOTO 100
        gdur0=gdur0+delt
C
C
C       INTAKE OF CLIMATE DATA
        CALL CLIMAT(IDAY,TEMP,RAIN,PGASS,ET0,E0,XIE)
        call daylen(iday,geop,dl)
        FINT=(1.-EXP(-1.*EXC*LAI))
C
C       IF(IWB.EQ.0) GOTO 164
C
C       ******  WATER BALANCE  ******
C
        SMFC=SM0*EXP(-1.*TSGF*ALOG(ZT+1.)**2)
        MH=EXP(SQRT(-1.*ALOG(SM/SM0)/TSGF))
        KMH=K0*EXP(-1.*TSC2*MH)
        IF(MH.GT.MHMAX) KMH=TSC1*MH**(-1.4)
C
```

```
C         CALCULATION OF XIMMAX
C
          S=S0*(1.-SM/SM0)/SQRT(DELT)
          XIMMAX=S+SOPE
C
C         CALCULATION OF THE ACTUAL EVAPORATION E
C
          E=AMAX1(0.,E0*EXP(-0.4*LAI)*(SM-SMW/3)/(SM0-SMW/3))
          if(iwb.eq.0) E=AMAX1(0.,E0*EXP(-0.4*LAI))
          E=AMIN1(E,E0*EXP(-0.4*LAI))
          EL=0.
          IF(SS.GT.0.) EL=E
          IF(SS.GT.0.) E=0.
C
C         INFILTRATION
C
          SS=SS+(RAIN+XIE-EL)*DELT
          SSR=SS/DELT
C          XIMMAX=XIMMAX*AFGEN(SALTB,SAL)*AFGEN(RNTB,RAIN))
          XIM=AMIN1(XIMMAX,SSR)
          SR=AMAX1(0.,SSR-SSMAX/DELT-XIM)
          tsr=tsr+sr
C
C         CALCULATION OF THE WATER FLOW THROUGH THE LOWER ROOT ZONE BOUNDA
C         -RY
          ZTMRD=ZT-RD
          IF(ZTMRD.LE.0.) GOTO 55
          HRD=MH-(ZTMRD)
          IF(HRD.GT.0) GOTO 60
          IF(HRD.LT.0) GOTO 70
   55     CRMD=0.
          GOTO 80
   60     CALL RIJTMA(RIJTEM,MH,ZTMRD,CRMD,KMH)
          CRMD=AMIN1(CRMD,(SMFC-SM)*RD/DELT)
          GOTO 80
   70     CRMD=AMAX1(-1.*KMH,-1.*(SM-SMFC)*RD)/DELT
   80     CONTINUE
C
C         CALCULATION OF THE ACTUAL TRANSPIRATION
C
          T0=ET0-0.1*E0
          TM=AMAX1(0.0001,(1.-EXP(-1.*EXC*LAI))*T0)
C
          if(iwb.eq.0) goto 130
          SWDEP=AFGEN(SWDPTB,T0)
          SMCR=(1.-SWDEP)*(SMFC-SMW)+SMW
          IF(SM.GE.(SM0-0.05).OR.SMW.GT.SM) GOTO 150
          IF((SM0-0.05).GT.SM.AND.SM.GE.SM10) GOTO 140
          IF(SM10.GT.SM.AND.SM.GE.SMCR) GOTO 130
          T=(SM-SMW)/(SMCR-SMW)*TM
          GOTO 160
  130     T=TM
          GOTO 160
  140     T=AMIN1(1.,((SM0-0.05)-SM)/((SM0-0.05)-SM10))*TM
          GOTO 160
  150     T=0
  160     CONTINUE
          IF(AIRDUC.EQ.0.) GOTO 162
          IF(SM.GE.SM10) T=TM
  162     CONTINUE
C
C         CALCULATION OF THE RATE OF CHANGE IN WATER CONTENT OF
C         THE ROOTED ZONE
          RSM=(XIM+CRMD-T-E)/RD
          if(iwb.eq.0) rsm=0.
C
C         CALCULATION OF DRAINAGE
C
          MT=2.*(DD-ZT)
          DMAX=AMAX1(0.,K0*3.14*MT/(3.14*MT+DSP*ALOG(DSP/(3.14*DRA))))
          IF(DRAINS.EQ.0.) DMAX=0.
```

```
C
C         CALCULATION OF THE CHANGE IN  GROUND WATER DEPTH
C
          if(iwb.eq.0) goto 165
          DZ=AMAX1(-1.*ZT,2.*(DMAX+CRMD)/(SM0-SM))
          IF(ZTMRD.LE.0.) DZ=AMAX1(-1.*ZT,(E+T-XIM)/(SM0-SMF))
          IF(ZTMRD.LE.0.) RSM=-1.*(SM0-SMF)*DZ/(ZTF+1.)
C
C
C
C
C
C
C
C         DEVELOPMENT STAGE
C
  165     DVRED=1.
          IF(DVS.LT.1.) DVRED=AMAX1(0.,AMIN1(1.,(DL-DLC)/(DLO-DLC)))
          DVRET=AFGEN(DVRETB,TEMP)
          IF(DVS.LT.1..AND.DSL.EQ.1.) DVRET=1.
          DVRC=DVRC1
          IF(DVS.GT.1.) DVRC=DVRC2
          DVR=DVRC*DVRED*DVRET
          SLA=AFGEN(SLATB,DVS)
C
C         PARTITIONING FACTORS
C
          FR=AFGEN(FRTB,DVS)
          FL=AFGEN(FLTB,DVS)
          FS=AFGEN(FSTB,DVS)
          FO=AFGEN(FOTB,DVS)
C
C         DAILY DRY MATTER INCREASE
C
          GASS=PGASS*FINT*T/TM
          RMRES=RMR*WRT+RML*WLV+RMS*WST+RMO*WSO
          TEFF=Q10**(0.1*TEMP-2.5)
          MRES=AMIN1(GASS,RMRES*TEFF)
          CVF=(CVL*FL+CVS*FS+CVO*FO)*(1.-FR)+CVR*FR
          ASRC=AMAX1(0.,GASS-MRES)
          DMI=CVF*ASRC
C
C         GROWTH RATE ROOTS
C
          GRRT=FR*DMI
          DRRT=0.
          IF(DVS.GT.1.5) DRRT=PERRT*WRT
          GWRT=GRRT-DRRT
C
C         GROWTH RATES OF THE AIRIAL PARTS
C
          ADMI=DMI-GRRT
C
C         GROWTH RATE LEAVES
C
          GRLV=FL*ADMI
          DSLV1=(1.-T/TM)*PERDL*WLV
          DSLV2=WLV*AMAX1(0.,1.-7./AMAX1(0.01,LAI))
          DSLV=AMAX1(DSLV1,DSLV2)
C
          LV(IL)=LV(IL)-DSLV
          IF(LV(IL)) 190,210,210
  190     IKIL=IL-1
          DO 200 IK=1,IKIL
          IIK=IKIL-IK+1
          if(iik.eq.0) goto 200
          LV(IIK)=LV(IIK)+LV(IIK+1)
          LV(IIK+1)=0.
          IF(LV(IIK)) 200,210,210
  200     CONTINUE
  210     IL=IL+1
```

```
      IF(IL.GT.ISPAN) IL=ISPAN
      DALV=LV(ISPAN)
      IF(IL.EQ.1) GOTO 230
      DO 220 IIK=2,IL
      IK=IL-IIK+2
      LA(IK)=LA(IK-1)
220   LV(IK)=LV(IK-1)
230   LV(1)=GRLV*DELT
      LA(1)=GRLV*SLA*DELT
      IF(IL.EQ.2) LV(1)=LV(1)+WLV
      IF(IL.EQ.2) LA(1)=LA(1)+WLV*SLA
      LASUM=0.
      DO 240 I=1,IL
      IF(LV(I).LE.0.) LA(I)=0.
240   LASUM=LASUM+LA(I)
C
      DRLV=DSLV+DALV
      GWLV=GRLV-DRLV
C
C     GROWTH RATE STEMS
C
      GRST=FS*ADMI
      DRST=0.
      IF(DVS.GT.1.5) DRST=PERST*WST
      GWST=GRST-DRST
C
C     GROWTH RATE STORAGE ORGANS
C
      GWSO=FO*ADMI
C
C     INTEGRALS OF THE CROP
C
      DVS=DVS+DVR*DELT
      WRT=WRT+GWRT*DELT
      WLV=WLV+GWLV*DELT
      WST=WST+GWST*DELT
      WSO=WSO+GWSO*DELT
C
      TADW=WLV+WST+WSO
      TDW=TADW+WRT
C
      DWRT=DWRT+DRRT*DELT
      DWLV=DWLV+DRLV*DELT
      DWST=DWST+DRST*DELT
      DWSO=DWSO+DRSO*DELT
C
      TWRT=WRT+DWRT
      TWLV=WLV+DWLV
      TWST=WST+DWST
      TWSO=WSO+DWSO
C
C     INTEGRALS OF THE WATER BALANCE
C
C
      SM=AMAX1(SMW/3.,AMIN1(SM+RSM*DELT,SM0-0.001))
      IF(ZTMRD.LE.0.) GOTO 777
      SMF=SM
777   CONTINUE
      SS=SS-SR*DELT-XIM*DELT
      ZT=ZT+DZ
      ZT=AMIN1(ZT,ZTMA)
      ZT=AMAX1(ZT,ZTMI)
      if(rd.ge.rdm) rr=0.
      IF(ZTMRD.LE.0.) GOTO 888
      ZTF=ZT
888   CONTINUE
      IF(AIRDUC.EQ.1.) GOTO 250
      if(rd.ge.(zt-10.)) rr=0.
      IF(ZTMRD.LE.0.) RR=0.1*(ZTMRD)/DELT
      IF(FR.LE.0.) RR=0.
      IF(RTDF.GE.5.and.iwb.eq.1) GOTO 245
```

```
      IF(RD.LT.10.) RTDF=RTDF+1.
      IF(RD.GE.10.) RTDF=0.
      GOTO 250
245   WRITE(ISP,3221)
3221  FORMAT(' Crop failure due to waterlogging !')
      RETURN
C
250   LAI=LASUM+SSA*WST+SPA*WSO
      RD=RD+RR*DELT
      wuse=wuse+t
C
C
C
C     *****  TEMPORARY OUTPUT  *****
C
      ID=ID+1
      if(wlv.le.1.and.dvs.gt.0.5) goto 100
      IF(MOD(ID,10).EQ.0)WRITE(ISP,1000)IDAY,WRT,WLV,WST,WSO,LAI,
     1ZT,SM,DVS
1000  FORMAT(1X,I3,4F9.0,2F9.3,2F7.3)
      IDAY=IDAY+1
      IF(IDAY.GT.365) IDAY=IDAY-365
C
      GOTO 1
C
100   CONTINUE
      WRITE(ISP,1000)IDAY,WRT,WLV,WST,WSO,LAI,ZT,SM,DVS
      if(iwb.eq.1) goto 101
      lvopt=twlv
      stopt=twst
      soopt=twso
      gdur=gdur0
      wuseo=wuse
101   continue
      wusew=wuse
C     *****  DATABASE OUTPUT  *****
C
      RETURN
      END
CC*****************************************************************
      SUBROUTINE CLIMAT(I,TEMP,RAIN,PGASS,ET0,E0,XIE)
CC*****************************************************************
      INCLUDE 'WOFOST.CMN'
C    DATA BLOCK OF DATABASE
     IMPLICIT REAL(A-H,J-Z),INTEGER(I)
     COMMON /PLANT/CVL,CVO,CVR,CVS,EXC
     COMMON /PLANT/PERDL,PERRT,PERST,Q10,RD
     COMMON /PLANT/RML,RMO,RMR,RMS,RR
     COMMON /PLANT/SLATB(30),SPA,SSA
     COMMON /PLANT/TBASE,IDW,CROPT,DVRC1,DLO,DLC,DVRC2,DSL
     COMMON /PLANT/FLTB(30),FRTB(30),FOTB(30),FSTB(30),SWDPTB(30)
     COMMON /PLANT/DVRETB(30),ISPAN,nperso,nperve,pperso,pperve
     COMMON /PLANT/AIRDUC,uptb(30)
C
     COMMON /SOIL/CEF,CLODAN,DD,DRA,DSP
     COMMON /SOIL/FAEF,FCEF,K0,MAMAX,RDM,S0
     COMMON /SOIL/SAL,SM0,SOPE,SURRO,ZTMI,ZTMA
     COMMON /SOIL/TSC1,TSC2,TSGF,ZT,TEXTCL
     COMMON /SOIL/SM,SMFC,SMFCF,SMW,SM10,SSMAX,RIJTEM(8,10)
     COMMON /SOIL/DRAINS,nbase,pbase,nrec,prec
C
     COMMON /CLMT/TEMPTB(30),RAINTB(12),ET0TB(30),E0TB(30),PGS4TB(30)
     COMMON /CLMT/PGS3TB(30),TBIRR(30),TCC4(30),TCC3T(30),TCC3M(30)
     COMMON /CLMT/GEOP,raind(12)
C
      common /div/manf1,manf2,twlv,twst,twso,lvbas1,stbas1,sobas1,
    1 lvbas2,stbas2,sobas2,nmax,pmax,gdur,lvopt,stopt,soopt,neo,new,
    2 wusew,wuseo,tsr
C
     LOGICAL*1 CRPNAM,CLMNAM
     COMMON /NAMES/CRPNAM(30),CLMNAM(30)
```

```
      RI40=1
      TEMP=AFGEN(TEMPTB,RI40)
      call rainda(raintb,raind,i,rain)
      ET0=AFGEN(ET0TB,RI40)/30.
      E0=AFGEN(E0TB,RI40)/30.
      IF(CROPT.EQ.1.)PGASS=AFGEN(PGS4TB,RI40)*AFGEN(TCC4,TEMP)*.682
      IF(CROPT.EQ.2.)PGASS=AFGEN(PGS3TB,RI40)*AFGEN(TCC3M,TEMP)*.682
      IF(CROPT.GE.3.)PGASS=AFGEN(PGS3TB,RI40)*AFGEN(TCC3T,TEMP)*.682
      XIE=AFGEN(TBIRR,RI40)/30.
      RETURN
      END
CC***************************************************************
      subroutine rainda(rtb,rda,idag,rn)
CC***************************************************************
      dimension rtb(12),rda(12),irand(30),idgn(12)
      data irand/7,6,13,9,8,21,25,12,23,1,4,28,16,20,29,17,27,3,15,
     1 2,11,18,26,5,19,14,28,30,22,10/
      data idgn/0,31,59,90,120,151,181,212,243,274,304,335/
      do 10 i=1,12
      if(idag.le.idgn(i)) goto 11
10    continue
      i=13
11    icd=idag-idgn(i-1)
      icm=i-1
C
      a=0.
      f=1.
      fact=0.01389*rda(icm)+0.672
      irada=rda(icm)
      do 100 ik=1,irada
      a=a+f
100   f=f*fact
      r=rtb(icm)/a
      do 200 i=1,irada
      if(irand(i).eq.icd) goto 150
      goto 200
150   rn=r*fact**(i-1)
      goto 300
200   continue
      rn=0.
300   return
      end
CC***************************************************************
      SUBROUTINE RIJTMA(R,RMH,D,L,KMH)
CC***************************************************************
      REAL KMH
      DIMENSION R(8,10),LMH(10),CCR(8),VMH(8)
      DATA CMH/20.,50.,100.,250.,500.,1000.,2000.,5000.,10000.,16000./
      DATA CCR/.5,.4,.3,.2,.15,.1,.06,.02/
      IF(RMH.LT.20.) GOTO 130
      IF(RMH.EQ.20.) GOTO 60
      IF(RMH.GE.16000.) GOTO 65
      DO 20 I20=2,10
      K20=I20
      J20=I20-1
      IF(RMH.LE.CMH(I20)) GOTO 30
20    CONTINUE
30    DO 50 I50=1,8
      SLOPE=(R(I50,J20)-R(I50,K20))/(LMH(J20)-LMH(K20))
50    VMH(I50)=SLOPE*RMH+R(I50,K20)-SLOPE*LMH(K20)
      GOTO 70
60    DO 61 I61=1,8
61    VMH(I61)=R(I61,1)
      GOTO 70
65    DO 66 I66=1,8
66    VMH(I66)=R(I66,10)
70    IF(D.LT.VMH(1)) GOTO 140
      IF(D.GT.VMH(8)) GOTO 120
      DO 100 I100=2,8
      K100=I100
      J100=I100-1
```

```
      IF(D.LE.VMH(I100)) GOTO 110
100   CONTINUE
110   SLOPE=(CCR(J100)-CCR(K100))/(VMH(J100)-VMH(K100))
      C=SLOPE*D+CCR(K100)-SLOPE*VMH(K100)
      RETURN
120   C=0.
      RETURN
130   C=KMH*(RMH/D-1.)
      RETURN
140   C=0.5
      RETURN
      END
C*********************************************************************
C
      subroutine daylen(i,gp,daglen)
C
C*********************************************************************
      DECL=-23.4*COS(0.0172055*(i+10))
      COSLD=COS(DECL*0.0174444)*COS(gp*0.0174444)
      SINLD=SIN(DECL*0.0174444)*SIN(gp*0.0174444)
      DAGLEN=3.82166*(3.14+2.*ASIN(sinld/cosld))
      return
      END
      function asin(x)
      asinhp=1./6.*x**3
      asin=x+asinhp
      do 10 i=3,101,3
      asinhp=asinhp*i**2*x**2/((i+1)*(i+2))
      asin=asin+asinhp
      if(abs(asin-asino).lt.1.E-8) goto 20
10    asino=asin
20    return
      end
C*********************************************************************
C
      subroutine nutri (ips)
C
C*********************************************************************
C
      include 'wofost.cmn'
C    DATA BLOCK OF DATABASE
     IMPLICIT REAL(A-H,J-Z),INTEGER(I)
     COMMON /PLANT/CVL,CVO,CVR,CVS,EXC
     COMMON /PLANT/PERDL,PERRT,PERST,Q10,RD
     COMMON /PLANT/RML,RMO,RMR,RMS,RR
     COMMON /PLANT/SLATB(30),SPA,SSA
     COMMON /PLANT/TBASE,TDW,CROPT,DVRC1,DLO,DLC,DVRC2,DSL
     COMMON /PLANT/FLTB(30),FRTB(30),FOTB(30),FSTB(30),SWDFTB(30)
     COMMON /PLANT/DVRETB(30),ISPAN,nperso,nperve,pperso,pperve
     COMMON /PLANT/AIRDUC,uptb(30)
C
     COMMON /SOIL/CEF,CLODAN,DD,DRA,DSP
     COMMON /SOIL/FAEF,FCEF,K0,MHMAX,RDM,S0
     COMMON /SOIL/SAL,SM0,SOPE,SURRO,ZTMI,ZTMA
     COMMON /SOIL/TSC1,TSC2,TSGF,ZT,TEXTCC
     COMMON /SOIL/SM,SMFC,SMFCF,SMW,SM10,SSMAX,RIJTEM(8,10)
     COMMON /SOIL/DRAINS,nbase,pbase,nrec,prec
C
     COMMON /CLMT/TEMPTB(30),RAINTB(12),ET0TB(30),E0TB(30),PGS4TB(30)
     COMMON /CLMT/PGS3TB(30),TBIRR(30),TCC4(30),TCC3T(30),TCC3M(30)
     COMMON /CLMT/GEOP,raind(12)
C
      common /div/manf1,manf2,twlv,twst,twso,lvbas1,stbas1,sobas1,
    1 lvbas2,stbas2,sobas2,nmax,pmax,gdur,lvopt,stopt,soopt,neo,new,
    2 wusew,wuseo,tsr
C
     LOGICAL*1 CRPNAM,CLMNAM
     COMMON /NAMES/CRPNAM(30),CLMNAM(30)
C
      pbas=pbase*afgen(uptb,gdur)*manf2
      nbas=nbase*afgen(uptb,gdur)*manf2
```

```
      if(cropt.eq.4) goto 10
C
C     non-leguminous crops
C
      rec=nrec*manf1
      nopt=nperve*(lvopt+stopt)+nperso*soopt
      nwab=nperve*(twlv+twst)+nperso*twso
      nmaxo=AMAX1(0.,(nopt-nbas)/rec)
      nmaxw=AMAX1(0.,(nwab-nbas)/rec)
C
      frbaso=amin1(1.,nbas/nopt)
      lvbas1=frbaso*lvopt
      stbas1=frbaso*stopt
      sobas1=frbaso*soopt
      lvbas2=amin1(lvbas1,twlv)
      stbas2=amin1(stbas1,twst)
      sobas2=amin1(sobas1,twso)
C
      neo=(nopt-nbas)/rec
      new=(nwab-nbas)/rec
      goto 50
C
C
C     leguminous crops
C
10    rec=prec*manf1
      popt=pperve*(lvopt+stopt)+pperso*soopt
      pwab=pperve*(twlv+twst)+pperso*twso
      pmaxo=amax1(0.,(popt-pbas)/rec)
      pmaxw=amax1(0.,(pwab-pbas)/rec)
C
      frbaso=amin1(1.,pbas/popt)
      lvbas1=frbaso*lvopt
      stbas1=frbaso*stopt
      sobas1=frbaso*soopt
      lvbas2=amin1(lvbas1,twlv)
      stbas2=amin1(stbas1,twst)
      sobas2=amin1(sobas1,twso)
C
      neo=(popt-pbas)/rec
      new=(pwab-pbas)/rec
C
50    continue
      write(ips,8754)nmaxo,pmaxo,nmaxw,pmaxw,nbase,pbase,nrec,
     1 prec,ztmi,ztma,neo,new,wuseo,wusew,tsr,
     1 LVOPT,LVBAS1,STOPT,STBAS1,
     1 SOOPT,SOBAS1,TWLV,LVBAS2,TWST,STBAS2,
     2 TWSO,SOBAS2
8754  format(//20x,'nmaxo= ',f4.0,'      pmaxo= ',f4.0,/
     1 20x,'nmaxw= ',f4.0,'      pmaxw= ',f4.0,/
     2 20x,'nbas  = ',f4.0,'      pbas  = ',f4.0,/
     2 20x,'nrec  = ',f5.2,'     prec  = ',f5.2,/
     2 20x,'ztmi  = ',f4.0,'      ztma  = ',f4.0,/
     2 20x,'npeo  = ',f4.0,'      npew  = ',f4.0,/
     2 20x,'wuseo= ',f4.1,'      wusew= ',f4.1,/
     2 31x,'       tsr   = ',f4.1,//
     1 18x,'LVOPT= ',f6.0,5x,'LVBAS1  = ',f6.0,5x,/
     2 18x,'STOPT= ',f6.0,5x,'STBAS1  = ',f6.0,5x,/
     3 18x,'SOOPT= ',f6.0,5x,'SOBAS1  = ',f6.0,5x,//
     1 18x,'TWLV  = ',f6.0,5x,'LVBAS2  = ',f6.0,5x,/
     2 18x,'TWST  = ',f6.0,5x,'STBAS2  = ',f6.0,5x,/
     3 18x,'TWSO  = ',f6.0,5x,'SOBAS2  = ',f6.0,5x,/'1'/)
      return
      end
```

Books already published in Simulation Monographs

C.T. de Wit and H. van Keulen. **Simulation of transport processes in soils** (2nd ed., revised). 1975. 109 pp. 90-220-0591-7. Price Dfl. 15.00

J. Beek and M.J. Frissel. **Simulation of nitrogen behaviour in soils.** 1973. 76 pp. ISBN 90-220-0440-6. (out of print)

H.G. Fransz. **The functional response to prey density in an acarine system.** 1974. 149 pp. ISBN 90-220-0509-7. (out of print)

M.J. Frissel and P. Reiniger. **Simulation of accumulation and leaching in soils.** 1974. 124 pp. ISBN 90-220-0530-5. (out of print)

H. van Keulen. **Simulation of water use and herbage growth in arid regions.** 1975. 184 pp. ISBN 90-220-0557-7. Price Dfl. 30.00

G.F. Makkink and H.D.J. van Heemst. **Simulation of the water balance of arable land and pastures.** 1975. 87 pp. ISBN 90-220-0566-6. (out of print)

G.W. Arnold and C.T. de Wit (Eds). **Critical evaluation of systems analysis in ecosystems research and management.** 1976. 114 pp. ISBN 90-220-0593-3. (out of print)

J. van den Bos and R. Rabbinge. **Simulation of the fluctuations of the grey larch bud moth.** 1976. 91 pp. ISBN 90-220-0589-5. (out of print)

R. Rabbinge. **Biological control of fruit-tree red spider mite.** 1976. 234 pp. ISBN 90-220-0590-8. (out of print)

J. Goudriaan. **Crop micrometeorology: a simulation study.** 1977. 257 pp. ISBN 90-220-0614-X. Price Dfl. 41.00

E. van Elderen. **Heuristic strategy for scheduling farm operations.** 1977. 233 pp. ISBN 90-220-0612-3. (out of print)

P. Kampmeijer and J.C. Zadoks. **EPIMUL, a simulator of foci and epidemics in mixtures of resistant and susceptible plants, mosaics and multilines.** 1977. 56 pp. ISBN 90-220-0636-0. (out of print)

T. Kozai, J. Goudriaan and M. Kimura. **Light transmission and photosynthesis in greenhouses.** 1978. 105 pp. ISBN 90-220-0646-8. Price Dfl. 17.00

K.R. Christian et al. **Simulation of grazing systems.** 1978. 121 pp. ISBN 90-220-0645-X. Price Dfl. 19.00

C.T. de Wit and J. Goudriaan. **Simulation of ecological processes.** (2nd ed. revised and extended). 1978. 183 pp. ISBN 90-220-0652-2. Price Dfl. 23.00

C.T. de Wit et al. **Simulation of assimilation, respiration and transpiration of crops.** 1978. 148 pp. ISBN 90-220-0601-8. Price Dfl. 23.00

R.A. Feddes, P.J. Kowalik and H. Zaradny. **Simulation of field water use and crop yield.** 1978. 195 pp. ISBN 90-220-0676-X. Price Dfl. 35.00

S.M. Welch and B.A. Croft. **The design of biological monitoring systems for pest management.** 1979. 84 pp. ISBN 90-220-0687-5. Price Dfl. 26.00

N.D. Barlow and A.F.G. Dixon. **Simulation of lime aphid population dynamics.** 1980. 171 pp. ISBN 90-220-0706-5. Price Dfl. 39.00

N. Carter, A.F.G. Dixon and R. Rabbinge. **Cereal aphid population biology, simulation and prediction.** 1982. 91 pp. ISBN 90-220-0804-5. Price Dfl. 27.00

F.W.T. Penning de Vries and H.H. van Laar. **Simulation of plant growth and crop production.** 1982. 320 pp. ISBN 90-220-0809-6. Price Dfl. 70.00

D.A. Haith, L.J. Tubbs and N.B. Pickering. **Simulation of pollution by soil erosion and soil nutrient loss.** 1984. 77 pp. ISBN 90-220-0842-8. Price Dfl. 25.00

E.Ng and R.S. Loomis. **Simulation of growth and yield of the potato crop.** 1984. 147 pp. ISBN 90-220-0843-6. Price Dfl. 34.00

S.J. Troester, F.A. Ress, A.S. Felsot and W.G. Ruesink. **Modelling of the persistence of pesticides applied to the soil.** 1984. 149 pp. ISBN 90-220-0844-4. Price Dfl. 43.00

H. van Keulen and J. Wolf (Eds). **Modelling of agricultural production: weather soils and crops.** 1986. 464 pp. ISBN 90-220-0858-4. Price Dfl. 110.00